Reclamation and Use
of Disturbed Land
in the Southwest

Reclamation and

Use of Disturbed Land in the Southwest

John L. Thames
Editor

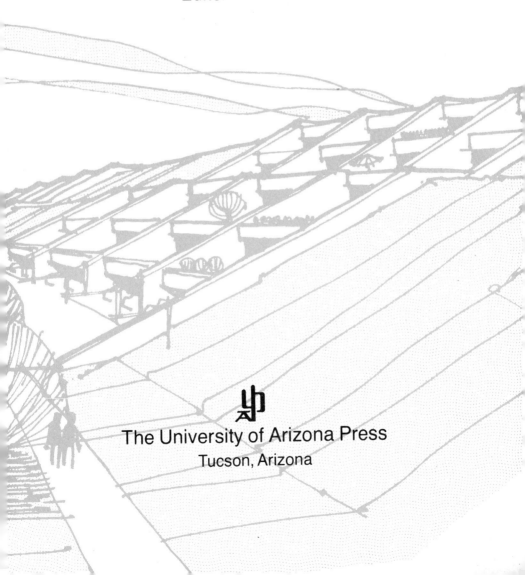

The University of Arizona Press
Tucson, Arizona

About the Editor ...

JOHN L. THAMES is professor and program chairman of watershed hydrology in the School of Renewable Natural Resources at the University of Arizona, Tucson. He received his Bachelor of Science degree from the University of Florida in 1950, Master of Science from the University of Mississippi in 1960, and a Ph.D. from the University of Arizona in 1966. During four years with the Corps of Engineers he engaged in research on off-road trafficability. Twelve years of research experience with the U.S. Forest Service includes land reclamation, flood and erosion control, groundwater recharge and watershed management. At the University of Arizona, he is active in hydrologic modeling as well as research on the hydrology and reclamation of surface mined lands.

THE UNIVERSITY OF ARIZONA PRESS

Copyright © 1977
The Arizona Board of Regents
All Rights Reserved
Manufactured in the U.S.A.

I. S. B. N. 0-8165-0535-7 paper
I. S. B. N. 0-8165-0598-5 cloth
L. C. No. 76-17133

Contributors

Aldon, Earl F., Principal Hydrologist, Rocky Mountain Forest and Range Experimental Station, USDA Forest Service, Albuquerque, New Mexico

Bach, Dan, Horticulturist and President, Dan Bach Associates, Tucson, Arizona

Bradley, Michael D., Assistant Professor of Hydrology and Water Resources, The University of Arizona, Tucson

Carter, Ralph, Director of Coal Extraction and Reclamation Program, Energy and Environmental Systems Division, Argonne National Laboratory, Argonne, Illinois

Daniel, Terry C., Associate Professor of Psychology, The University of Arizona, Tucson

Dean, K. D., Metallurgist, Salt Lake City Metallurgy Research Center, Utah

DeRemer, Dale, Consultant and General Manager, Agronomics, Inc., Avondale, Arizona

Ermak, Donald L., Regional Studies Program, Biomedical Division, Lawrence Livermore Laboratory, University of California, Livermore, California

Grandt, Alten F., Director, Land Use and Reclamation, Peabody Coal Company, St. Louis, Missouri

Guilbert, John M., Associate Professor of Geosciences, The University of Arizona, Tucson

Haggard, Jerry L., Attorney, Evans, Kitchel and Jenks, P.C., Phoenix, Arizona

Halpenny, Leonard C., Consulting Hydrologist, Water Development Corporation, Tucson, Arizona

Hassell, Wendell G., Manager, Tucson Plant Materials Center, Soil Conservation Service, Tucson

Hodder, Richard L., Montana Agricultural Experiment Station, Montana State University, Bozeman

Kay, Burgess L., Wildland Seeding Specialist, University of California, Davis

Kennedy, Allen S., Assistant Division Director, Energy and Environmental Systems Division, Argonne National Laboratory, Argonne, Illinois

Kercher, J. R., Regional Studies Program, Biomedical Division, Lawrence Livermore Laboratory, University of California, Livermore, California

Kesten, S. Norman, Assistant to the Vice President, Environmental Affairs, American Smelting and Refining Company, New York

Kilburn, Paul D., Chief, Ecological Group, Woodward-Clyde Consultants, Denver, Colorado

King, David A., Professor of Resource Economics, The University of Arizona, Tucson

Kollman, Alden L., Manager, Project RECLAMATION, The University of North Dakota, Grand Forks

LaFevers, James R., Energy and Environmental Systems Division, Argonne National Laboratory, Argonne, Illinois

Leaming, George F., Director, Southwest Economic Information Center, Marana, Arizona

Ludeke, Kenneth, Agronomist, Pima Mining Company, Tucson, Arizona

Matter, Fred S., Associate Professor of Architecture, The University of Arizona, Tucson

Nord, Eamor C., Range Scientist-Plant Ecologist, Forest Fire Laboratory, Pacific Southwest Forest and Range Experiment Station, Riverside, California

Ogden, Phil, Professor of Range Management, The University of Arizona, Tucson

O'Neil, Thomas J., Associate Professor of Mining and Geological Engineering, The University of Arizona, Tucson

Plummer, A. Perry, Project Leader and Range Scientist, Intermountain Forest and Range Experiment Station, USDA Forest Service, Ephraim, Utah

Richardson, James K., President, Arizona Mining Association, Phoenix, Arizona

Ritschard, Ronald L., Manager, Regional Studies Program, Biomedical Division, Lawrence Livermore Laboratory, University of California, Livermore, California

Royer, Lawrence, Assistant Professor of Outdoor Recreation, Utah State University, Logan

Shirts, M. B., Metallurgist, Salt Lake City Metallurgy Research Center, Utah

Springfield, H. W., Range Scientist, Rocky Mountain Forest and Range Experiment Station, USDA Forest Service, Albuquerque, New Mexico

Stairs, Gerald R., Dean, College of Agriculture, and Professor of Watershed Management, The University of Arizona, Tucson

Verma, Tika R., Assistant Research Professor, School of Renewable Natural Resources, The University of Arizona, Tucson

Vivian, R. Gwinn, State Archaeologist, The Arizona State Museum, Tucson

Wali, Mohan K., Associate Professor, Biology, and Principal Investigator, Project RECLAMATION, The University of North Dakota, Grand Forks

Weissenborn, Kenneth, Supervisor, Coronado National Forest, USDA Forest Service, Tucson, Arizona

Widman, Gary L., General Counsel, Executive Office of the President, Council on Environmental Quality, Washington, D.C.

Zimmerman, R. E., Assistant Chief Engineer, Soil Testing Services, Inc., Northbrook, Illinois

Contents

FIGURES

TABLES

Preface

THIS VOLUME EVOLVED from the symposium, "Disturbed Land Use and Reclamation in the Southwest," given at the University of Arizona in January 1975. Its purpose is to contribute to an understanding of the constraints, alternatives and techniques in the reclamation of lands disturbed by mining, and to present the latest results of major research efforts in disturbed land reclamation in the southwest. It is designed to aid those involved in the planning, operation, use and management of reclamation efforts on disturbed lands in the arid and semiarid system. It also aims to present the problems of industry, government, and conservationists. These groups, frequently at odds, actually have the same problem in common: to better the lives of our citizens. Controversy has arisen only in the interpretation of the problem and in the solutions given; interpretations that led to the bitter confrontations and power struggles of the 1960s, and are now shifting toward the sobering energy concerns of the 1970s.

This is a period of rapid change. Controversies will continue, but there seems to be a general awareness, possibly a growing enthusiasm, that minerals and fossil fuels must be extracted for the national welfare, but that extraction and reclamation must be done in a manner that minimizes social and environmental damage. The pessimism of just five years ago that surrounded the impact of mining and the prospects of reclamation in the arid environment is now yielding, thanks to a growing base of knowledge from imaginative research, to an optimism which holds that even the most fragile and refractory site disturbed by mining can be returned to productive use, perhaps even greater than that before mining.

Much has been learned. Much more remains to be learned. The problems are complex and touch sensitive areas of our social, political, economic and technological structure. Solutions will not be easy, but they can be obtained when the best minds endeavor to find objective solutions suitable to the development of wise operational policies that are also acceptable to the public.

Grateful appreciation is expressed to Bob Crist of ASARCO for his invaluable help. Sincere thanks are also due to the authors for their cooperation in the tedious job of manuscript preparation. The editor is deeply grateful for sponsorship by ASARCO, Inc., Argonne National Laboratory, Lawrence Livermore Laboratory, and SEAM (U.S. Forest Service); and for the cooperation of AMAX — Arizona Incorporated; Anaconda; Anamax Mining Company; Cities Service Company; Miami Operation; Cyprus Pima Mining Company; Duval Corporation; Inspiration Consolidated Copper Company; Kennecott Copper Company; Magma Copper Company; Peabody Coal Company (a subsidiary of Kennecott Corporation); Phelps-Dodge Corporation; Ranchers Exploration and Development Corporation; and the Arizona Mining Association.

JOHN L. THAMES

PART I

MINING RECLAMATION
and
LAND USE PLANNING

THERE WAS A TIME in this country when extreme exploitation of natural resources was not only accepted but enthusiastically encouraged as a means of pushing back the frontier and creating capital for a developing nation. Inevitably, land was misused either deliberately for quick returns or through unwitting human error. The costs were deferred to later generations. Nevertheless, through trial and error, an effective type of land planning system did evolve, regulated by natural economic forces and the environment. This concept is explored in chapter 1 by Gerald R. Stairs, and extended to include mine land reclamation in chapter 8.

Paradoxically, we are a nation of both conservationists and developers. Conservation movements appear to gain their greatest momentum at times slightly lagging behind those of greatest industrial development. The conservation movement that originated in the 1960s reflected a change in national priorities from development toward conservation. We are also a highly legal-political society, shifting away from natural economic constraints in land use toward more federal and state control. Mining, once a favored if not a pampered institution, is now slipping into public disfavor and toward greater legislative control. Jerry L. Haggard discusses the likely effects that land planning systems, developed under this shifting attitude, will have on the mining industry.

Some of the major problems in the new mining and reclamation legislation are discussed by James R. La Fevers. One important problem is the lack of flexibility in reclamation laws to allow adjustment for technological advancements. This is particularly true in the southwest where reclamation is only in the very early stages of experimentation. Perhaps because of the lack of specific knowledge, the wording of reclamation laws is frequently vague and subject to different interpretation, a task passed on to the regulatory agencies.

In the southwest, a greater percentage of the land is in federal or state stewardship. Thus, much of the land-use planning that affects mining and reclamation in the region is the direct decision of the government agencies which control resource use. The decision process which such an agency undergoes in planning and regulating land use, under the broad legislative acts, is outlined by Kenneth Weissenborn. In chapter 5, A. S. Kennedy, R. E. Zimmerman and R. Carter present a comprehensive method for planning reclamation. The methods developed for mid-western strip mines, are also applicable to many of the problems in the southwest.

1. Land Use Planning – An Overview

Gerald R. Stairs

THE LITERATURE OF LAND USE PLANNING is proliferating. Although varying in specific content, its commentary has a similar ring. The impact of the environmental decade of the 1960s is usually noted, along with contemporary comment about the interaction between economics and environmental quality. Another familiar bit of commentary relates to the purported fact that the United States has never had a formal commitment to land use planning and that state or local level efforts have been no better. Because of these and related issues, the usual conclusion is that the time has come when we must immediately legislate and regulate land use planning at all levels. Perhaps in a formal sense, land use planning has been scant in the U.S. But we have had, in fact, a rather intensive set of efforts directed at this issue, and these efforts have been effective.

A look at history and at the current state of land use will be valuable as a means for suggesting the next move in land use efforts. The subset of land use planning histories includes basically three components, two of which have been much more significant than the third. The first of these can be categorized as the trial and error method established during the development of our country at a time when manifest destiny rode high and land resources were in the initial stages of development. The second relates to the establishment of our public lands and their subsequent management or use. Finally, the third may be listed as zoning — a topic less important historically in the management and development of large, natural resources, albeit an important one in the urban development scheme.

One means of comfortably surveying the state of land use in the U.S. might be to imagine a vacation trip through various regions of the country. Beginning in the northeast, one might choose the countryside of New England or upstate New York as a sampling site. One becomes aware of a countryside made somewhat more attractive by lines of stone fences across the hills and through the otherwise virginal appearing forests. These fences stand as silent testimony, not to a desire on the part of early settlers to make the landscape more rustic or attractive, but rather to the fact that early attempts to farm the land met with significant soil structure problems. As a convenience, the rocks which hindered the plow were hauled away and integrated

[3]

into fences at property lines. They became permanent fences. The fact that much of this region is no longer intensively cultivated and has reverted to forest production as a land use is clear from even a cursory view of the area. It is not as clear whether suitable effort went into the attempts to farm the area, but a search of the region's history would show many generations of hard-working families who attempted to make a living on this relatively poor soil under conditions of short growing seasons and variable weather systems. The fact that the area is no longer intensively farmed provides a signal example for a type of land use planning that would be difficult to duplicate in any modern, decision-making laboratory — however computerized or sophisticated the data system.

A similar view of evolved land use can be gained by continuing our journey into the southeastern U.S. Here we can trace the evolution from natural forest land to intensive cotton farming. In the 1970s, in many regions, there is a return to forest production on an intensive scale. Farming on a large scale is still to be found in this region, just as is true in the northeast, but the location is closely correlated with better soil types. Land remaining in productive agriculture in this region has stood the test of trial and error over time and has provided for a substantial integration of soil quality, environment and economics.

The north-central region was also included in that great forest area that covered the eastern one-third of the United States at the start of colonization. Only a brief period was required to harvest the immense white pine forests of this region and to quickly plow the prairie ecotype into productive agriculture. Farming effort followed the timber harvest, with better soils remaining in agricultural production today, but with the bulk of the northern region yielding back to forest production. Lands to the south yielded prairie vegetation to cultivated crops. The region continues in a productive mode where better soils and climate provide an economically sound environment for agriculture.

Continuing west, one travels into the Great Plains where early settlers tore the sod from some of the best grazing land in the world and in a short time changed it into a dust bowl that reflected the misery suffered by nature and man.

The far western United States provides an end to our scenario. It also underwent a similar pattern of trial and error settling and development. Examples of overgrazing and excessive timber cutting, attempts at farming where soil was poor or unsuited, blended chronologically with evolving improvement of resource management and, finally, the establishment of agriculture on areas where it was economically sound to do so.

In summation, this period of national development, played out over a hundred years or so, provided substance to the manifest destiny concept that provided a rallying point for the early history of land settlement. The sweep of land occupation preceded or took place concurrently with minerals exploitation, industrial development, and urbanization as a way of life. To call all of this land use planning in the formal sense would be a mistake, and would overlook the chronicle of errors made in the process. The price of this system was unusual exploitation of nature, the native Americans, the blacks and other minority groups. Nevertheless, it was a system that has delivered to us today a land use pattern that is clearly in keeping with maximum productivity identifications in the generic sense. Certainly, specific improvements can still be made and modest adjustments or transfers of land use are available; yet in the main, we have well identified the land areas of our country based upon the reinforcement of real world economics. It seems particularly important that this history be acknowledged and, further, that a land use sense which provides a clear understanding of the significance of these events be instilled in today's planners. It is particularly important to understand that well defined land use patterns are not easily converted without serious economic consequence. That this issue is of concern can be seen from the example of farmland loss to urbanization, utility corridors or transportation systems. About 1¼ million acres are removed from production each year in the United States. The new interstate highway system, when completed, will have required a land area larger than the State of Delaware. Productive land cannot continue to be lost at the rates extant in New Jersey, or Maryland where one-half of the farmland will be lost in a period shorter than that just passed in land settlement. In short, it is imperative to go back to the resource base to define land use planning.

The second type of land use planning deals with the matter of federal "set-aside" or the public domain. The sale or reserve of public domain has been a principal factor in our land use development pattern. In the first forty or fifty years of our nation's history, about 90 million acres of land were sold to establish a national treasury. Later, in the 1850s, enormous grants were made to the private sector for the construction of railways. These giveaways were followed or were concurrent with settlement grants such as home-steading and the set-aside of large national reserves in the form of national forests or parks. In the process of federal set-aside or sales, the U.S. devel-oped the beginning of a strong conservation ethic. Although latter-day conservationists are usually somewhat loathe to admit it, this nation has not done badly in terms of land reserve, a factor that most western states can attest to. It was consistent with the stage and timing of settlement in the U.S. that the early conservation movement was forced to focus most heavily upon the then less accessible and less settled western U.S. Today many western states remain over one-half public domain. The State of Arizona is one of the leading examples, with over 80 percent public land. This legacy of public land derives, primarily, from the national forests and national parks domain, from reverted grant lands, and from Indian reserva-tions that represent a unique type of land, public only in relation to its native American inhabitants.

In the southwest, it is apparent that any broad question of land or natural resource utilization must begin with an understanding of the inter-action between public and private land. This requires that one be conversant with the legislation that created this type of reserve, with current manage-ment policy, and with any broad policy review studies providing guidance or recommendation for management of the public lands. While the former are best illustrated through contact with an appropriate agency, the latter are typified in the recent report of the Land Law Review Commission. This report, entitled "One-Third of the Nation's Land," represents an abstract of a multimillion dollar study of public land management and policy. It began as a letter from Congressman Wayne Aspinall to President John F. Kennedy in 1962 and culminated in the establishment and report of the Land Law Review Commission. Many significant recommendations are embodied in the report, certainly more than could be discussed here. However, three areas of concern can be focused upon.

One clear call for action in the report was the suggestion that activity among various federal agencies should be closely coordinated. This topic is often expressed by those considering the natural resource sector and seems at first to be reasonable and easily accomplished. That it is in reality difficult to obtain is no surprise to anyone experienced in the problems of agency management. With various federal department lineages, the problem of coordination is amplified. Always a topic of apparent agreement at the public relations level in Washington, D.C., the effort at local levels remained a problem until the 1970s. Since then, however, increased effort has been put forth to insure better interagency cooperation. Hopefully, the facility for good private sector relations has also been increased. One type of organizational effort that offers promise in most states is the State Rural Development Committee (SRDC). Composed of numerous federal and state agencies with membership from political subdivisions and from the private sector, the SRDC provides a forum for local community contact with an integrated array of public officials and agency personnel. It also offers the private sector, be it mining, transportation, agriculture, or tourism, an opportunity to focus the services and capabilities of diverse agencies upon problem areas in resource development.

A second recommendation called for state and local government to be given an effective role in federal agency land use planning and that such planning be made to conform where possible with state and local planning. This suggestion is heartily endorsed at the local level and is most sensible in relation to development of resources by the commercial sector. The economic well-being of a majority of western states depends to a large degree upon such effective interaction and cooperation. Again, the SRDC can be a useful vehicle to accomplish planning of this type — as can various local government or citizen participation sessions conducted in open hearings where private and public views can be discussed.

A third recommendation called for the establishment of regional commissions. This concept, analagous to the older river basin commissions, calls for both intra- and inter-state regions based upon resource utilization lineage rather than specific political subdivision. Again, the integration of public and private land was called for in the development of planning units.

Taken together, the three illustrations from the Land Law Review Commission report provide a reasonable frame of reference for resource planning in the southwestern U.S. Acknowledgment by all that the federal set-aside is an important bit of land use planning history with significant impact is necessary if the western U.S. is to deal with many problems of land use and resource extraction in the next few decades. In so doing, Americans must learn to echo the recommendation often put forward that "users can be conservationists too"; in fact, this attitude must be adopted if we are to integrate properly the public domain and the private sector.

The final broad classification of land use planning with historical and present-day importance is that of zoning or classification of land. Although never of great significance outside of the urban area, the concept of zoning represents a precedent of importance. Zoning began to gain momentum in U.S. cities in the 1920s and 30s, was duly tested in courts, was found constitutional, and became a mainstay in the planning of cities and towns. Generally considered a restrictive concept, it is often referred to as a "negative" form of planning. The impact in rural areas has been modest, and where installed, easily overturned by various development efforts. Still, the precedent is important as is the historical action that followed assignment of this authority to the state level. Almost without exception, the states decentralized the zoning authority and provided an early example of planning authority at the local level. Today attempts to reinstate authority at the state, regional or federal level have usually met with significant resistance at the local level.

In overview, three major types of land use planning have been considered: a trial and error system; federal set-asides; and local zoning. All of these have had significant impact upon our current state of land and resource identification.

The question now posed is: if these factors add up to suggest that current land use patterns are reasonably consistent with land capability, how can it be assured that new planning thrusts will not overlook the lessons of history? Obviously, the answer is land use planning based upon true resource values. The problem facing society is to define these values. It would be inconsistent to attempt a specific value definition for various resources in this chapter. Instead, it seems more reasonable to suggest an outline of land use planning and a frame of reference for public action in setting forth future decisions.

A simple way of looking at the broad aspects of land use planning is to consider three functional areas on a continuous, interactive scale. The first area would be an array of individual resources — forests, agriculture, wetlands, mineral lands, flood plains, etc. — each with a significant body of information, professional specialists, and supporting public or private sectors. The second area would be to consider the interaction potential between each or all of the resource areas. Many of the land uses are interchangeable; all can interact with others in significant ways. For example, the discovery of mineral deposits on an area already in use as a recreation area creates immediate potential conflict. Finally, the third area to be considered in broad view is that of political decision making, regulation and policy setting. It is in this final phase that public expression of concern is finally molded into a plan for management of the resource base.

The foregoing outline suggests an approach long utilized but one that is growing in sophistication and intensity today. Perhaps the most important aspect of this entire process is to relate it to the current ruling attitude of society-at-large. A difficult task, and always somewhat speculative, it is nonetheless necessary to characterize broad parameters that may play determinate roles in future resource decision making. At the usual risk level for any prophetic pronouncement, two major topics are worthy of consideration. The first is that the world seems about to enter a resource era that might be best characterized as the management of scarcity. The second, intimately tied to the first, is that we must seek a change from a demand-oriented economy to a supply oriented economy if we are to find answers in this era of limited resources. Final definition of accuracy for such sweeping generalizations must await the future, but it would seem that acceptance of these two tenets might at least allow society to err on the more desirable side of conservation.

Certainly the foregoing historical outline must be referenced to a consideration of individual resources, their interaction and, finally, the decision or policy setting mode. It remains the purpose of this overview to consider briefly the mechanisms for organizing the functional processes necessary to accomplish desired levels of planning. In the 1970s, operational modes in most states revolve around a cluster of mechanisms ranging from governor-appointed commissions to local town hall groups. A new wave of citizen "listening sessions" has been attempted, in most cases with rather little significant numerical representation from the body public.

A note of concern in much of the present organization for public participation in the decision making process relates to the involvement of professional or industry groups. A typical state planning format might include a governor's commission of leading citizens, several decentralized regional sessions and finally presentation to the legislature with additional citizen input possible during legislative hearings. During the entire process there is often little opportunity for formal presentation by professional groups, and unfortunately, the professional representation at various hearings is often lacking. It is unfortunate that this is so, and it would seem that correction of the problem should proceed on two fronts. Firstly, it would be consistent with logic to consider appointment of a "companion council" of key professional members to assist the "leading citizen" type of commission approach. Testimony by the professional group could be vital in early identification of problems and alternative solutions. Such inputs would broaden and reinforce the validity of hearings conducted at the public level. Secondly, the professional resource managers need to exert more personal initiative in participating at the decision or policy making level. Certainly, there is little to complain about if a decision is lost by default. The need for agency and industry recognition of this problem may offend some policy statements of non-advocacy, but it seems consistent with the real needs of society to speak up and out for sound basic professional concepts in resource management. To do less will be to invite a delegation to secondary involvement and provide a constant reactionary rather than leadership posture.

2. Land Use and the Mining Industry

Jerry L. Haggard

THE BACKGROUND for the subject of land planning and the mining industry may be set by reviewing the position which the use of land for mining has occupied in the United States in the past and the changes which are occurring in that position. Then, I will describe the evolution of land planning nationally in the legislative and administrative branches of the federal government, and what effects the usual types of land planning systems can be expected to have on the mining industry. Finally, I will review the land planning system which has most recently been proposed for Arizona.

Although state and federal legislatures have been flooded with land planning legislation during the past several years, land planning is not new in the United States. Almost since the beginning of our nation, cities, counties, states and the federal government have designated lands for special uses and have regulated uses on other lands. Our state and national parks, monuments, forests, wilderness areas, other similar categories of land, and city and county zoning ordinances are all familiar examples of land planning. What is new are the expanded concepts of comprehensive land planning which have been generated in the 1970s by the environmental movement.

MINING AS A LAND USE

Not to be nostalgic, but rather to set the stage for the situations which the mining industry faces today, the position of esteem which the mining industry once occupied in "the good old days" should be examined. At one time, and in fewer cases continuing now, mining was considered to be so important that it was exempted from statutes granting zoning authority to local governments. For example, the Arizona statutes which grant zoning authority to counties specifically prohibit counties from using that authority to restrict major mining and metallurgical uses as well as land uses by the other basic industries of grazing, agriculture and railroads (ARS § 11–830, Supp. 1975).

Further, a basic premise of common law continues to be, at least in theory, that the mineral estate is the dominant estate, carrying with it the right to use, damage or destroy as much of the surface as is reasonably necessary to remove the minerals from the property. This right of the mineral estate owner is subject only to agreements with the surface estate owner, to the rights of lateral and subjacent support of the surface, and to the obligations of the mineral owner to compensate the surface owner for damages to his surface estate. The obvious rationale for this is that unless the mineral estate is dominant and there is at least a conditional right to use and damage the surface, the mineral estate could be made worthless.[1]

Continuing with this history of fair treatment for mining, a very interesting case decided in Illinois held a surface mining reclamation statute to be unconstitutional. This statute provided that the operator must return the surface "to approximately the original contour which existed prior to mining." That phrase will be familiar to those who have followed federal surface mining legislation. The Illinois Supreme Court held that the statute was unconstitutional because it invaded the private property rights of the mineral estate owner. The Court said that the state has no authority, under either the guise of a conservation theory or the guise of regulation for the preservation of health, to compel an owner at his own expense, to convert his private property to what the legislature considers to be a higher or better use.[2]

Mining has been considered to be such a necessary public purpose that the industry in several states is granted the power of private eminent domain to condemn rights of way for mining operations.[3] Then came the environmental revolution of the 1960s, giving birth to the Wilderness Act of 1964, the Water Quality Act of 1965, the Clear Water Restoration Act of 1966, the Federal Clean Air Act of 1967, the Wild and Scenic Rivers Act of 1968, climaxed by the National Environmental Policy Act of 1969.

The extent to which we have moved (forward or backward, whichever one may view it) from a position of mining as a favored land use and toward a position of mineral activities being most unfavored land uses, may be exemplified by the startling decision reached in 1973 by the United States District Court for the District of Minnesota in *Izaak Walton League v. St. Clair* (353 F. Supp. 698 [D. Minn., 1973]). In that case, the owner of a reserved mineral interest in the Boundary Waters Canoe Area applied to the Forest Service for a permit to conduct mineral exploration. The Izaak Walton League and others sued to enjoin the exploration. The court found that mineral activities are incompatible with wilderness and, therefore, it "took judicial notice" that Congress could not have intended to allow mineral activities to take place in the wilderness area. This holding is, of course, entirely contrary to the express provisions of the Wilderness Act which permit mineral activities to be conducted in Wilderness Areas until 1984.[4]

---

[1] Ferguson, *Severed Surface and Mineral Estates,* 19 Rocky Mt. Mineral Law Inst. 411, 414 (1974).

[2] Northern Illinois Coal Corp. v. Medill, 397 Ill. 98, 72 NE2d 844 (1947). This continues to be the law in Illinois and the statute has been upheld only in situations where surface mining would have caused an actual threat to the physical health and safety of a community. See Village of Spillertown v. Prewitt, 21 Ill.2d 228, 171 NE2d 582 (1961).

[3] Arizona — Art. II, § 17; Colorado — Art. II, § 14; Idaho — Art. I, § 14; Illinois — Art. IV, § 31.

[4] 16 U.S.C. § 1133(d)(3). Other special provisions of the Wilderness Act, however, apply to the Boundary Waters Canoe Area 16 U.S.C. § 1133(d)(5). Therefore, a similar situation arising in another wilderness area might be distinguished from the result reached in the Izaak Walton League case.

However, the 8th Circuit Court of Appeals reversed and remanded the case, but only to allow the Forest Service to determine whether a permit should be granted.[5] Therefore, the merits of the District Court's decision with respect to mining in the Boundary Waters Canoe Wilderness Area have yet to be determined.

Another case decided in 1974 gives cause for additional concern in the mining industry. In *Bureau of Mines of Maryland v. George's Creek Coal Company*, 321 A.2d 748 (1974) the Maryland Court of Appeals held:

> We think the statutory prohibition of the open-pit or strip method of mining coal constitutes reasonable regulation under the State's police power calculated to protect the environment and to preserve State-owned land for public use for present and future generations of citizens.

This case was remanded to the trial court to allow the owner of the reserved mineral interest to establish his right to compensation by satisfying the burden of showing that the ban against strip mining of coal amounted to a prohibition of all usefulness of the property.

These are some of the indications that we are experiencing a profound change from mineral development being a preferred land use to a most unpreferred land use. Environmental groups, encouraged but not satisfied with their successes of the '60s, began to feel that legislation dealing with individual resources — water, air, wilderness and solid waste disposal — was not quite enough. The single control which would encompass all of these elements and more was control of the land.

FEDERAL LAND USE PLANNING

Although land use planning as such began as early as 1961 in Hawaii and in some of the northeastern states, the national movement began in earnest in 1970 with the introduction of S. 3354 which would have established a national land use policy. Various similar bills were introduced and considered during succeeding years. In June, 1973, by a vote of 65 to 21, the United States Senate passed S. 268 which would have established the Land Use Policy and Planning Assistance Act. The House version of that bill (H.R. 10294) enjoyed no difficulty until it left the House Interior and Insular Affairs Committee in February, 1974. From that point the land planners' castle commenced to crumble in the House Rules Committee and finally collapsed in the House of Representatives on June 11, 1974 by a vote defeating the rule for the bill. However, rather than the defeat of federal land planning, we must realistically consider this action to be only a delay, as its proponents promised high priority attention to federal land planning in the more liberal 94th Congress.

The evolution from air, water, and solid waste regulation to total land planning is illustrated by the growth of the Environmental Protection Agency (EPA) programs. These would make an interesting case study of the creeping spread of federal agency power. The extension of EPA's control

---

[5] Izaak Walton League v. St. Clair, 497 F.2d 849 (8th Cir., 1974), U.S. Cert. denied 419 U.S. 1009.

from regulating these individual matters to regulating land use may be marked by the decision in *Sierra Club v. Ruckelshaus* (344 F. Supp. 253 [D.C., 1972]) which gave birth to the nondegradation of air standards and control of indirect sources of emission programs of EPA.

Regulations to control indirect sources of emissions, published in the Federal Register on October 30, 1973 (38 Fed. Reg. 29893) would require EPA or state approval for the construction of such facilities as airports, parking lots and industrial developments. The Arizona Environmental Planning Commission[6] presented a statement opposing the regulations, saying that EPA was exceeding its authority by getting into the business of land planning. Then on August 27, 1974, EPA published proposed regulations to prevent significant air quality deterioration.[7] Again, the Arizona Environmental Planning Commission, the Arizona Office of Environmental Planning, the Department of Health Services and the Governor of Arizona objected to the proposed regulations for the reason that EPA was getting into the business of land planning.

Then, almost as if Arizona's protests had illuminated the opportunities for EPA, in November, 1974, EPA announced that it was indeed getting into the business of land planning. EPA released a proposed policy statement setting forth the land use implications and requirements of EPA programs which will wrap up its control over air, water, and other environmental media into a program of "structural controls" and "nonstructural controls." This bureaucratic language means that EPA proposes to regulate land use as a means of controlling all pollution elements. By a resolution dated September 18, 1974, the Arizona Environmental Planning Commission again opposed this policy of EPA to involve itself in land use planning under the guise of EPA's pollution abatement activities.

GENERAL EFFECTS OF LAND PLANNING ON MINING

As with the goals of clean air and clean water, few people can be opposed to the abstract goal of using land in such a manner that will benefit our society. The opposition arises with the question of who will determine what that best use will be, and with the determination that such use will be other than what the landowner desires.

The typical forms of the state and federal land planning programs present two major threats to mining. The first arises from the usual requirement that "areas of critical environmental concern" be designated. The ill-fated federal legislation would have required that such areas include all

---

[6] The Arizona Environmental Planning Commission was created in 1972 to establish a state land use planning program for Arizona. A.R.S. §§ 37–161, *et seq.*

[7] 39 Fed. Reg. 3100. The regulations to prevent significant air quality deterioration became effective on January 6, 1975 (39 Fed. Reg. 42510) and provide for the establishment of three classes of areas. Class I would be established for areas in which practically no further development affecting air quality will be allowed. Class II areas would permit only moderate well-controlled growth. Class III areas are those in which the already excessively restrictive national ambient air quality standards of EPA would apply. EPA arbitrarily designated the entire United States as Class II upon the promulgation of the regulation.

lands on which "incompatible development could result in damage to the environment, . . . or the long-term public interest. . . ." Such areas would also have included fragile or historic lands, scenic areas, valuable ecosystems and agriculture, forest, watershed and grazing lands. (§ 412, H.R. 10294, 93rd Congress). Obviously, with such a broad definition, any area of land could be designated an area of critical environmental concern. In the areas so designated, no use and development may substantially impair the values for which they were designated (*Id.* § 105 b). This would create the authority, or the requirement, to prohibit mining activities in the broadly defined areas of critical environmental concern.

The other major threat to mining in the typical forms of land planning legislation is the requirement to control the impact, and perhaps to prohibit adverse impacts, of large-scale development upon the environment wherever such development may take place (*Id.* § 105 e). We can be certain that mining activities will be included in the term "large-scale development." The purpose of identifying large-scale developments is to study, plan, and regulate them. Each of us knows that any project can be studied, planned and regulated to death.

LAND PLANNING IN ARIZONA

An effort was made in 1973 to enact a comprehensive land use planning system for Arizona. The concern generated by this proposal, and the resulting opposition, soon made it apparent that another approach would be necessary. This took the form of S.B. 1014, which was passed by the Legislature and approved by the Governor in 1973. This bill established the Arizona Environmental Planning Commission and gave it the job of studying the question of land planning, conducting hearings throughout the state, and preparing a land use plan by January, 1975.

The Commission was formed with six members from the Arizona Legislature and nine members, appointed by the Governor, representing various land using interests in the state. Commencing in November, 1973, the Commission held 24 public hearings throughout the state and many additional meetings with local, state and federal agencies, Indian tribes and user organizations. In December, 1974, the Commission approved a proposed bill which was introduced in the Arizona House of Representatives on January 15, 1975 (H.B. 2028, 32nd Ariz. Leg., 1st Reg. Sess.). Even before the bill was introduced, it had been criticized by some groups as being a pro-growth bill. I suspect the real objection of these groups is that it is not an anti-growth bill. At the same time it was being criticized by others as being excessively restrictive. Perhaps this is an indication that its real effect would be somewhere in the middle.

One of the first decisions faced by the Commission was whether the state land planning system would be of a comprehensive type, regulating all uses on all lands in the state, or whether it would be more limited in scope and regulate only certain uses and certain areas. The latter more limited approach was selected and the bill may be characterized generally as establishing a land planning program which would regulate certain land use activities in all areas and all land use activities in certain areas.

Those areas in which most land use activities would be regulated would be designated by the state as areas of critical state concern and could be selected from the following categories:

1. Areas which contain or have a significant impact upon environmental, historical, natural or archaeological resources of intercounty, statewide, interstate or international importance. Because of the broadness of these area categories and because they would be somewhat analogous to the establishment of state parks, recreation areas and similar areas, the Commission decided that it should be only the State Legislature which could designate such areas.
2. Areas significantly affected by a land use activity of state concern.
3. Areas necessary for the long-range production of certain resources in which incompatible major development is imminent.
4. Areas containing natural hazard conditions which could unreasonably endanger life or property (*Id.* § 41–116).

Areas in categories 2–4 would be designated by the State Office of Resource Administration to be established by the bill. Most land uses within such designated areas would be regulated under guidelines established by local governments (counties and cities) and approved by the State Office.

The Commission considered placing a ceiling on the total amount of land within Arizona which could be designated as areas of state concern. Florida has limited such areas not to exceed 5% of the entire state. However, the majority of the Commission elected not to establish such a limit.

The second subject of regulation which would be established by the legislation is referred to as activities of state concern. Those land use activities would be:

1. Major airports.
2. Highways.
3. Major facilities of a public utility.
4. Facilities for the development, use or treatment of water resources or for flood control.
5. Development, extraction or processing of resources within the state.
6. New communities.
7. Disposal sites of hazardous materials.
8. Nuclear detonations.
9. Sources of major air pollution (*Id.* § 41–118).

The development of a mine and the extraction and processing of minerals would be classified as activities of state concern and the effects of these activities on the land would be subject to regulation by the Office of Resources Administration. In regulating these activities, the bill requires the Office to follow certain standards and to consider both beneficial and detrimental effects of the proposed activity and regulation.

A most important provision of the bill is the requirement that no restriction or prohibition of a use of real property which results in a significant reduction in the value of the property may be carried out unless the state compensates the owners of any interests in that property (*Id.* §41–119.01). One of the subtle vices of the federal and some state land planning legislation is a provision that actions taken under those land planning systems shall not diminish the rights of property owners contrary to the provisions of the federal or state constitutions. This is misleading and somewhat hypocritical because, while it sounds like a protection, it avoids the problem that the federal and some state constitutions do allow the rights of private property owners to be severely diminished without compensation. The constitutions of many of the states and of the United States provide only that private property shall not be *taken* for public use without just compensation. But, generally, the restriction of property uses by zoning actions does not entitle the landowner to compensation even though the value of his property has been seriously reduced by the restriction. This requirement for the State of Arizona to pay compensation for reductions in private property value caused by land planning restrictions was approved by the Commission without dissent.

The Arizona land planning system would create three administrative bodies. One would be the Office of Resource Administration which would administer the system (*Id.* § 41–113). There would also be a Board of Appeals to which any affected local government or private citizen could appeal regulations, actions or orders of the Office. The Board would be made up of seven members, selected by the Governor, who would represent groups interested in resource use and conservation and who would have knowledge of land use or land use planning (*Id.* § 41–115). Appeal *de novo* from the Board to the state courts is expressly provided for (*Id.* § 41–115.01). The third arm of the land planning administrative structure would consist of a Coordinating Council on which state agencies having jurisdiction over land and resource uses would be represented (*Id.* § 41–114).

One of the features of this bill which is significant to the mining industry is the exemption of subsurface resource exploration from regulation. We can expect that the exemption of mineral exploration will be questioned as the bill is considered by the Legislature. The answers to such questions, I believe, are these.

Most land surface resources, in contrast to subsurface resources, are identifiable and their relative values can be approximated and compared. These include, for example, forests, forage, recreation values, and surface water. However, it is self-apparent that subsurface resources are hidden and a meaningful determination of their existence and value can be made only with very sophisticated exploration equipment, methods and personnel. It is doubtful that any planning office could develop sufficient mineral information in an area to properly compare the subsurface and surface values and to determine the best uses. Such information can be obtained only by leaving the area open to exploration by industry.

We can expect the argument, however, that once an area has been explored, then the continued availability of the area for exploration should be determined by the land planners. But this ignores two fundamental truths about the mining industry. The first is that the technology of mineral exploration is constantly advancing. Improved exploration methods can discover ore deposits not discoverable with earlier methods. This position also ignores the facts that market values change and mineral recovery methods improve. Mineral deposits which were not recoverable at one time frequently become so at a future time. If land planning decisions based on prior exploration conclude that an area is unsuitable for mining, it may cause such an area to be unavailable for future mineral exploration and production when changes in markets and improved recovery methods could make mineral development feasible. Finally, the Arizona Environmental Planning Commission recognized that exploration for subsurface resources is a temporary and minimal use of the land surface, and that bringing the land effects of a mine development and the mining and processing operations within the control of the legislation is all that is necessary.

We can expect that the exemption of mineral exploration from Arizona's land planning system will be a major controversial issue and some compromise may be required. The only way in which a valid determination can be made that exploration should be restricted in an area is to assume that the area does contain an ore deposit, and to attach a value to the assumed deposit. Then, using that assumption, if the surface values are determined to be greater than the value of the assumed mineral deposit, there may be some basis to conclude that the area should not be explored or developed. Unless this determination can be made, the area should be left open to exploration by industry to determine its true subsurface values.

One of the questions which will concern the mining industry (once the industry has recovered from the prospect of having the land effects of mining operations regulated as an activity of state concern) is the level of government at which its activity will be regulated. The bill provides initially that local governments (in most cases, counties) will have the opportunity to propose and have approved their plans for regulating such activities. The ultimate authority, however, for the approval, disapproval or modification of those local plans and for their implementation will be at the state level (*Id.* § 41–118). The mining industry should consider whether it might expect more reasonable treatment at the county level or the state level.

This is a difficult question and it might be answered differently depending upon the particular area in which a mine development will be proposed. If it is in an area under the jurisdiction of a local government which desires to have the economic benefits created by the development, it may be that the final authority being left at that local level would be more desirable to the industry. On the other hand, if a development is proposed, for example, near Arizona metropolitan areas, whose residents believe they already have a sufficient economic base so as not to need additional development, then the development might receive more reasonable treatment at the state level.

The system provided for in the bill will be the best approach. The attitudes and feelings of the local governments should be reflected by their initial proposal of the plan and regulations to the Office. The Office must consider and be guided by that local proposal. However, if it is determined that state revenues and state development are of sufficient statewide importance, the state could overrule local opposition.

Some suggestion has been made that the regulation of the major activities of state concern and of all activities in areas of state concern could establish a state environmental impact statement review system. It is likely that some features resembling such a procedure will be found in this land planning system. However, the time limits and the requirements to consider the economic and materials supply, as well as the environmental, aspects of development proposals will preclude the prolonged delays and nonobjective studies presently being experienced under the National Environmental Policy Act and under some state environmental policy acts.

There is a major concern respecting the power, and checks on this power, which will be held by the Office of Resource Administration. Under this bill, this Office would certainly have the potential of becoming the most powerful agency in the State of Arizona. This concern caused the Commission to consider placing the land planning and the enforcement functions of the Office into separate and independent divisions under the direct authority of the Governor. This should have been done, but the majority of the Commission disagreed. One of the major controls on the power of the Office will be the requirement for the state to pay compensation for reductions in private property values. The power of the Office will also be checked by the Appeals Board. Although there was a strong feeling that the operations of the Office should not be encumbered by the superior authority of an Appeals Board, the Commission decided that this was a necessary protection along with the right of appeal from the Board to the courts.

CONCLUSION

The land use planning systems which have been proposed by the federal government and by Arizona may be regarded as limited in their approach. However, to avoid complacency which such an initial limited program may create, the following illuminating paragraph from "The Quiet Revolution In Land Use Control," a study prepared for the Council on Environmental Quality (1971), should be heeded:

> As a political matter probably the most feasible method of moving towards a well-planned system of state land use regulation is to begin with a regulatory system that concentrates on a few goals that are generally perceived as important, and then to gradually expand the system by adding more comprehensive planning elements, as is being done in Vermont.

This statement is a surprisingly candid expression of the land planners' strategy in accepting what may appear to be limited land planning systems. Therefore, the mining industry should be watchful of both state and federal land planning legislation, should be active in its consideration in the legislatures, and should beware of this strategy to gradually expand the systems by adding more comprehensive planning elements.

3. Effect of Legislative Change on Reclamation

James R. LaFevers

ALTHOUGH THERE IS SOMETHING OF A CONSENSUS among industry representatives and legislators alike that legislation is a necessity to ensure
consistent practices by all operators and to assure them of an acceptable
framework for reclamation planning, there is little agreement as to the
precise form reclamation laws should be given. Since Arizona officials
reviewed the texts of several other states' reclamation laws during the early
1970s, it is likely that the law that will be devised for Arizona will contain
many of the elements already tested elsewhere. This is a commendable
approach, and one that has been taken in other states preparing similar
legislation. It is partly for this reason, therefore, that although each state
tailors its reclamation law to suit its own particular needs and conditions,
a certain set of similarities often appears between state laws. Important
differences are also built into the laws to reflect local physical conditions
and public attitudes concerning mining, reclamation, and specialty problems.
Important differences in reclamation result from the way in which the laws
are administered rather than from significant differences in the laws themselves. In order for Arizona legislators to enact a law that will be equitable
to all concerned parties and have the greatest beneficial impact, it is important
that they be apprised not only of the language of other states' laws but also
of the complexity of effects that can be brought about by subtle differences
in interpretation. It is beyond the scope of this chapter to examine each
state law, or to attempt to discuss all the effects of legislation on reclamation
practices, and only a few examples are given to illustrate the broad scope
of possible effects.

SIMILARITIES AMONG LAWS

Of the twenty-five states in which coal was produced by surface methods
in 1974, twenty had operational reclamation laws, and most of the others
were considering such legislation. Most of these laws are quite similar in
the imprecision of their language and the definition of their goals. Rarely
are such terms as "reclamation," "rehabilitation," and "higher land use"
defined in a workable way, even though they often constitute the central
theme of the law. The 1974 Indiana law states, for example, that "Reclama-

[18]

tion shall mean rehabilitation of the area of land affected," and like most reclamation laws, it provides for "proper reclamation" and "improved land use practices," while stating that it shall "prevent or minimize injurious effects to the people and the natural resources of the State," "protect our lakes and streams from pollution," "decrease soil erosion and hazards of fire," "improve the aesthetic value of the landscape," and "increase the economic contributions of the affected areas." Although these are undoubtedly admirable qualities in a reclamation law, they are too imprecisely defined to be of much utility in administering such legislation. In such cases, the interpretation of the applicable meaning of such phraseology as "improve" aesthetic value and land use practices is often the responsibility of a single individual. In Indiana, the Director of the Division of Reclamation is charged with the burden of applying these imprecise regulations equitably to thirty separate operators under hundreds of variations in conditions. Although he is a highly qualified and competent individual, the Director admitted that personal bias can influence the total reclamation picture. The law is administered, therefore, as in other states, according to a separate set of "guidelines," which can be modified and updated periodically without legislative changes. These guidelines define the precise reclamation requirements and restrictions, but are also open to considerable variation in interpretation.

The reclamation laws of most other states are similarly imprecise in language. In Ohio, for instance, "reclamation" means "backfilling, grading, resoiling, planting, and other work to restore an area of land . . . so that it may be used for forest growth, grazing, agricultural, recreational, or wildlife purpose, or some other useful purpose of equal or greater value." The Missouri law states that "at least 75% of all land strip mined be restored to a rolling topography suitable for farming. The other 25% can be graded for use as a park or wildlife habitat." One wonders what the criteria should be for determining useful purposes "of equal or greater value," and why land use should be restricted to farming, parks and wildlife habitat. The major point of interest in these laws, however, is in the use of the term "restore" instead of "rehabilitate," since restoration implies concern for returning to pre-existing conditions rather than creating a landscape with a high land use potential.

The latitude allowed in administering the reclamation laws can be demonstrated by any of a number of examples. The reclamation requirements in Illinois, for instance, are quite similar to those in Indiana. Both states allow for separate grading practices for land reclaimed for row crops, pasture, forest, recreation and other uses. Furthermore, both require grading and revegetation for aesthetic purposes along highways, the construction of dams to form lakes or ponds in final cuts, and the covering of acid forming material. Many other stipulations in the Illinois and Indiana laws are also similar, and although differences exist, mining areas in both states are fairly similar in terms of the physical setting. One would expect, therefore, that the reclamation patterns in these two adjacent states would also be similar, especially considering that in some cases the same companies are responsible for the reclamation in both states. The case is, however, that in Illinois 86% of the land reclaimed under the law through 1974 had been graded for row crops or pasture, whereas in Indiana 90% had been reclaimed as forest and range (maximum slope of 33⅓%) with only 9% graded for pasture and hay (25% maximum slope) and less than 1% was in row crops (less than 8% slope). Beginning in the early 1970s there has been a trend toward lessening the distinction between these two states, with Illinois recognizing the need for recreation areas and allowing more land to be reclaimed for forest and recreation sites, and with Indiana requiring a larger percentage of the land (22%) to be rehabilitated for use as farm land. This further exemplifies the fact that it is often the interpretation of the law rather than the law itself that dictates reclamation procedures. This can be extremely beneficial in that it is a major factor giving the laws the flexibility necessary to be adaptable to changes in technology and local land use needs. It should, however, be guarded with such features as Indiana's Natural Resources Advisory Committee and Natural Resources Commission which act as review boards to dilute some of the Director's responsibility in interpreting the law.

TRENDS IN RECLAMATION LEGISLATION

The predominant trend in reclamation laws since the early 1940s has been toward requiring increasing inputs to the reclamation programs to ensure eventual re-use of the mined areas. Of those states where area stripping, as opposed to contour mining, is the dominant method of coal extraction, as it is in the southwest, Indiana has the oldest reclamation law and one which has endured several modifications. Prior to Indiana's first reclamation law, some attempts at revegetation were undertaken by those companies interested in postponing or reducing the clamor for proposed legislation. This reclamation consisted of planting tree seedlings on ungraded spoils, and became the foundation for the first regulatory law, which became effective in 1941. Since this law was actually drafted in part by coal company officials, the requirements were rather meager by today's standards. The major objectives were to establish vegetation on mined areas and to reduce the amount of orphaned land in the state. This was accomplished by requiring that 101% of the acreage mined be revegetated each year. By 1951, dissatisfaction with the resulting topography resulted in an amendment to the law which required certain lands be graded so they could be traversed with farm machinery. In practice, however, nearly all the land reclaimed up to about 1963 exhibits a typical washboard, ridge-and-valley topography. In 1953, the requirement to reclaim 1% additional land was revoked because it was becoming increasingly difficult for companies to gain access to land they had not mined. This marks what is probably the only significant reduction in the stringency of the reclamation law requirements.

In 1963, the Indiana law first required top grading along public roads for aesthetic purposes. Also included in this amendment were requirements to grade areas for cultivation, cover acid forming materials, and build dams for water impoundment. This law was rewritten in 1967 to require better preplanning, more extensive grading, and to ensure that vegetative cover would be permanent. In addition, bonding and fee requirements were raised again as they had been on several previous occasions.

APPLICATION TO SOUTHWESTERN PROBLEMS

To determine the nature of the shortcomings of reclamation laws, discussions were held with those officials responsible for the planning and administration of the laws in the states of Arkansas, Indiana, Illinois, Missouri, and North Dakota, and with persons involved in drafting legislation in Arizona and elsewhere. Although there are numerous areas of disagreement, several points applicable to western lands have been brought out. In order for laws to be efficiently administered, they must be flexible. There should be mechanisms by which the law can evolve in response to changes in technology and other factors without the necessity of legislative action. This flexibility should also extend to certain reclamation requirements. It must be realized that mined land reclamation is still in the experimental stages. This is true even in the eastern states where trial and error tactics have been a necessity for more than fifty years. In some cases, however, there is a tendency for new laws to restrict experimentation by the rigidity of their requirements. Such might be the case with a time limitation on the establishment of vegetation. Although laws require reclamation to be completed within 24 months after mining is completed in Iowa, and within 12 months of the expiration of the permit in Kentucky and Kansas, such requirements in Arizona might encourage the use of rapid growth vegetation which lacked the stamina necessary for permanent endurance. A necessary requisite of effective reclamation in the southwest will be to allow ample time for successful types of vegetation to become established.

Legislating the types of vegetation that can be used to reclaim a mined area is a restrictive practice which should be avoided as much as possible. One stipulation in the federal strip mine regulations proposed in 1974, as well as the Montana bill, that has caused some controversy because of its rigidity is the requirement to revegetate with self-regenerating vegetation. The usually stated objection to this is that it could be interpreted to completely eliminate such agricultural crops as wheat and barley in the west and corn in the midwest. Similarly, a requirement to restore the vegetation that existed prior to mining, or the elusive "climax vegetation," would greatly restrict the development of many areas. The Black Mesa of Arizona is an example of an area that could benefit from a well planned mining and reclamation program. Overgrazing and poor land management practices have effectively destroyed much of the land use potential of the mesa. It would be a considerable misappropriation of resources to reestablish vegetation that is only there now because sheep won't eat it when economically superior species will be available.

Topsoil replacement requirements can also be self defeating if their wording is so restrictive as to preclude experimentation. In some cases, the spoil material may be higher in nutrients than was the surface material. It is also possible that an underlying stratum will be higher in some needed minerals than the surface matter, in which case the opportunity should exist to choose the most promising alternative. Furthermore, even in areas where the surface matter appears to be the best choice of material to be returned to the surface, stockpiling can destroy many of the micro-organisms which originally made it attractive.

   This discussion is not meant to give the impression that reclamation laws should be made less stringent. On the contrary, as research continues to discover hazards not previously realized, the laws need to be revised to ensure an increasing degree of environmental protection. However, the laws are subject to considerable variation in interpretation, and great care should be taken to assure that the language of the law clearly expresses its intent.

4. The Forest Service and Land Use Planning

Kenneth Weissenborn

THE FEDERAL GOVERNMENT HAS STEWARDSHIP of 755 million acres of land in the United States — just slightly more than one-third of the total land area. The U.S. Forest Service administers one-quarter of this public land, about 187 million acres. The Bureau of Land Management (BLM) administers another 451 million acres, most of it in the western United States. Most of the lands administered by these two agencies are managed for multiple use and are also available for mining activities.

The Forest Service planning system is based on the same ecological and environmental concepts as the systems of the other land management agencies. Under the Forest Service Organic Act, the Forest Service is charged with furnishing a continuous supply of water and timber to meet the nation's needs. Every year the eleven national forests in the southwest produce about 375 million board feet of lumber, enough to build a city the size of Tucson, Arizona, every two to three years. Nationwide, national forests produce about 11 billion board feet of timber. Thus, there is concern about activities which may take land out of production. The Multiple Use Act of 1960 and other later laws expanded the charge under the Organic Act to give legal recognition to the multiple use practices which were, in fact, already being followed.

National forest lands provide forage for the production of livestock, habitat for big game, such as deer and bighorn sheep, small game of all kinds, and a wide range of recreational opportunities including camping, fishing, hunting, and wilderness travel. Mining activities are permitted on most national forest and BLM lands, in addition to oil, gas, and geothermal leasing. Many of these activities take place on the same land areas. As can be imagined, the problem of conflicting use and demand is real. Whatever is done in one part of the system impacts favorably or unfavorably on every other part. Planning helps allocate land, resources and uses to meet the nation's needs in the best way. It deals with interrelationships of resources, uses and impacts.

The term renewable natural resources needs to be stressed. The fact that some resources are renewable opens many options to decision makers for consideration of acceptable combinations in their use. The renewable aspect also complicates planning because it means there is often a wide array of alternatives to develop and analyze. There are no "go — no go" decisions as is often the case in other resource fields.

In *The Closing Circle,* Barry Commoner (1971) summarized the laws of ecology as 1) Everything is connected to everything else; 2) Everything goes somewhere (often to places and in ways we have not anticipated); 3) Nature knows best (our alternatives cannot circumvent natural processes); and 4) There is no such thing as a free lunch (somewhere, somehow, someday we must pay the price). The Forest Service relies heavily on input data from many disciplines, but more knowledgeable outside input from subject matter specialists is needed.

There is a fairly good inventory on timber, soils, range, and wildlife for the nation. Quite a bit is known about the effects of many kinds of land treatments through experience on the ground and application of research results. In many areas, the answers are still unknown, such as how to rehabilitate some particular kinds of disturbed lands. The Forest Service Surface Environment and Mining Project is an effort to improve performance in this one area.

The basic elements in the planning system are soil, water, vegetation and climate. These are the building blocks for resource management. The way in which these elements are combined determines the suitability of land for various resource uses. For example, climate limits the range of vegetative species. Land form also determines suitability. Steep mountain slopes limit access and are not suitable in many cases for development. Bottom lands may be suited for a variety of uses — many of them on the same acre. Canyon lands, alluvial fans, and all of the many other land forms influence the planning process. In many cases, land suitability limits its use.

Availability is a second determinant in land use planning. Some lands and resources are not available for some kinds of uses. For example, timber is not available for harvest within a wilderness area. Minerals are not available in areas withdrawn from mineral entry for other uses. There may be other reasons why lands or resources may not be available, such as archaeological sites, historic sites, or areas occupied by endangered species. In planning, attempts are made to identify these areas and values and to protect or preserve them. Determination of availability is an administrative, and in the broadest sense, a political determination in that politics means allocation. What are the tradeoffs? What are the values foregone? Which lands will be made available for timber, range, recreation, wildlife, and when may they be suitable for all of these?

Manageability is another determination. Who owns the land? What are the topographic considerations — can a wilderness boundary be managed? Can the public identify delineated areas? Can we, in fact, do what we say we want to do? Much of this is an administrative decision. If poorly made, it can be undone by an uncooperative public.

Perhaps the most important determinant is need — where does this fit national, regional, or local priorities? What are the real objectives for the planning area? This determination of need is, and should be, a political process. The Forest Service role in this is one of advocacy (where other advocates are lacking). The role is also one of facilitating a collaborative decision between conflicting demand groups. In this regard, public land resource planners might do the following:

1. State the problem as clearly as possible
2. Provide the facts completely and accurately
3. Acquire, aggregate, analyze and interpret data
4. Offer reasonable alternatives, with anticipated outputs and consequences
5. Provide counsel to the limit of their knowledge.

In this scheme, planners do not make decisions. This point could be emphasized by looking at how, and by whom, decisions are made. Resource management decisions are made in these ways:

1. Legislative decisions — made in the form of law. An example of such decision is the creation of a new wilderness area through the legislative process, with all of the sanctions and constraints such action implies.
2. Judicial decisions, which affect our management actions by legal actions. An example of such a situation is the Monongahela Forest timber sale case, where a court order specifies the conditions under which timber can be sold on that particular forest.

These two types of decision processes are becoming more and more common. In spite of this, the vast majority of resource management decisions are:

3. Administrative or agency decisions, with varying degrees of public involvement in the entire planning and decision making process.

These decisions range from the details of daily operations to the longer ones of allocating lands and/or resources between conflicting demands. The objective in the Coronado National Forest, for example, is to encourage the latter process and to invite the involvement of all who feel strongly about their public lands, and to encourage participation. Many of the questions concerning land management decisions cannot be resolved by science, even though they are generally classified as scientific. Most environmental decisions transcend the scientific. Many natural science professionals are frequently unprepared to analyze the social, economic, and political elements. Some are not in touch with society's true values. The same may be true of many in the mining industry.

Land use planning is not an exact science. Much of it is an art. Perhaps with the help of planners and a concerned public, resource managers can arrive at realistic conclusions and decisions.

5. Surface-mined Land Reclamation Methods

A. S. Kennedy, R. E. Zimmerman, and R. Carter

IN 1973, THE ENERGY AND ENVIRONMENTAL SYSTEMS DIVISION at Argonne National Laboratory entered into a cooperative project with the Illinois Institute for Environmental Quality to develop plans for the reclamation of 109,000 acres of land that were strip-mined in Illinois prior to 1962 — the year the first state reclamation law was passed. The objectives of the project were to develop cost-effective alternative plans for the reclamation of these "pre-law" lands; to design and develop demonstration projects that would establish and verify such important inputs as reclamation costs and technological feasibility; to establish and develop a base of practical experience with alternative methods of reclaiming the land; and to develop and apply data-collection and management systems that would allow the large quantities of information required to be assembled and analyzed efficiently.

This comprehensive planning was a necessary first step in the development of a pre-law strip-mined land reclamation program for Illinois. It viewed the problem from a statewide planning perspective and provided a basis for establishing a comprehensive, long-range reclamation program plan. Data derived from a comprehensive survey of Illinois strip-mined lands were combined in a computerized data management system with a detailed assessment of the unit costs associated with alternative reclamation techniques. This information was supplemented by laboratory studies of revegetation processes. The feasibility of achieving alternate end-uses of reclaimed land was evaluated, and cost-effectiveness measures were developed that enabled the project team to design a step-by-step, long-range, public investment and private incentive reclamation plan for the pre-law lands. This chapter describes briefly the methods and approach used in the analyses of the pre-law land inventory and in the development of the reclamation program plan. The final recommendations and program plan are described in detail in a report entitled, *Strip Mine Land Reclamation in Illinois* (Carter, Zimmerman, and Kennedy, Illinois Institute for Environmental Quality, December 1973).

STUDY APPROACH

The approach used by the study team to analyze the surface mined land problem in Illinois involved the following steps:

[26]

1. Obtain a data base in a computer-readable format, describing current condition and utilization (if any) of pre-law lands. (A data base prepared by Southern Illinois University's Cooperative Wildlife Research Laboratory was used for this purpose.)
2. Develop a general data management system to retrieve specified information from the land condition data base as required.
3. Categorize the land in terms of current condition as either environmentally degraded or non-utilized.
4. Prepare reclamation cost estimates as functions of land condition descriptors and desired end-use categories.
5. For analysis purposes compute and display cost functions for specific current-condition/end-use combinations.

An existing commercial data management system (MARK IV), which was already operational at Argonne, was tested and deemed satisfactory for use as the retrieval program. The data base was coded in MARK IV format and stored in the system. The basic data element in the system was the affected land parcel as determined from ownership records. Characteristics of each parcel were stored as the number of acres having a particular condition. For example, one entry might describe a 300-acre parcel as having 100 acres of spoil with a slope of $5°–10°$, 100 acres of $10°–20°$, and 100 acres with a slope $20°–30°$. Each parcel was given a record of over 200 entries (or characteristics); over 1,000 parcels were stored for the Illinois analysis.

Criteria for lands whose environment was affected by mining were established and a report of such lands was prepared using the MARK IV. Data retrieved from the land condition file were also fed to computational cost routines for the purpose of calculating reclamation cost functions. The operation of the data management and analysis system is diagrammed schematically in Figure 5.1.

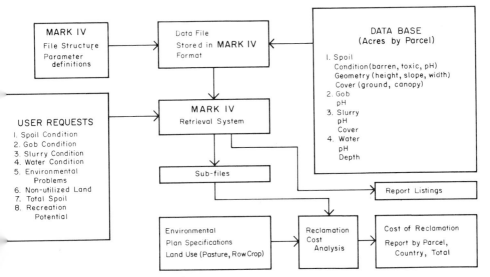

Fig. 5.1.   Environmental data system.

PRESENT LAND CONDITION

Because Illinois strip-mined areas differ from place to place due to their age, the equipment and procedures used in the mining process, soil chemistry and topographic factors, reclamation costs vary as well. Some areas have undergone a degree of natural reclamation, thanks to natural seeding and watering, while others are so contaminated that they are completely barren. Reclamation costs also vary depending on the type of land being reclaimed — spoil, water, gob, or slurry. Therefore, for the purpose of conducting a cost analysis, affected lands were categorized into environmentally affected (as defined by the criteria in Table 5.1) and non-environmentally affected lands. The breakdown of environmentally affected lands and the criteria for selecting sites with environmental problems are shown in Table 5.1.

TABLE 5.1
Categories of Land Affected by Mining

Category	Environmental Problems (Acres)	Non-Environmental Problems (Acres)	Total
SPOIL Utilized*	2,317	47,584	49,901
Non-utilized	8,707	29,077	37,784
Subtotal	11,024	76,661	87,685
WATER†	767	11,072	11,839
GOB & SLURRY‡	5,077	205	5,282
Total	16,868	87,938	104,806

*Toxic (pH < 5.0) or Ground Cover < 25% and Canopy Cover < 10%.
†Acid (pH < 5.0).
‡Toxic (pH < 5.0) Cover < 75%.

A large fraction of the affected lands has already been put back to some productive use. If these lands are not environmentally affected, they were omitted from further consideration in this analysis. Non-environmentally affected lands that are still idle were retained as part of the existing affected lands "problem." The breakdown of these lands is also shown in Table 5.1.

LAND USE ALTERNATIVES

Reclamation costs also vary as a function of desired end use. Since a majority of surface mining in Illinois has been done in rural areas, it is reasonable to expect that most reclaimed lands will be returned to agricultural production — forest, pasture, or cropland. It was assumed that isolated instances where affected lands may be particularly suited for recreation or residential development would not significantly alter the results of the analysis. For purposes of this study, two types of pastureland and two types of cropland were designated as possibilities for agricultural end use:

1. Strike-off pasture: Spoil land where the tops of the spoil piles have been graded, where acid conditions (if any) have been neutralized, and to which fertilizer, mulch, and seed have been applied.
2. General grade pasture: Spoil land that has been graded to rolling contours (slope $\leq 10°$), on which acid conditions (if any) have been neutralized, and to which fertilizer, mulch, and seed have been applied.
3. Neutralized row crop: Spoils have been graded to fine tolerance (slope $\leq 5°$), acid conditions (if any) have been neutralized, and fertilizer, mulch, and seed have been applied.
4. Topsoil row crop: Spoils have been graded to fine tolerance, two feet of topsoil have been put in place, and fertilizer, mulch, and seed have been applied.

RECLAMATION ALTERNATIVES — COST ANALYSIS

Each of the existing land condition categories described above was evaluated in terms of the total expenditures required to reclaim pre-law lands under each of the desired land-use categories (less acquisition costs). On the basis of data items from the master file that describe the current condition of each parcel, a set of computer programs was developed that calculates reclamation costs as a function of these conditions. Costs that were considered include vegetation removal, recontouring, soil neutralization, covering, revegetation, and water treatment. These cost calculations are described briefly below.

The cost of shrub and small tree removal is small when compared to tree removal and earthmoving. Since this growth can more or less be worked into the spoils as part of the recontouring operation, only a slight additional cost will result from the particularly large "small" tree that requires special handling. The costs were assumed to be:

% Cover Small Trees and Shrubs	Cost of Removal
More than 75%	$60/acre
50% to 75%	$45/acre
25% to 50%	$30/acre
Less than 25%	None

Tree removal is much more expensive and is assumed to vary with the size of trees as well as cover density as shown in Table 5.2. It is assumed

TABLE 5.2
Cost of Tree Removal in Dollars per Acre

Tree Height (ft.)	$/Acre in Terms of % of Tree Cover		
	1%–10%	10%–75%	75%–100%
1–20 + Heterogeneous	$100	$300	$ 500
21–40	200	400	700
41–60	250	600	1,000
Greater than 60	300	800	1,000

that vegetation removal for row crop land is 25% higher than these costs to account for root removal. Costs of vegetation removal for strike-off pasture are assumed to be 50% of these costs. Estimates of acres of cover by type and density for each parcel were obtained from the mined land data bank.

Recontouring costs were assumed to depend on the geometry, stoniness, and aggregation of existing spoil levels as well as the desired finished contour. While the primary considerations in the cost of earth-moving are volume and distance, published cost figures for earth-moving by the manufacturing industry often do not include cost items other than direct operating cost, and are for optimum working conditions with the best available operators. Thus these figures usually are much lower than those actually encountered in projects. The figures presented in Figure 5.2 and Table 5.3 were the result of consultation with manufacturing and mining-industry people, as well as consulting engineers, and represent a "best estimate" for conditions likely to be encountered.

The first conclusion to be drawn from this list is that either someone is wrong or else the figures have a different cost basis. In effect, both are right. The earth-moving equipment industry costs are based upon ideal conditions

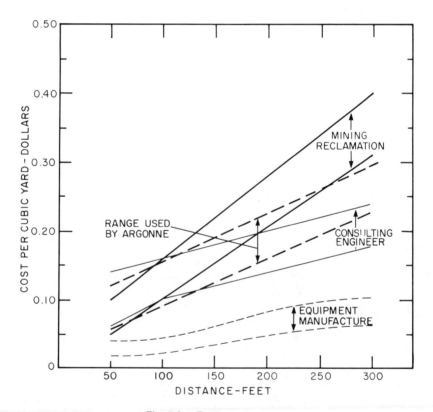

Fig. 5.2.   Earth moving costs.

TABLE 5.3
Range of Cost for Earthmoving (per cubic yard)

Source	Effort	Distance Moved (ft.)					
		50	100	150	200	250	300
Consulting	Easy	6¢	10¢	12¢	14¢	16¢	18¢
Engineer	Difficult	14¢	16¢	18¢	20¢	22¢	24¢
Mining	Easy	5¢	10¢	—	—	—	30¢
Industry	Difficult	10¢	15¢	—	—	—	40¢
Equipment Manufacturing	Easy	2¢	2.3¢	3.4¢	4.6¢	6.0¢	6.8¢
Industry	Difficult	4¢	4.6¢	6.3¢	8.2¢	9.8¢	10.0¢

and do not include all cost items, while the other figures are based upon actual operating experience.

To take account of these differences, an empirical cost formulation was developed as follows. First, it was assumed that some effort will go into the initial cut, independent of the material characteristics, and that this effort is about equal to that of a forty- to fifty-foot push. The effort involved in the set-up in the more difficult material will be about twice as high. Second, the very stony material will call for an effort nearly equal to an easy cut, and the extremely stony will equal the more difficult cut. Third, based on manufacturers' data and formulas, the distance factor will be equal to the distance the spoil is moved divided by 9.6. The resulting point system that best fit the data was:

Condition	Points
Distance	Distance in feet/9.6
Aggregation	
Compact	7
Loose	4
Stoniness	
Extremely	7
Very	3
Moderately	0
None	0

In tabulating the unit cost, a base cost of 8¢ per cu. yd. for an 80-ft. push in easy material was assumed. The points for such a condition are:

$$\text{Distance} = 80/9.6 = 8.3$$
$$\text{Aggregation} = 4$$
$$\text{Stoniness} = 4$$
$$\overline{\text{Total} \qquad 12.3}$$

Therefore, the cost per yd. (in cents) of material was computed as:

$$\text{¢/yd.} = \left(\frac{\text{aggregation points} + \text{stoniness points} + \text{distance points}}{12.3} \right) \times 8$$

The volume of material in cubic yards to be moved was determined from geometric considerations — initial spoil slope, height, and strike-off width to final contour — and the total recontouring cost was then easily obtained. Geometry parameters as well as aggregation and stoniness characteristics were obtained from the mined-land data bank.

Soil neutralization is required for acid spoils and costs will vary depending on the degree of acidity present. For extremely acid spoil (pH less than 3), it is assumed that 2 ft. of topsoil will be required to re-establish vegetation at an in-place cost of $1.50/cu. yd. For moderately acid spoils (pH between 3 and 4) it is assumed that 10 tons of limestone will be required per acre at an installed cost of $10/ton, while slightly acid spoil (pH between 4 and 5) will require only 5 tons/acre. Installed fertilizer, mulch and seed costs are assumed to average $180 per acre.

Water treatment costs vary greatly depending on the type of treatment equipment selected. An efficient method is to react lime with the contaminated water within the pond itself rather than in a treatment facility. A hydroseeder along with pumps can be used to do this if hydrated lime is used as the base. In addition to reducing the time required, the capital equipment can be used for other projects totally unrelated to mine reclamation.

The water treatment costs used in this analysis were based on the assumption that a hydroseeder and pumps would be used to react hydrated lime in the pond itself. The data available from the data bank for these calculations were:

1) the area of the pond
2) the depth — less than or more than five feet
3) the pH

This limited information did not lend itself to an accurate cost analysis, because neither the makeup of the water nor the volume was known. However, several assumptions were made to estimate these two items. The assumed makeup and hydrated lime required for water treatment are:

pH	Acidity	Iron	Sulfate	$Ca(OH)_2$ Tons/Acre ft.
3.0	5,000 ppm	2,000 ppm	500 ppm	10.0
3.0–4.5	500 ppm	200 ppm	50 ppm	1.0
4.6–5.0	100 ppm	100 ppm	0	0.33

The composition of mine drainage is extremely variable — water with a pH less than 3 may have very little iron or up to 15,000 ppm. However, for the purpose intended it was felt that the numbers listed above would provide some basis for the cost estimate. Volumes were calculated by assuming that the depth of the water over five feet is two-thirds the height of the overburden. This is also a general figure. In many areas within the state the water is very low or very high; however, as with the makeup assumptions, some numbers were needed.

Using these assumptions, the volume of lime entering the pond was calculated. The required time was found by assuming an application rate of five tons of lime per hour. Pumping and hydromulcher costs, including labor, were assumed to be thirty dollars per hour each. In addition, a mobilization cost of $200 for the first 200 tons of lime, and $100 for each additional 200 tons, was used to calculate the total cost of adding lime to the water.

Finally, mining wastes (gob and slurry) are expensive to reclaim and are of little value after revegetation. Gob acre reclamation costs are given in Table 5.4.

Slurry area reclamation costs were assumed to vary with pH and vegetation cover as shown in Table 5.5.

The results of the cost calculations applied to pre-law lands in Illinois are displayed in Table 5.6. Several conclusions are immediately evident from this table:

1. Gob and slurry areas are a significant part of the environmentally affected lands problem and a significant part of the cost.
2. Virtually none of the disturbed lands can be economically returned to row crop land.
3. Some small percentage of non-utilized lands may be economically returned to good pastureland, but the more likely end use is strike-off pasture.
4. It may be possible to economically return some environmentally affected spoil lands to good pasture.

TABLE 5.4

Costs of Gob and Slurry Reclamation (in dollars per acre)

Percentage of Vegetation Cover Before Reclamation	Surface Covered Before Reclamation	Material Not Covered Before Reclamation
< 10	2,600	4,200
10–25	1,800	4,200
25–50	1,000	4,200
50–75	500	3,100
75–100	0	2,400

TABLE 5.5

Assumed Costs of Slurry Area Reclamation (in dollars per acre)

Percentage of Vegetation Cover Before Reclamation	Slurry pH	
	< 4.5	> 4.5
< 10	3,100	1,000
10–25	3,100	700
25–50	1,500	500
50–75	1,000	300
75–100	300	0

TABLE 5.6
Costs of Reclaiming Mined Land in Illinois

Alternative Set	Type of Affected Land and Condition	Reclamation and End Use	Current Total Cost (M$)	Total Acres	Average Cost ($/Acre)
Environmental	a. Gob and Slurry	a. Vegetation	13.6	5,077	2,674
	b1. Spoil b2. Gob, Slurry and Water	b1. Pasture b2. Vegetation	29.7	16,868	1,759
	c1. Spoil c2. Gob, Slurry and Water	c1. Row Crop (neutralization only) c2. Vegetation	49.2	16,868	2,916
	d1. Spoil d2. Gob, Slurry and Water	d1. Row Crop (with 2 ft. topsoil) d2. Vegetation	101.1	16,868	5,993
Non-Environmental Non-Utilized	a. Spoil	a. Strike Off Pasture	13.0	29,077	446
	b. Spoil	b. Good Pasture	21.2	29,077	731
	c. Spoil	c. Row Crop (neutralization only)	66.5	29,077	2,456
	d. Spoil	d. Row Crop (with 2 ft. topsoil)	221.3	29,077	7,611
Non-Environmental Utilized	a. Spoil	a. Strike Off Pasture	30.7	76,661	401
	b. Spoil	b. Good Pasture	50.5	76,661	658
	c. Spoil	c. Row Crop (neutralization only)	164.6	76,661	2,147
	d. Spoil	d. Row Crop (with 2 ft. topsoil)	572.7	76,661	7,471

COST FUNCTION ANALYSIS

In order to evaluate the effectiveness of alternative financing and sub-sidization policies on the timing and extent of reclamation activity, cost functions were developed that displayed the acres that could be reclaimed for a given average acre expenditure. These functions were generated by applying the above cost analyses to each ownership parcel and then ranking these parcels from the least to the highest average reclamation cost per acre. These curves then revealed the acreage that could be expected to be reclaimed under various market assumptions, subsidy policies, or state reclamation programs. For illustration purposes, only the cost function analysis for envi-ronmentally affected lands will be considered. The complete analysis appears in *Strip Mine Reclamation in Illinois,* referred to earlier in this chapter.

Before acid spoils can be returned to pastureland, they must be graded to rolling contour and neutralized. Thus, the only feasible and potentially economical agricultural use for these lands is low to medium grade ($300/ acre) pastureland. Some environmentally affected sites also contain acid water, which must be neutralized, as well as gob and slurry areas, which must be covered and vegetated. Because it was originally felt that areas with potential surface water problems should be given a high priority in the reclamation program, cost functions for environmentally affected lands were calculated separately for sites with acid surface water and those without acid surface water. The unit-cost functions for each of these site categories are shown in Figs. 5.3 and 5.4, respectively. It can be concluded from Fig. 5.3 that virtually no sites with acid water will be reclaimed by the private sector, assuming a current market value for pasture of $300/acre. As illustrated in Fig. 5.4, a few sites without acid water have the potential for being reclaimed. However, for this latter land category, it is also evident that the unit cost curve is extremely elastic in the area from $400 per acre to $900 per acre, indicating a strong latent market potential for private reclamation efforts with a minimal incentive program. This is particularly true if the price of pastureland can be expected to rise in the next few years. Thus, if the neutralization costs borne by the private owner could be subsidized, a substantial incentive toward reclamation might result in the private sector. This form of subsidy is particularly efficient, since there would be no incentive to encourage the reclaimer to over-neutralize.

Unit-cost curves were generated, assuming that such a subsidy would be in effect, and the results are also displayed in Figs. 5.3 and 5.4. Some benefit is derived from this subsidy for lands without acid water, but a dramatic improvement can be noted for lands with acid waters. It can be concluded that if neutralization costs are subsidized, at least at the lower ends of each cost curve, a strong market potential for private reclamation exists for lands both with and without acid water. Therefore, as long as water runoff is contained on site, both types of land can be treated equally and an effective rate of reclamation can be attained with a combination of a neutralization subsidy and increased land prices relative to reclamation costs.

It remains to consider the high ends of each cost function. It will be noted that each cost curve displays a "kink" at approximately $1,000 per acre, at which the slope of the cost curve increases dramatically. In each case, these "tails" are due to sites containing gob and slurry areas that are

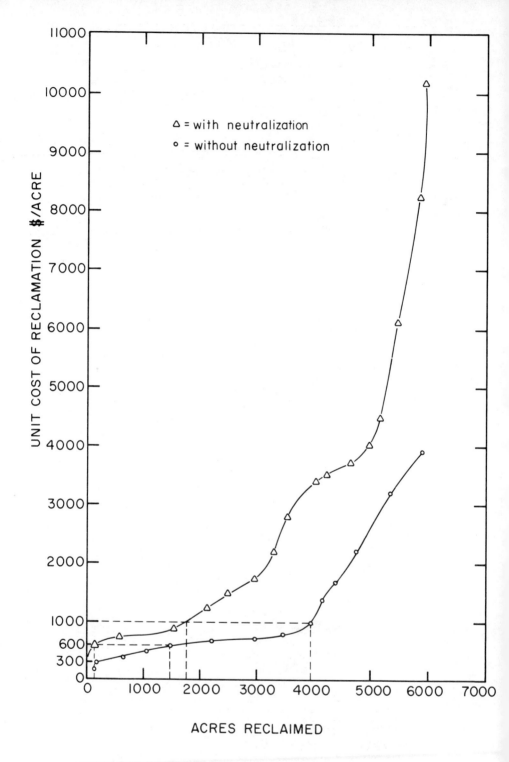

Fig. 5.3.   Environmental lands with affected surface water.

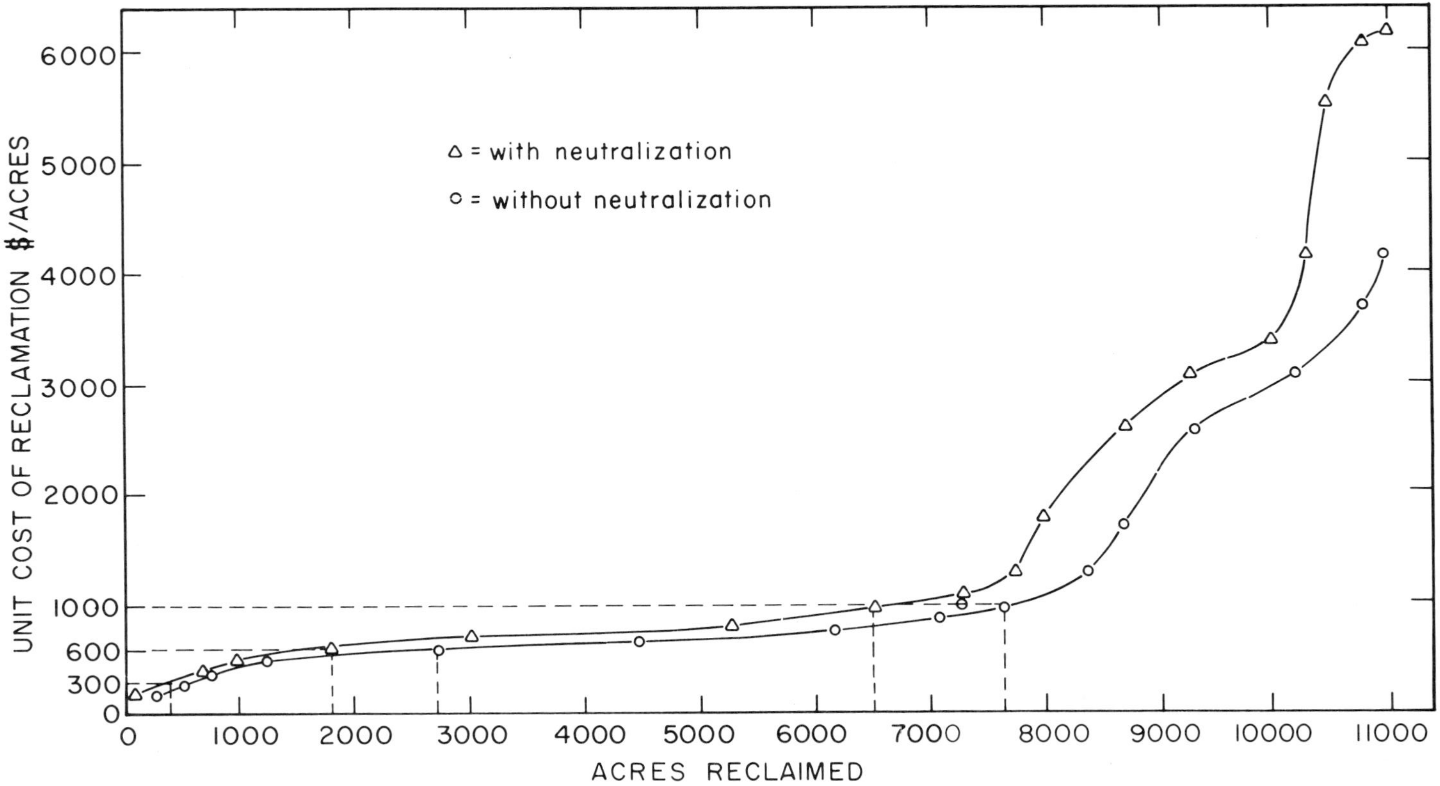

Fig. 5.4. Environmentally affected lands without affected surface water.

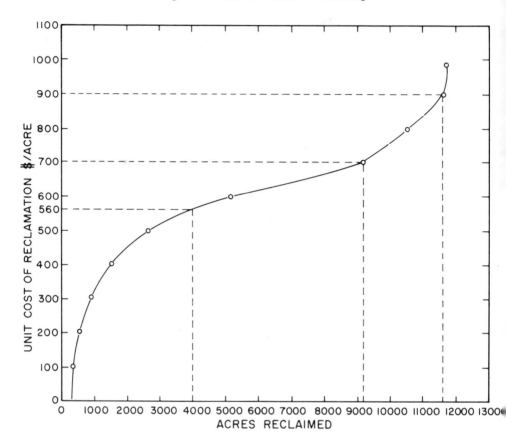

Fig. 5.5.   Environmental lands without gob and slurry, neutralization not included.

extremely costly to reclaim and which dramatically increase the average cost per acre for the entire site. If reclamation of these areas were also subsidized or if they were reclaimed by the state, in addition to the neutralization subsidy, a strong market potential would exist to reclaim most of the environmentally affected lands, provided that land prices continue to increase in the next five to ten years relative to reclamation costs, as shown in Fig. 5.5.

This analysis leads to the conclusion that an effective and efficient reclamation program for environmentally affected lands need not include the purchase or condemnation of these private properties, except possibly for those sites that cause off-site damages and for which the owner refuses to take voluntary corrective action. Rather, direct subsidies, in the form of neutralization-cost reimbursement and gob and slurry covering and vegetation, would, in a reasonable period of time (10–12 years), be sufficient to reclaim most of these lands, with the private sector bearing the costs of tree removal, earthmoving, fertilizing, seeding, and project management.

CONCLUSIONS

Some of the conclusions that were derived from this analysis are listed below:

1. Although 109,000 acres of pre-law lands exist in the state of Illinois, the reclamation problems associated with these lands are not as extensive as might be expected. The following breakdown summarizes the current condition and utilization of pre-law lands:

Spoil lands currently in use (acres)	49,901
Environmentally affected lands	16,868
a. Gob and Slurry	5,077
b. Acid Water	767
c. Acid Spoil	11,024
Non-utilized spoil lands	37,784

2. Gob and slurry areas are extremely expensive (average $2,674/ acre) to reclaim, and may have only marginal value after 10 to 15 years. It is consequently unlikely that they will be reclaimed by private interests.

3. Acid water can be neutralized, but the real cause of this condition — runoff from acid spoil, gob, and slurry — must be treated simultaneously in order to correct the condition permanently.

4. If acid spoil is to be neutralized, it must be graded to rolling contour. The latter can be done at reasonable prices (average $585/acre).

5. Non-utilized, non-acid spoil land can be returned to low-grade pastureland in a relatively inexpensive manner (average $444/acre) by use of the so-called "strike-off" grading technique.

6. Returning pre-law lands to acceptable row-crop lands with topsoil replacement techniques is very expensive (average $2,287/acre without topsoil replacement for non-utilized lands).

7. The potential exists in the private sector to reclaim a significant fraction of the remaining idle affected lands (assuming a steadily increasing difference between land values and reclamation costs) if the state would subsidize spoil and water neutralization and gob and slurry stabilization costs. This is an efficient and least costly solution for the state since it would not have to acquire, reclaim, and resell the land.

IMPLICATIONS FOR SURFACE MINING AND LAND RECLAMATION IN ILLINOIS

Illinois is rich in strippable coal reserves. The energy shortage and the resurgence of coal as a prime energy source in the near- and mid-term time horizons means that, in all probability, strip mining will continue in Illinois. The state is also rich in prime farmland, much of which lies above coal fields. The shortage of grains and row-crops, as well as meat, has increased the value of these lands for agricultural use. Thus, a major conflict has arisen that must be resolved.

This analysis was retrospective in that it involved an examination of the practices for purposes of assessing what can be done to mined land after the fact, but some of the lessons learned are applicable to decisions in the 1970s. Among the more important are:

1. Mining wastes are toxic and expensive to reclaim, unless their disposal is properly planned before mining commences.
2. Row-crop land cannot be restored economically from mined land. Until methods of saving and restoring topsoil during mining become feasible and economical, prime row-crop land should be mined only after poorer quality land has been mined.
3. Existing pastureland and marginal row-crop land can be restored economically to pastureland after mining. Efficient mining practice may further reduce these costs.
4. Neutralization of acid spoils causing runoff and surface water pollution can be expensive and costs for correcting these conditions should be borne by the mine operator, coincident with the mining operation. Again, improved mining techniques will help to reduce these costs.
5. Marginal agricultural lands can be converted into alternative recreational and open space facilities or natural preserves after mining if the demand exists. The acquisition of these areas should be planned before mining, so that the mining operation can be geared to minimize earthmoving costs after mining.

PART II

CONSTRAINTS IN
DISTURBED LAND RECLAMATION

THE EXTENT TO WHICH land disturbed by surface mining can be reclaimed depends upon the economic, social, technological, geological and environmental constraints which bound the system. As an academic exercise, if the constraints on a system are given and an objective function is defined, an optimum solution follows. The problem in the real world resolves in defining a generally accepted objective. But invariably the objective is confounded by two opposing philosophies. One philosophy maintains that it is essential to occupy the land and make it productive for man's needs and wants and for his economic gain. The other considers it an obligation to set aside or restrict human activity on what remains of the available land. The challenge to meet material needs guides the first, and the fear of consequences, the other. Both are based on the assumption that the human population and its wants and needs will continue to increase. They each have in common the belief that their objective will provide the greatest benefit to future generations.

The mineral industry is in a period of rapid change and apparently recognizes its social obligation to produce more than just material wealth. The greatest fear is that public policy may crystallize prematurely on an objective without considering all of the constraints and their ramifications in the total system.

Thomas J. O'Neil provides an introduction to mining operations and the considerations in developing mine lands. He emphasizes that environmental programs should be subjected to the same cost/benefit analyses as other capital investments. The economics of mining and reclamation are further elucidated by George F. Leaming, who also suggests that a broader concept of land conversion be adopted rather than the more restrictive ideas of reclamation, restoration, and rehabilitation. He presents several case studies of economically viable mine land conversions that illustrate this concept.

James K. Richardson points up the dilemma of the engineer faced with the necessity to meet stringent regulatory standards with a yet undeveloped technology. Similarly, John M. Guilbert discusses technological and geological constraints, particularly as applied to minerals exploration in planning and regulating mining operations.

The social considerations of mining and reclamation are discussed by Michael D. Bradley, who suggests that future social turbulence can be met by residuals management, ecological and landscape planning, and long-range national resource planning; but he cautions that the road may well be long and difficult.

6. Operating Considerations

Thomas J. O'Neil

MINE OPERATORS ARE ATTEMPTING TO COMPLY with a bewildering and conflicting assortment of pressures — legislation, coercion and not-so-gentle persuasion — under the general heading of social welfare. For example, a serious move developed in the 1970s to ban surface mining and permit only underground mining in the future. This would definitely reduce the amount of disturbed land created, but would cause an inevitable deterioration in mine safety, to say nothing of the economic penalty to the consumer. Subsidence from underground mines may also become illegal, although to comply with this apparently reasonable request, operators will need to leave larger support pillars, thereby reducing mineral recovery. With current technology, an underground room-and-pillar coal mine may recover only 50% of the coal; with larger pillars this could easily drop to 35%. In a time of energy scarcity, this loss of coal resources seems to be contrary to the national interest. The advantages of surface mining are simply too great to consider prohibition as a reasonable alternative. That is why rehabilitation of mine lands is such a vitally important topic. It is nonetheless a topic that should be subjected to the same type of rigorous benefit/cost analysis as other capital investments. Studies of this type must have engineering input from the mining community to yield useful results.

In assessing the redevelopment potential of a particular surface mining area, a number of factors are relevant. Some of these are:

1. Magnitude and geometry of the disturbed lands.
2. Chemical properties of the mine waste materials.
3. Physical properties of the mine waste materials, such as structural stability — both static and dynamic, particle size distribution, permeability and porosity.
4. Economic analysis of the mine operation and of the redevelopment plans.
5. Climatic factors.

Although the miner may have little to add to a discussion of climatic factors, he obviously is in a position to contribute a great deal with respect to the other four items. In general, mine land redevelopment can be considered as roughly equal parts of analysis and design — analysis of the

[43]

existing waste areas and how they can be modified (mining engineers and geological engineers).

It is necessary to focus on some of the capabilities and knowledge mining engineers bring to the discussion of mine reclamation alternatives. The presentation is concerned more with breadth than depth, as the details of specific mine operations vary to such an extent that generalization would be futile. One aspect that should be discussed is the operational differences between strip coal mining, where most reclamation effort to date has been invested, and open pit copper mining, which is so important in southern Arizona.

STRIP COAL MINING

Bituminous coal operations have accounted for fully 40% of all lands disturbed by mining operations in the United States through 1971, mostly in strip mines (Fig. 6.1). Furthermore, individual strip coal mines are typically high production installations covering considerable area, and the industry has been heavily concentrated in six states which totaled 83% of coal mine lands. In comparison, although the industrial minerals industry (e.g., sand, gravel, stone, and so forth) accounts for nearly another 40% of mine lands, the average installations are small and the mines are distributed throughout the entire fifty states. Thus, coal strip mining is highly visible, and the bulk of reclamation activity has been directed toward this branch of the mining industry.

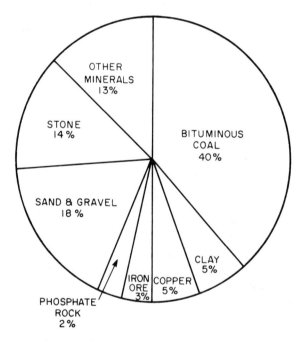

Fig. 6.1.   Land utilized by mining, by commodity. 1930–71 and 1971. (from Paone, et al. 1974)

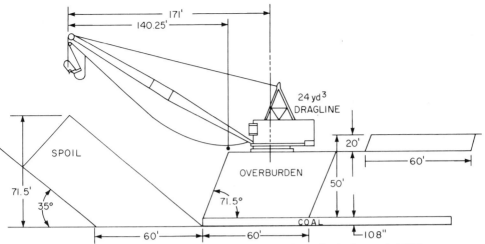

Fig. 6.2.   Typical Bench Cut Dragline Pit Layout (after Stefanko, et al. 1973).

Coal is a stratified carbonaceous rock of widely varying composition. The coal series begins with peat and ends with graphite, with the most important members being lignite, bituminous coal, and anthracite. Being confined to particular strata, coal provides a classic definition of structural ore limits. That is, there is seldom any difficulty in determining what should be mined and what shouldn't be in a strip coal mine. The coal ends abruptly at the overlying and underlying strata which are normally of a totally different composition. The boundaries are also visually distinct. Thus, in Fig. 6.2, whereas production economics and product prices will define the horizontal limits of mining — particularly where the topographical relief is moderate to severe — it is clearly obvious how deep to mine. In multiple seam mining, the decision becomes somewhat more complicated, as shown in Fig. 6.3, but in most strip coal operations — unlike metal mines — it seldom is economically attractive to reactivate a closed mine.

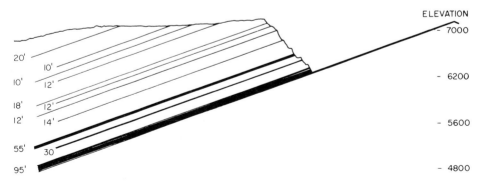

Fig. 6.3.   Proposed Cross Section for Multiple Seam Strip Mine in Wyoming. (from Stefanko, et al. 1973)

Surface topography determines whether contour mining or furrow mining is to be followed. In hilly areas contour mining is practiced where the coal is mined from the cropline into the hill until the economic stripping ratio is exceeded. The resulting excavation is a long sinuous trench that circumscribes the hill much as an elevation contour. Furrow mining, commonly known simply as strip mining, is employed in flat areas and will usually extend a considerable distance in all horizontal directions. A large strip mine will ultimately disturb thousands of acres, although at any one time mining operations may be restricted to a relatively small area. During a year's time a moderately large coal stripping operation might mine 500 acres, but once mined the 500 acres are seldom needed for subsequent production operations. Thus, from the standpoint of efficient mining, only enough active mining land is required to maintain an adequate buffer against interrupted operations. Since it is critically important that electrical generating plants not run out of coal, a fairly large mine inventory of coal must be carried in situ. Consequently, as a rough estimate, active mining requires the areal equivalent of one year's production of coal. If coal blending is required from two or more mining areas, this land may not be concentrated in one area. Further, if the excavation is used to bury waste material such as fly ash, additional open land will also be needed. However, the main point is that whereas a strip coal mine will cumulatively disturb a large amount of land during the life of the operation, a relatively small amount needs to be disturbed at any one time. A given plot of coal land may, therefore, be in a disturbed state for only a year under ideal conditions. This is in marked contrast to a major open pit porphyry copper mine which may work the same lands for as long as 75 years.

Coal strip mining methods involve some of the largest — and most costly — land-based machines ever constructed by man. Similar to other capital intensive industries, strip coal mining relies on continuous 24-hour, 7-day operation to remain profitable. When one large stripping shovel can cost more than $13 million, operators can ill-afford to let the machine remain idle. Thus, although strip mining methods are technically fairly straightforward, mine planning and scheduling are vitally important and quite complex.

Both overburden and ore (coal) are relatively light and easily mined in comparison to other types of deposits. In fact, in some mines it is unnecessary to drill and blast either the overburden or the coal as the materials are fractured well enough naturally to permit efficient mining. Breakage costs are minor in most strip coal operations; overburden stripping accounts for most of the cost of mining.

Mammoth stripping shovels and draglines can produce extremely low unit costs under favorable operating conditions. Unfortunately, these required conditions are quite restrictive and inflexible. Thick, uniform, easy digging overburden is essential as these machines are economically sound only when they are actually excavating, not when they are moving, positioning, or grappling with blocky ground. One requirement that discourages the use of the large strippers is overburden segregation as required for post-mining reclamation. There is an important trade-off involved here as each additional soil or rock type segregated requires an additional step in the extractive process.

This, in turn, reduces the maximum economic size of equipment used, causes a greater mining area to be held open at any one time, and reduces capital and labor productivity. The inevitable result is higher prices to the consumer. This implies that the maximum social good can be attained by minimizing the degree of segregation necessary, consistent with redevelopment objectives. As a general rule, the public (through higher prices caused by higher production costs) should not be required to pay for improving the productivity of the land over that which existed prior to mining.

Finally, with thin seam coal having fairly thick overburden such as encountered in much of the midwest, the disturbed lands can be pretty much restored as they were prior to mining. Providing that the public is willing to pay the bill, the lands can be reclaimed to the point where there is no apparent difference between virgin lands and reclaimed lands. With the thick seam western coals, sometimes there is not enough spoil to return the lands to their original elevation, but even here the post-mining appearance and productivity of the land need not be seriously impaired. As discussed below this is simply not the case with most open pit copper mines.

OPEN PIT COPPER MINING

In many respects open pit copper mining and strip coal mining represent the extremes of the spectrum of surface mining operations. Coal and copper mining and processing methods are vastly different due not only to the characteristics of the commodity produced but also to the type of ore deposit.

Most primary domestic copper is mined from a group of loosely-defined deposits called the porphyry coppers. For the purpose of this discussion a porphyry copper is a massive, low grade deposit of magmatic origin where the economic mineralization is thoroughly disseminated in a rock mass in three dimensions. This last point is of crucial significance from a land rehabilitation standpoint. Because a porphyry copper has a deep-seated igneous source, the vertical dimension of such deposits is usually large. This gives rise to the classic cone-shaped open pit copper mine where the mine becomes progressively deeper with time and develops laterally mainly to provide stable slopes on the sides of the pit. This presents two formidable problems to the land reclaimer: (1) open pit porphyry copper mines stay in operation for a long period — usually more than 20 years and occasionally more than 75 years; and (2) the excavation doesn't move, but simply keeps getting larger and deeper. There are about 18 open pit copper operations in the state of Arizona, some of which have been producing for over 40 years. Only about four mines of any significance in the state have ever stopped producing — having apparently been "worked out." One of these mines — Cities Service's Pinto Valley Mine — recently reopened and will probably be in operation for at least another 25 years, a development that could well be repeated at the other dormant operations.

Finally, although full reclamation of strip coal mines is an accepted operating constraint, restoration of an open pit copper mine is a mind-boggling problem. Backfilling these gigantic holes is not a feasible suggestion. Phelps Dodge Corporation recently suspended operation in its Lavender Pit in Bisbee, Arizona, after having mined over 364 million tons of material.

It has been estimated that it would cost approximately 100 million dollars to backfill the mine — a colossal waste of financial and energy resources that would prevent reentering the mine if copper prices rise high enough to again sustain a profitable operation. As Zahary, et al. (1973) note, "The only practical approach to reclamation (of large open pit mines) is to find an alternative use for the site which incorporates the permanent features." With copper pits, this is not easy, although with clearly exhausted ore deposits, pump storage power generation and urban refuse disposal might be considered.

Another important distinction between coal deposits and porphyry coppers is that while coal observes definite structural (stratigraphical) limits to the ore, a porphyry copper provides an excellent example of assay limits because copper values in a porphyry usually decline gradually from a higher grade core to background values over a considerable distance. Ore is defined almost entirely by current production costs and prevailing market prices. The ore reserve is, therefore, a dynamic concept. If the price of copper rises (or production costs drop) lower grade mineral can be profitably mined and the ore reserve increases. Obviously, the process is reversible in the face of falling prices or rising costs. Every consumer is well aware that other prices rose 12% to 15% in 1974, so with both declining prices and rising costs copper producers were also experiencing a severe profit squeeze. The net effect of the cost-price squeeze is to lower ore reserves, "ore" being defined as mineral which can be economically recovered.

A corollary to this characteristic has to do with final pit limits. A common question to copper mines is, "How big will the pit be when you're finished?" When the questioner is told that it's impossible to say, he is usually skeptical. Surely a competent miner must know something as basic as the ultimate size of his excavation. However, due to assay limits a final pit size and shape cannot be determined unless sales price and production costs are specified first — and there obviously are an infinite number of combinations of future costs and prices. Fig. 6.4 is a cross-sectional drawing of a hypothetical, but fairly typical, porphyry copper deposit. Final pit limits are plotted for different cut-off grades to illustrate the above concept. In recent years rising copper prices have permitted the mining of lower grade materials which has led to expanding final pit limits for nearly all mines as shown in the drawing. Time and again what had been mineralized waste suddenly became ore through improved technology, higher prices, or both. Sometimes the magnitude of these changes was unanticipated by the mine operators who proceeded to construct surface facilities over what appeared to be waste at the time, but later could be mined at a profit. There is a saying in mining that the best place to explore for ore is beneath the mill, which is, of course, remotely related to Murphy's Law.

Every open pit copper mine has experienced the phenomenon discussed above: changing economics creates or destroys ore so that it is rarely possible to say with any degree of certainty how long the mine will be in production or what the final size and configuration of the pit will be.

Fig. 6.5 is a simplified flow sheet for conventional copper production through the smelting stage. It shows that the mining of one ton of ore from a low grade porphyry copper open pit yields approximately 10 lbs. of metal,

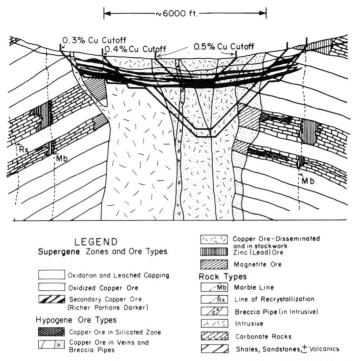

LEGEND
Supergene Zones and Ore Types

☐ Oxidation and Leached Capping
☐ Oxidized Copper Ore
▨ Secondary Copper Ore
(Richer Portions Darker)

Hypogene Ore Types

▨ Copper Ore in Silicated Zone
▨ Copper Ore in Veins and
Breccia Pipes

▨ Copper Ore-Disseminated
and in stockwork
▨ Zinc (Lead) Ore
▨ Magnetite Ore

Rock Types

☐ Mb Marble Line
☐ Rx Line of Recrystallization
☐ Breccia Pipe (in Intrusive)
☐ Intrusive
▨ Carbonate Rocks
▨ Shales, Sandstones,± Volcanics

Fig. 6.4. Generalized Porphyry Copper Deposit Showing Pit Limits for Various Cut-off Grades. (after Jerome 1966)

but generates nearly 8,000 lbs. of waste products. Considering that domestic primary copper production in 1973 totaled nearly 3.45 billion lbs., a truly staggering amount of waste material was generated. There are several types of waste products, however, each with its unique characteristics and each posing unique problems to the redeveloper.

Mine overburden. Overburden waste can consist of every type of material from loose, unconsolidated soil and sands to barren, massive, competent rock to shattered and mineralized, but economically submarginal rock. Operationally, there is no reason to segregate various types of barren materials. Whether these are soils (such as desert alluvium) or rock is irrelevant. From the standpoint of rehabilitation, however, it makes a great deal of difference. Soils generally support vegetation much better than rock so that a benefit/cost analysis of soil segregation is usually advisable.

Mineralized waste rock presents a different problem. Just as changing economics can make ore out of waste (or vice versa) in the mine, waste dumps having copper values may very well be tomorrow's ore. Historically, many mining fortunes have been made by reworking old waste dumps, and most operations today separate low grade materials for possible future processing. One Arizona producer maintains several such dumps according

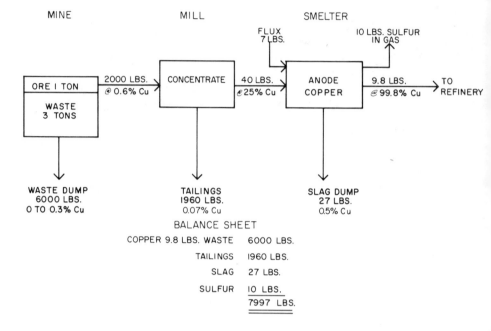

Fig. 6.5.   Flow sheet for a Typical Open Pit Porphyry Copper Operation.

to grade, with the highest grade material stockpiled closest to the crusher. Clearly, it would be foolish to adopt costly and/or permanent redevelopment schemes on such dumps, thereby tying up an important national resource. Unfortunately, the same degree of uncertainty as to time and extent of utilization surrounds this material as it does in-place rock.

Under certain conditions copper-bearing waste dumps can be leached with sulfuric acid solutions thereby providing extremely low-cost, low-energy consumption copper. Dump leaching is a relatively slow process typically taking several years to complete.

Tailings. Tailings are a finely-ground rock slurry from which most of the copper has been removed. The copper separation process requires a basic environment so that the pH of tailings initially ranges between 10 and 11. If pyrite is a major constituent of the tailings, however, its subsequent oxidation will lower the pH greatly, as low as about 3 in some cases. The revegetation problems are obvious. Tailings have excessive amounts of heavy metals and dissolved salts and are void of organic matter. Furthermore, the extremely fine size of tailings (frequently 50 percent minus 200 mesh) and the large amount of contained slimes give copper tailings a low permeability. Tailings hold their moisture for years, and dewatering is difficult. Consequently tailings dams may liquefy and fail under strong earthquake loading, and poor consolidation makes such impondments potentially hazardous for any construction activity. Tailings are simultaneously the least attractive and most vegetatively inhospitable mine waste material. Finally, with an assay

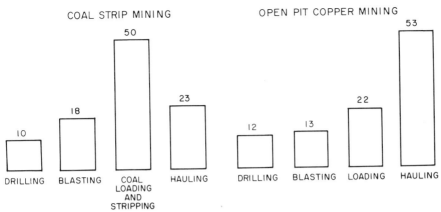

Fig. 6.6. Average Distribution of Operating Costs. (General Mine Costs are Allocated to Unit Operations.)

of about 0.1% copper, tailings also represent a potential future resource which it may not be prudent to destroy with permanent redevelopment.

Slag. As solidified, molten rock, slag does not offer much opportunity for redevelopment. Slag has some industrial uses, such as railroad ballast. The relatively small size of slag dumps indicates that reclamation efforts should probably be minimal.

PRODUCTION COST CONSIDERATIONS

In considering redevelopment schemes on any mining lands the economics of the operation should be analyzed in detail. Rehandling previously mined rock is a costly operation, and any plans which involve rearranging dumps and waste areas should receive a careful engineering economics analysis. Fig. 6.6 shows average cost distributions for the unit operations of both coal and copper surface mining. In strip mining, most of the cost is incurred in overburden stripping and coal loading. The industry has responded by adopting mammoth excavating equipment to reduce unit costs as shown in Fig. 6.7. Overburden segregation is a selective mining practice which tends to limit the effective utilization of such equipment. This, in turn, leads directly to significant cost increases. Certainly, some overburden segregation is nearly always advisable, but the benefit/cost analysis should be as explicit as possible.

In open pit copper mining, truck haulage accounts for most of the operating cost. Therefore, redevelopment plans that call for increased haulage distances or moving old dumps would be particularly damaging to the economics of the operation. The mining engineer should, therefore, be an early participant in planning the redevelopment of waste areas.

WEIGHING THE COSTS OF ENVIRONMENTAL EXPENDITURES

Teaching mathematical programming is a task that is satisfying to the scientific mind, for the subject matter is bound into tight, albeit fairly complex, bundles of logic. Given a generalized objective function and appropriate constraints there always seems to be an optimal answer. Real life problems, of course, are much nastier because frequently it is impossible to quantify an objective function that is generally acceptable. And so it is with environmental expenditures. It is reasonable in concept to demand that industry pay the total costs of production, including the external costs now borne by the public. The higher production cost, of course, ultimately becomes higher prices to the consumer, but still it is only fair that each consumer, not the public at large, pay the total costs of a product. The difficulty arises in quantifying the objective function, the benefit/cost model for environmental expenditures. No rational person would, for example, advocate maximum air purity as a national objective function to the exclusion of all other objectives. This would require, among other actions, the elimination of all automobiles and coal-fired power generating plants — 45% of the U.S. capacity. The cost — public and private — would simply be too high. A referendum would indicate that the public would prefer a little dirt

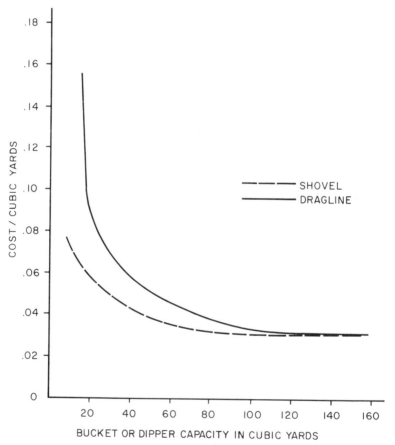

Fig. 6.7. Shovel and Dragline Costs per Cubic
Yard for Various Bucket Sizes. (from Cummings and Given 1973)

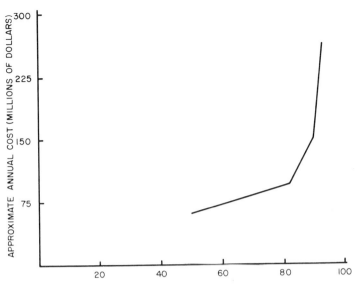

Fig. 6.8 Cost vs. percent Sulfur Removal for Domestic Copper Smelters. (Adapted from Arthur D. Little, Inc. 1973)

in the air in exchange for motorized transportation and electric power. Naturally, this extreme case is absurd, but it does point out the inevitability of trade-offs and the necessity for sound benefit/cost analyses.

In the context of benefit/cost appraisals, two points seem to be frequently overlooked in the public zeal to clean up the environment. The first is that old friend, the law of diminishing returns, stating that return is not a linear function of expended effort. For example, with SO_2 emission control from copper smelters, it has been estimated (see Fig. 6.8) that 80% effectiveness can be achieved for less than 25% of the cost. The incremental cost for each additional unit of control is greater than the previous one such that the cost of 90% control is about twice that of 80% control. In 1975 dollars, over $250 million would be required to achieve the extra 10 point reduction. Is it worth it? Many would say yes, that the smelting companies should make up for all those years of neglect. The decision might be different, however, after the consumer learns that he will pay this cost in the long run. Furthermore, even if the consumer is willing to accept a higher cost of living, $250 million spent for a marginal improvement in smelter emission control is $250 million that is not spent elsewhere. This opportunity cost is the second point mentioned above. Melman (1971, pp. 229–30) gives some idea of the magnitude of $250 million in relation to various items in the 1972 federal budget:

Program	Millions of Dollars
Vocational Education	500
Indian Economic Development	200
All Federal Crime Prevention Programs	1,000
Water Quality Control/Sewage Treatment	1,000
Urban Mass Transportation	500

$250 million would also provide over 16,000 units of low-income housing.

The point of this discussion is that environmental expenditures should be subjected to careful benefit/cost analyses, just as all other capital investments are. The concept of the trade-off will then be plainly evident, and the economic welfare of the firm and the local resident will not be considered lightly. Such analyses that affect a highly technical, capital intensive mining and processing operation must have mining engineering input. Certainly some benefits — particularly aesthetics — cannot be quantified, in which case the opportunity cost concept may provide some guidance. It is, nonetheless, clear that operating considerations must be the main component in planning for the redevelopment of mined lands.

REFERENCES

Cummings, A. B., and Given, I. A., eds. 1973. *SME mining engineering handbook,* Soc. of Mining Engr. Salt Lake City.

Jerome, S. E. 1966. Some Features Pertinent in Exploration of Porphyry Copper Deposits. In *Geology of the porphyry copper deposits.* S. R. Titley and C. L. Hicks, eds. pp. 75–85. Tucson: University of Arizona Press.

Little, Arthur D., Inc. 1972. Economic impact of anticipated pollution abatement costs — primary copper industry. Report to the Environmental Protection Agency. Cambridge, Mass.

Melman, Seymour. 1971. *The war economy of the United States.* St. Martin's Press, New York.

Paone, James, Morning, John L. and Giorgetti, Leo. 1974. *Land utilization and reclamation in the mining industry, 1930–71.* U.S. Bureau of Mines Information Circular 8642.

Stefanko, Robert, Ramani, R. V. and Ferko, M. R. 1973. *An analysis of strip mining methods and equipment selection,* final report to the Office of Coal Research on Contract No. 14-01-0001-390.

Zahary, G., Murray, D. and Hoare, B. 1974. "Reclamation — a challenge to open-pit mining operations in Canada," paper presented to the World Mining Congress, Lima, Peru.

7. Geologic Considerations

John M. Guilbert

IF EVERYTHING WERE KNOWN about the distribution, timing, economics, and occurrence in the Earth's crust of minerals and other geological phenomena that might eventually be exploited or affected by man, this text would be unnecessary. With the universal desire of the public to reap the benefits of mining but not pay the cost, we must learn to take materials from the land cheaply and inconspicuously and still be able to use the land for other purposes. We must learn to occupy safe areas of the terrain, leaving for other uses those areas that are geologically hazardous. We must also establish priorities, recognizing that some land is best used only for mining. Clearly, here is where land use planning and geology enter.

Minerals exploration is probably the most complex aspect of the interface between geology and land use planning. Mining itself is generally misunderstood, but there is almost no public understanding of exploration. The general public is by and large totally unaware of predrilling activities in mineral exploration, and quite unaware of the nature of drilling as part of the discovery process. As was evident during an industry-public clash in Tucson in 1974 and 1975, the public equates drilling with the presence of an ore body and impending rape of the land. There was no willingness to consider the fact that drilling indicated only that information exists that the odds are good that a valid mineral deposit is in the area. It was not considered that if a mineral deposit were discovered, the value of the land would then be reconsidered in the light of various imperatives, which would not result in a heedless headlong demolition of the entire terrain. Intrinsic characteristics of minerals exploration that make it difficult to deal with in terms of land use planning are: (1) it does not necessarily lead to ore; in fact, most projects terminate unsuccessfully; (2) its successful continuation is essential to national economic-industrial health; (3) it involves a high ratio of area which must be covered to target size. Hundreds or thousands of square miles are normally explored in hopes of finding an ore body that may occupy only tenths of a square mile; (4) it is in a constant state of change, with new techniques and new geologic understanding (such as with the recent elucidation of plate tectonics); (5) there is continued development of new imperatives from the market place; (6) it indeed may lead to ore bodies and mining in scenic or otherwise valuable areas. Minerals exploration is clearly not a static phenomenon to be handled with static declaratives.

I will return to some of the problems of minerals exploration with direct regard to land use planning. The following sections cover four subject areas that impinge upon the use of lands, namely geologic hazards, geohydrology, mining, and minerals exploration.

GEOLOGIC HAZARDS AND LAND USE PLANNING

Geologic hazards involve people and land use on the one hand, and dangers inherent in natural or induced rapid local change of geology, topography, and hydrologic phenomena on the other. Geologic hazards are best handled legislatively at a local level since conditions vary widely from place to place. For example, Tucson, Arizona, has very real geologic hazard problems, but they by no means typify those of other cities. Geologic hazards in the Tucson area can be divided into four general categories: 1) sheet flow flooding and floodplain problems; 2) subsidence fissuring due to groundwater withdrawal; 3) slope instability, or landslides and mud flows; and 4) boulder falls near mountain fronts (Peirce, 1972). A fifth category would include faulting and more far-reaching geologic phenomena. In the 1970s, there has been a meritorious effort by the Pima County Planning and Zoning Commission to prepare legislation concerning geologic hazards. County supervisors are talking in terms of one hundred year events, over-bank water velocities, and river reaches. They are aware that most geologic hazard areas can be defined by professionals and set aside. This growing awareness that geologic hazards are real, but definable and manageable, has led to the Flood Plain Management Act of 1973 at the state level. The Pima County Board of Supervisors is now working on a local floodplain management ordinance. The Planning and Zoning Commission is working on Article 45, which involves ordinances concerning active fracture zones (primarily groundwater subsidence fissuring); Article 46, a slope instability zone measure; and Article 47, a sheet flow, flooding zone ordinance. Such far-sightedness in planning is by no means general among the cities and counties of the southwest and reflects commendable cooperation between legislators and scholars of the land. Quite properly, floodplain zoning, sheet flow zoning, active fracture zoning, and boulder-fall zoning should override virtually all other zoning configurations.

State codes should impel local governments to consider local geological problems. State offices should also be concerned with more general hazards involving large river floodplains, mass wasting and erosion, earthquake problems, and zoning involving geological engineering aspects of siting, foundation selection, and highway construction. Geophysical and geological aspects of nuclear plant site selection should be considered at the state or federal level. The perhaps less obvious geologic hazards, such as the results of sand and gravel mining from river beds on the downstream side of bridge abutments, should also fall to state government.

It is debatable that the federal government should reach to the local level in any great detail, but federal impetus is needed for an overview of geologic hazards problems for long-term or interstate action, and for coordination of effort between local and state agencies. At all levels, administrators and legislators should avail themselves of trained earth scientists, hydrologists, and demographers.

GEOHYDROLOGY AND LAND USE PLANNING

Incredibly there is presently no involvement of geohydrology in zoning legislation. Legislation is needed that, at least, considers groundwater depletion, pollution, and recharge. There has been a dramatic change in the concept and status of water ownership in the last 50 years from adamantly private to much more public ownership (with minerals generally going into the private sector). Along with this trend, zoning problems and land use problems continue to develop. At the moment, the only pertinent acts are the Critical Groundwater Basin Acts of 1948 and 1952, which involve most of southern Arizona's intermontane basins. They were designed as conservation acts to prevent rapid or abusive development of land in heavy overdraft areas. They provide that water pumpage may not proceed from one basin to another or from a basin to surrounding high land without retirement of appropriate acreage within the valley.

Mining companies are now buying basin lands with thoughts of retiring them as they pump water to mining properties in surrounding hills. The City of Tucson has, for example, bought substantial retireable acreage in the adjacent Avra Valley in order that it might pump water from the Avra Valley basin to the Tucson basin. As many as 40,000 acres may ultimately be retired.

This activity raises several questions. To what uses will retired Critical Groundwater Basin Act land be placed? What reclamation measures will maintain land already cultivated that becomes retired? How long is groundwater irrigation of crops wise and feasible for the common good in a given basin? What are the real ramifications of the Critical Groundwater Basin Act? What indeed will be the land use implications of the Central Arizona Project regulations? These are all problems that should be immediately confronted. As with geologic hazards, planners and administrators should use the services of professionals such as hydrologists, agronomists, and agricultural economists.

MINING AND LAND USE PLANNING

It is conceivable that society might ultimately be able to regulate itself so effectively that mining activities could be carried on inconspicuously and away from population centers. Such an eventuality is possible only in a totally primitive society or in one of extraordinarily cheap energy in which people and materials flow with infinitesimal transportation charges. However, this is unlikely, and in fact, transportation and energy costs will escalate in coming years.

Many American cities, especially those of the west, are in effect mining towns conveniently situated to supply labor, capital, and materials to mining operations. Some mines create population centers, particularly mines involving metals of high unit value such as copper, gold, silver, lead, and zinc. Another type of mine is created by the population centers themselves. These mines typically involve low-unit-value, high-place-value commodities such as sand and gravel, aggregate, and cement rock. A sand and gravel pit is as much a mine as one of the large open-pit copper properties, and it must be considered assiduously in land use planning. But clearly, complete separation of mining and population centers is not feasible, since mining attracts people, and people in turn may create local mining.

Both surface and subsurface mines should be considered in land use planning. Surface, open-pit operations generally produce large tonnages of ore and waste material and are therefore conspicuous. Subsurface mines are underground operations whose actual excavations are inconspicuous but whose surface facilities (head frames, mine buildings, waste dumps, and slag piles) may be conspicuous. The latter require nearby towns; the former are somewhat less dependent upon them. Mines could be categorized as large, involving considerable acreage like the typical open-pit copper mine, or small, requiring perhaps one or two head frames or a gravel pit, and a commensurately low profile of visibility.

Geologic constraints upon zoning, land use planning, and mines stem almost totally from the fact that ore bodies, a special class of mineral deposit that can be worked by man at a profit, are distributed according to geological constraints. That fact is critical to the understanding of land use planning, natural resources, and mining. With rare exceptions, widely distributed deposits involve materials of low value and mildly restrictive specifications such as sand, gravel, or brick clays. Ore bodies of the base and precious metals generally are associated with the same geologic forces that build mountains. It is therefore no surprise that many of the mining towns of the west are in the mountains, that much of the metallic mining activity of the western states is in mountainous areas, and that the greatest confrontation between mining interests, minerals explorationists, and nature conservationists has involved mountainous areas. Serious problems have arisen with the concomitant upsurge in environmental concern and the realizations that many of our minerals are in short supply. With the basic tenet that the distribution of mineral deposits is primarily a geological phenomenon, decisions concerning use of particular parcels of land must change as the value of minerals to society changes, as discoveries occur in new areas, and as the whole natural resource picture evolves. For example, if geologic phenomena have placed a copper orebody at the southern end of the Tucson mountains, society must decide whether that land is more valuable as residential, as tourist-scenic, as a national monument, or as a copper mine — a decision which is neither easy to make nor necessarily permanent. But it should be remembered that national and local welfare must be served.

LAND USED BY MINING

What has been the record of mining with respect to land use? Two of many pieces of evidence suggest that the percentage of land affected by mining is very small. First is the evidence from the series of space craft photos beginning with those of Gemini IV. All of the space imagery shows mining activity to involve far less than one percent of the total terrain which is orders of magnitude less than the land used for agriculture or residences, and far less than has been consumed by interstate highway programs in the western United States.

Secondly, an article by Ernest L. Ohle (1975) presents a map of the western United States with dots of proportionate sizes to scale representing land affected by mining. When the spots are coalesced, the total area per state still constitutes only 0.13 percent of the total land surface. But, since the populace lives close to many centers of mining activity, it quite naturally thinks that a far greater percentage of the surface has been affected.

It should be pointed out that even the largest single mineral deposits seldom disturb more than 10 to 20 square miles of land, including the mines themselves, buildings and concentration facilities, waste disposal, refining, and environmental restoration facilities. The largest single underground copper mine in the world at San Manuel, Arizona, is also the second largest copper mine of any type with production of 150,000 tons of copper a year, yet it involves less than 10,000 acres of desert. Only 100,000 out of 73 million acres of land in Arizona are affected by mining, a scant 0.14%. Bingham Canyon, Utah, and Butte, Montana, two of the largest ore bodies in the world, each produce from 7,500 acres of disturbed land enough copper for 11 million cars per year or for five years of home construction. One acre of land in a porphyry copper deposit of the type so common in Arizona produces from a 300 foot depth over 10 million dollars worth of copper. Assuming that our present standard of living is desirable, as Ohle (1975) comments, to what better use can this land be put? It is an interesting paradox that while most people seek to avoid mines and mining, they also like the hard goods, jobs, and the comforts that technology provides.

Zoning around existing mines is being considered in Arizona, with generally very sensible use of adjacent land to maximize extraction of the commodity sought but to minimize exposure of mine sites to the public through separation, greenbelts, etc. A point to be stressed, however, is the importance of minerals and mining to society. The dangers of too restrictive zoning in advance of discovery are alarming, and the great desirability of keeping decision options open for continual or periodic review should be clear. In Arizona, Pima County's proposed ordinances to regulate zoning in areas of moderate potential for mineralization (Article 43) is exemplary in that it recognizes the need for repeated exploration entry to land and the need for review of net value in addition to offering needed protection to society. Land use and land use planning appear to be in harmony here, but considerable concern is expressed by natural resource producers elsewhere that both exploration and mining as land uses may become prohibitively regulated and restricted.

MINERALS EXPLORATION AND LAND USE PLANNING

Minerals exploration constitutes an area of land use that is of great concern to minerals resource planners. It is vital to the nation's welfare. The continuation of supplies of commodities upon which our standard of living and our commercial sector depend hinge upon minerals exploration. The situation is aggravated by poor understanding of exploration by the public and even by many policy makers. Some wish to exclude geological exploration from various areas in the fear that "a disfiguring mine" will be discovered; others wish to avoid the surface damage done by exploration itself. However, there are imperatives that require that minerals exploration not be prevented from reasonable and effective function.

For example, the average automobile is composed of

2,775 pounds iron and steel (frame, engine, body)
100 pounds aluminum (castings, engine components)
50 pounds copper (radiator, wiring, electronics)
25 pounds lead (battery, solder)
50 pounds zinc (castings, galvanized parts, tires)
100 pounds glass (windows, lights, ceramics)
250 pounds rubber and plastics (tires, seats, trim)
150 pounds miscellaneous (fabrics, insulation, materials).

Implicit in the analysis, although many of the trace elements, the rare earth elements, and the manufactured plastics are omitted, is our dependence upon an astonishing amount and variety of minerals products. Figures released by the Department of Commerce and the Department of the Interior show that 15 percent of the United States gross national product and 30 percent of the Canadian gross national product are mineral based. One and three quarter billion dollars of mineral products are mined in the United States per year. Forty percent of our transportation capacity involves mineral commodities. Harder to depict is the increasing number of elements and minerals involved in our daily activities. For example, the entire product line from the Mountain Pass carbonatite deposit in California had virtually no market ten years ago, but it has since become a remarkable part of our commerce. The rare earths produced from the mine are vital to color television, computer technology, satellite communications and telemetry, the ceramics industry, and several others. Had this area been removed from exploration, the entire rare earth technology might not have developed.

A more explosive aspect of minerals use stems from the Malthusian problem of population explosion combined with a growing consumption per capita. Total consumption is still increasing in this country but, it is increasing substantially faster in the developing nations. Thus there is increasing competition for the purchase of foreign materials. As our command of the world market diminishes, it becomes even more necessary that we develop realistic national attitudes toward exploration and mineral production.

The minerals industries have not fared particularly well in the early 1970s through all of this. Costs have risen substantially faster than profits. Compared to the Standard and Poor 500 stock average for a decade ending in 1974, the cement business has been a disaster, the copper business has been poor to very poor, and the steel industry has fared less well than the 500 stock average. As a result, the mineral industries of this country are somewhat capital impoverished. They do not have the money to invest in enormous mineral ventures and start-up costs of the type that are required to meet the demands of exponential growth. The minerals industry is entering a period of uncertainty and difficulty in meeting its obligations to the manufacturers. It is a time when land use planners need more than ever to consider the ultimate posture of the country in terms of its traditional mining base, the importance of its natural resource and mining base today, and its predictably important base in decades to come, and to provide reasonable availability and access to lands for exploration purposes.

There are geological constraints on land use in mineral exploration just as there are in mining. Mineral deposit geology and ore distribution patterns are now known much better than they were ten years ago. A number of federal agencies such as the USGS and the Resource and Land Investigation (RALI) program are directed toward an understanding of where mineral deposits are located, and how access can be assured in the future.

What is the present status of land use in the context of geologic exploration? The Multiple Surface Use Act of 1953, the Multiple Purpose Act of 1954, and the National Environmental Policy Act (NEPA) of 1963 all will have tremendous impact on mining and land use planning. The Wilderness Act of 1964 includes in its provisions that the United States Geological Survey and related professional organizations are charged with the function of reassessing and reevaluating land in wilderness preserves. In 1983, most of the original wilderness preservation lands will be closed to mineral access. However, the government is charged with continually reassessing that land in terms of advances in geology and exploration techniques, presumably with thoughts of how land serves the greatest number with the greatest good. Few are naive enough to say that there are not some land uses that are more important to society than most single mineralized areas, but mineral explorationists are concerned that huge tracts of land will be shut off such that only "barefoot" access will be permitted and that the key (in fact) will be thrown away. But the Act provides that the key is not to be thrown away. The largest problems have been misunderstanding, undermanning, and underfinancing of the federal agencies that are responsible for updating the mineral inventories of wilderness lands, national forests, and roadless areas. The minerals industry is in danger of having to stand in line with reasonable operating plans while government agencies, as benevolent as they are, simply cannot handle the tremendous volume of business. USGS staffing has not been improved to handle the supposed evaluation of the 9.1 million acres of land that are about to be put away as wilderness areas. More lands have been added to the original withdrawal list by requests from various individuals, organizations, agencies, and societies. The total is now about 30 million acres. Although there has been an extension of the 1983 deadline, there has been no expansion of the survey and other staffs that are supposed to evaluate this land. There are cases on record where the land was "put away" in a wilderness category before the USGS could finish its report. When completed, the report concluded that the areas were indeed promising, potentially mineralized, and should be kept open. It is important that the agencies responsible for keeping the key in the lock are suitably financed and staffed. One way might be for professionals to volunteer their services to various city, county, state, and federal agencies. It is likely that there will be a trend of increasing responsibility given to local governing bodies, which means that citizens must be informed as to what mineral resources are and how they can best serve the nation.

Land use planners interested in the local and national welfare must think not only of local pressures, which are genuine and poignant, but also of the conservation and development of mineral resources in those singular geologic situations where they occur. Planners must recognize that: (1) certain areas are more likely to be endowed with developable resources than others and allowance for production from those areas must be entertained; (2) the variable mineralization potential of different tracts must be recognized; (3) the immobility of an ore body indicates that mobile land uses should be directed toward areas of low potential mineral endowment, like areas of thick valley fill or thick volcanic or sedimentary rock sections; (4) access by responsible explorationists willing to "cover their tracks" be permitted to as much public land as possible; and (5) discovery of new mineral resources requires realistic reassessment of the worth to the commonwealth of a particular parcel of land. The situation appears good in 1975, with all interest groups in reasonably salubrious equilibrium, but there are difficult times ahead. Careful planning with regard to the totality of public welfare — with regard to both wilderness and mineral conservation — must continue.

Land use planners have an awesome duty to us all, and they are only as well prepared to perform it as they are informed and aware. Professionals owe it to those who have the legislative-administrative-planning timetables to assist them in acquiring enough wisdom to make the most satisfactory decisions for the public.

Acknowledgment. Materials in this chapter were drawn from conversations with a number of western North American exploration managers, explorationists, and consultants; from colleagues at the University of Arizona, especially Drs. Spencer Titley and Brent Cluff; from the staff of the Arizona Bureau of Mines, especially Drs. H. Wesley Peirce and Richard T. Moore; from personnel in the U.S. Department of the Interior, the U.S. Department of Agriculture, and from a number of federal agencies. I am especially indebted to Kenyon Richard, a Tucson geological consultant and author, and Ernest L. Ohle, past President of the Society of Economic Geologists and an exploration manager in Salt Lake City.

REFERENCES

Ohle, Ernest L. 1975. Economic geologists, the Society of Economic Geology and the future. *Econ. Geol.* 70:612–22.

Peirce, H. W. 1972. Geologic hazards and land use planning. *Fieldnotes.* Arizona Bureau of Mines, vol. 2, no. 3.

8. Engineering Constraints

James K. Richardson

THE HISTORY OF MAN'S SPAN ON EARTH is a story of his alteration of the face of that earth in the evolution of a continually improving living standard. This driving force has become as instinctive as those other thrusts which have led to man's survival. There is no reason to believe that edicts from society or social systems will materially alter this drive any more than they will make survival less attractive. Suffice it to say at this point, however, that with the accumulation of knowledge, there is a developing awareness that our nest need not be fouled (indeed, must not) if man's span is to be perpetuated. Hopefully, as a result of the symposium on which this text is based and other similar exchanges of ideas, the scars of many activities may be partially healed or at least minimized. It is unlikely they can be eliminated, for a living compromise dedicated to man's survival is as important as pristine purity.

Mining and agriculture are the fundamental industries that convert natural resources into usable forms. The United States has developed a living standard that is the envy of the world. There is no question that this nation has been blessed with a bounty denied, or perhaps postponed, men elsewhere. As free men, operating under a free enterprise system, we have brought to and shared with the world at large a bounty of knowledge and resource conversion. Those who abjectly criticize our "profligate" use of resources have neglected to allocate to the balance of the world their fair share of our costs. This nation has repeatedly demonstrated a concept of sharing which has never been, and probably never shall be repeated. Despite this Judeo-Christian ethic, much evidence is developing that we must take care and use our resources wisely, for tomorrow they become more difficult to take from the earth.

In man's early beginnings, waters teemed with fish; fields swarmed with animal and vegetable life; mountains were rich with relatively easily available metals; and man's presence was hardly suspected. Today, almost all of these situations have been reversed as man's presence begins to dominate the earth.

The purpose of this chapter is to discuss the engineering constraints to obtaining those mineral resources which provide man with his tools and the energy needed to effectively use them. This concept is difficult to bring clearly to the public, which often forgets that the entire basis of civilization as we know it is embraced within man's intelligent use of tools and energy. A prelude of that picture was brought to the public's attention as the petroleum shortage emerged and developed. We have all seen a curtailment of employment, limitations upon transportation, and our whole national well-being severely strained because of one single item in our mineral picture. Many strongly believe that we are today operating on the basis of expediency in this field and that no long-term solution has emerged. Many also strongly believe that the potential total mineral shortfalls can conceivably make the oil exercise appear simple. It is, therefore, essential to give every consideration to our emerging mineral problem, which is every day crying out for recognition. There is no easy way to come home free.

The impact of the oil problem struck the American people, while only 30% of our oil was being imported to meet national needs. Most Americans thought of oil only in connection with their day to day activity, recognizing its need in autos, but failing, completely, to appreciate it as the basis of energy for their jobs, for their food, for their clothes, for their transportation, and so on. How then can anyone understand fully the significance of the following list of elements — of which over 50% of our needs are imported: aluminum, antimony, chromium, cobalt, mica, strontium, fluorine, titanium, columbium, manganese, nickel, asbestos, potassium, mercury, tin, zinc, platinum, gold and silver? Few of us know of the existence of cartels in some of these elements, whose object was and is exactly the same as that of the oil producing Arab nations. The United States must, therefore, quickly become as self-sufficient as possible, if our living standards are to be secure. The price may be high.

Despite any allegations to the contrary, any mineral company must be able to demonstrate the feasibility of recovering the tremendous sums required to assure self-sufficiency during some projected period. This is the criterion required by the private investor. The money-lender refuses to make such capital available where there exist so many unknowns that evaluation of projects goes beyond normal "risk taking" by the addition of human vagaries and shortcomings. The emerging nations, too, are seeking to develop their resources concurrently and the risk taker's money goes where the return is greatest and the risk least. There is reason to wonder whether the enormous sums required to meet all national needs and aspirations exist. There can be little or no question that this availability of capital represents a major constraint upon the developing of national mineral self-sufficiency.

Before there is even the thought of getting a mineral project underway, the economic mineral source must be found. Exploration has been on the decline in our nation for many years. Legislative restrictions in the form of taxation, recreation, environmental and land policies have materially contributed to this decline. And where do they look for these economic mineral deposits? There is an old adage, "When hunting for elephants, one goes to elephant country."

Arizona, insofar as copper is concerned, is "elephant country." This unquestionably accounts for the presence in this area of one of the largest assemblages of exploration professionals in the world. These groups represent aspiring copper producers as well as representatives of all major copper producing companies. The cost of exploration for elusive mineral reserves has become so large that the risks of exploration are becoming increasingly shared.

Vast areas of the public domain have been so restricted, and their status so uncertain, that the risk of exploration cannot be economically justified.

Without entry to these lands for exploration, and without assurance of tenure in the event of a discovery, no engineer can calculate the relative costs and benefits which permit determination of the most beneficial use of land. Arizona is replete with examples of the most beneficial use principal, for only about 17% of her land mass is in private hands.

Congress should be very chary of closing land to mineral entry, prior to competent mineral examination. A classic example of what could happen is demonstrated by a letter dated September 26, 1922, from the Commissioner of the General Land Office in Washington, to the Phoenix office, advising that a certain half-section of land had been classified as "non-mineral in character." Fifteen years later, an aerial survey, embracing 525 square miles, included the half-section within it. There was no evidence of mining activity in the area. In 1974, in the area of which the half-section is the hub, almost 400,000 tons of copper were extracted. The junction of the Pima and the Mission mines is included within it.

When this evaluation was made, it was a perfectly reasonable and honest classification. There was no surface expression of mineralization. Indeed, had there been any reason for examining beneath the surface in 1922 and discovering the existing mineralization, it still would not have been classified as "ore" for its average copper content was, and is, only about 7/10 of one percent copper, which at that time would have possibly classified it as waste. If, however, it had been withdrawn from entry, it is questionable whether these valuable deposits would ever have been discovered.

The area, after development, provides employment to almost 8,000 people, each earning an average of about $12,000 per year. From property taxes alone, almost 9 million dollars went into state and local government revenues. Prior to mineral development, the area supported some 800 cattle.

Obviously, the advances in mining technology converted this non-mineral land into the hub of the modern Arizona copper industry. Geophysical prospecting techniques pinpointed the Pima mine first, and then the others. Modern technology and equipment made it possible to mine the properties economically, while metallurgical advances made possible higher recoveries of available copper and by-product minerals earlier lost.

Other areas have similar history, and undoubtedly there are many sections which today qualify as non-mineral, but which in a few years will experience the transformation from "non-mineral" to "mineral" in character. The constraint that non-accessibility of land for evaluation imposes on our mineral posture should be apparent.

The time lag between finding, evaluating, and placing an ore body in production may be on the order of 10 to 12 years. One ore body on record was 27 years in the making.

Capital needs to place our nation in a self-sufficiency posture, and the competition existing for capital funds was mentioned above. In Arizona today it costs about $5,000 per ton of ore mined to provide the facilities needed to produce copper concentrates. In order to economically mine low grade ore bodies, a 25,000–40,000 ton daily ore production is needed. This in turn indicates a risk venture involving $125 million to $200 million. For example, the new Metcalf mine near Morenci, Arizona, has announced a total development cost of $200 million. This is only the beginning. The new Hidalgo smelter of Phelps Dodge, with an initial capacity of 100,000 tons of copper annually, smelting Tyrone concentrates, is expected to cost another $200 million. Without a smelter, copper-bearing concentrates contribute little to our mineral position.

The prime interest rate on the total $400 million investment in 1975 was 10.5%. This is the dilemma of the private investor and compounds the problem facing the engineer who is called upon, in today's uncertain world, to factually justify projects of this nature. One also appreciates why the minerals industry is considered highly "capital intensive."

Following the 1974 meeting of the American Mining Congress, a *New York Times* story reported the industry as laying all its ills at the door of environmentalists. There is no point in doing this here, but it might be useful for the public to become aware of the costs being forced upon it by well-intentioned, but often uninformed, organized minority groups.

The environment is both the source of resources and the sink for the waste. At no step in the flow system do resources and the environment fail to interact. No stage of the system can be treated in isolation — a change at one point invariably affects many others. Few other subjects can be found in which so many hard-core adherents are on both sides — little wonder that the public is confused by its leadership, which from day to day vacillates as a vocal electorate dictates.

The smelter in Hidalgo County, New Mexico, cost $200 million of which $70 million is attributable to anti-pollution facilities. Hopefully, they will work and thus meet the standards set. These standards, backed by deadlines, facing an industry where control technology is minimal, pose continuing cost-operational problems that exasperate both proponents and opponents. Both sides become suspect as changed restrictions are added helter-skelter.

The operator of another smelter in Arizona recently reported that the trial and error method of attempting to meet EPA requirements had left him with a "newly constructed plant which was already 20 years old." His problem is still unsolved 50 million dollars later. Each smelter could perhaps tell a similar story. How can the engineer evaluate a project when technology is non-existent and the requirements so fluid?

This area, it would appear, is one wherein government should have developed the technology prior to the establishment of standards. Industry, alone, should not be faced with the task of making technological breakthroughs in order to meet theoretical standards and deadlines. A cooperative balance between viewpoints must be reached or we shall destroy one another.

Land reclamation and its attendant cost implication represents a new unknown in the engineer's formula. Hopefully, common sense will somehow prevail and no stampede occur as did in air pollution control. Here too, the engineer must use his own judgment, having no guidelines as to what will be acceptable. The University of Arizona has studies underway relative to the ultimate reclamation of lands used in mining south of Tucson. The concept is interesting and worthy of consideration. The major short-coming immediately apparent is a lack of economic data, but the assurance that such creative and cooperative processes are underway is encouraging.

It is generally recognized that while operational facilities can be zoned, no way yet has been discovered to zone where a mineral deposit can be mined. Such a deposit cannot be moved from one location to another. However, even facility zoning creates another constraint, for it injects conditions which may add or detract from costs.

Before the engineer can recommend a project anywhere in the world, he needs a minimum of unknowns. The mineral industry is exposed to sufficient risk without adding to his problem. In any land, a stable tax policy is essential. Proposals continue to emanate from special interest groups, which would change the rules in the middle of the game. Were some of these to become law, not only would nationalization of our industry become a real possibility, but much of that which is ore today would become waste tomorrow. That nebulous "cut-off point" which is so susceptible to uncontrolled tax costs, either extends the life of a mine or materially shortens it.

Not the least disconcerting of these "unknowns" is the engineering distraction of reading each morning of corporate profits being termed "obscene." This term, attributed to Senator Henry Jackson, is indicative of the continuing Congressional sniping to which the minerals industry has been subjected. What is an "obscene" profit in view of national well-being? From these profits spring employment and those goods which permit industry to give Americans that which they have termed "the good life." It is those profits that industries plough back into their companies to fill the charter under which they were organized. Those profits assure the presence of material reserves that give stability to our national economy. How can the engineer visualize that return which will be termed "obscene" tomorrow?

A recent interview held by Daniel Moynihan relative to India could be used as a conclusion. When India won independence from Britain in 1947, it was producing 1.2 million tons of steel a year. Vanquished Japan was producing 800,000 tons. Twenty-five years later, India's output had risen to 6.6 million tons. But Japan, which had to import the iron ore and coal that India had in abundance, surged to 106.8 million tons. "This isn't a question of luck," noted Ambassador Moynihan, "and it isn't because the Japanese are any smarter or harder working than the Indians. It is simply the result of a conscious decision made by both countries."

Japan's decision was to encourage a private enterprise economy, while India opted for a planned economic system. This choice was made because India felt internal stability more important than economic growth. One of the most important drawbacks of such a choice is that it breeds stagnation. Moreover, he felt that unproductive economic arrangements eventually undermine the very stability they aimed at preserving.

The Wall Street Journal, in editorializing on the reports, says: "The one lesson that stands out today amid our economic gloom is that, troublesome as the problems of affluence are, the problems of poverty and stagnation are far worse. Everything from vital social services to a country's ability to feed itself depends on a strong, healthy economy. Merely drifting along at about the same rate may avoid some temporary economic headaches, but it also makes it just about impossible to do much more about serious social problems than to give up in despair."

Hopefully, our national leadership, with the guidance and at the behest of the electorate, will recognize the essentiality of a posture reasonably compromising the needs of environment and America's like needs for mineral products. Such an attitude, seemingly, promises stability to the desires of both with minimal detriment to either. Admonitions must no longer be blithely thrown aside by the uninformed. America's strength, well-being and welfare depend upon mineral production, based upon a conscious, balanced, national policy, designed to make our national heritage secure.

9. Economic Constraints

George F. Leaming

THE RELEVANCE OF ECONOMIC CONSIDERATIONS

STAKING A CLAIM is an old and honored, albeit at times abused, practice in mining. Before too much more is done in the political and legal arenas regarding reclamation for mined lands it might be best to stake a claim of relevance for economics in discussing mining's relationship to the physical environment. If economics can be defined, and it has been, as the study of the allocation of limited resources among alternative uses, then economics is obviously concerned with the full range of problems and situations involved in the place of mining in, and its relationship to, both physical and human environments. Problems of mined land reclamation are basically problems of allocating resources, both human and non-human, among alternative and often competitive uses. In other words, decisions made in reclaiming disturbed lands are economic ones, whether the resources involved be measured in dollars or not, or in fact, if they be measured at all.

The use of mineral resources has always involved the use of one or more other natural resources — land, air, and water — as well as capital and labor. Mining and related activities have also had their impact on the human environment as well as on non-human biological resources, normally acting through direct impacts on air and water supplies as well as through occupancy of the land.

In alleviating the adverse effects of mining upon any one aspect of the environment it has usually been necessary to increase the impact upon another. The prevention of undue contamination of water resources that results from returning liquid waste to the environment by substituting evaporation for seepage or open discharge to streams requires the use of either greater land area for evaporative surfaces, or greater amounts of energy to produce that evaporation. The prevention of air pollution by converting noxious gases to usable or more readily disposable liquids normally requires the greater use of water resources. Where resources are limited, as they all are, the achievement of any one objective can only be made at the expense of a competing goal. Nothing can be accomplished at zero cost. This is particularly apparent in the case of the varied impacts of mining upon the environment, and especially upon land, air, and water resources.

Defining Some Terms

If disturbed land reclamation is to be examined in an economic context, then it would probably be well to substitute the term, used land conversion. "Reclamation" implies that land was in its optimum condition prior to mining and that return to that condition consequently assures optimum future use of the land. This is not necessarily so, especially in the semiarid areas of the mountain west. Examples of better use of land after mining than before mining are numerous, and will not be presented in detail at this point.[1]

"Disturbed land" implies that mining and other excavations are fundamentally different from other uses which occupy or use the earth's surface. In reality, most of the basic economic principles and constraints that affect mining's use of the land are also applicable to other land uses which either occupy or disturb the surface, even if only to rearrange the vegetation. Most of the basic principles and constraints that are discussed in the following are applicable, either as is or with only slight modification, to used land conversion problems encountered in agriculture, construction and every other human activity that involves land use whether or not it involves some rearrangement of the lithosphere. As a result, the broader concept, used land conversion, has been used here with the associated implication that any land that has been occupied by human activity or changed to a useful form may be converted to another beneficial state once the original utilization is finished, and that this land conversion is subject to essentially the same economic constraints that govern the mining industry.

The Guiding Question

In applying economic principles to used land conversion, one guiding question should always be asked: Will the finished product of land conversion be worth what has been put into it in terms of other resources to change it from what it is to what it will be? If the answer to this question is no, then there has been or will be a misallocation of other resources, labor, capital, water and air.

For example, the expenditure of $2,000 per acre to put land into a condition in which it is worth only $500 per acre on the land market, is a misallocation of $1,500 worth of other resources just as much as spending $2,000 to make a $500 motorcycle. That $1,500 per acre misallocation could represent a month's medical services for impoverished Papago families, or dental work for 100 minority children, or 2,500 loaves of bread for unemployed automobile workers. The fact that it may come directly from mining company earnings makes it no less of a misallocation.

Granted, other costs and values besides the sales price of land must be reckoned with in converting used land. But what is the value of reduced siltation from waste dump erosion or the cost of dust blown from unconsolidated tailings dumps? And what is the value of an abandoned and unreclaimed Lavender Pit? Or an old head frame in Tombstone? Or the ruins of a 100-year-old smelter in the Coronado National Forest?

---

[1] See, for example, Ralph P. Carter, et. al., *Surface Mined Land in the Midwest: A Regional Perspective for Reclamation Planning,* Chicago: Argonne National Laboratory, 1974, 675 pp., and George F. Leaming and Willard C. Lacy, *Non-fuel Mineral Resources and the Public Lands,* vol. III, *Minerals and the Environment,* Tucson: University of Arizona, Division of Economic and Business Research. 1969, 235 pp.

The guiding economic principle of used land conversion can be expressed mathematically as

$$\sum_{i=1}^{i=n} V_1 = \sum_{i=1}^{i=n} C + \sum_{i=1}^{i=n} V_0 + R$$

where V_1 represents value after conversion, C represents cost of conversion, V_0 represents value before conversion, and R is a reclamation differential. Where R is zero or positive, conversion of the land from its existing state to a subsequent "reclaimed" state is economically desirable. Where R is negative, however, such a conversion represents a misallocation of resources and the destruction, rather than the creation, of value.

The conversion of land to another condition after any use must be paid for somehow. If the sale or subsequent use of the converted land cannot pay for the conversion, then either the original activity must pay for it or the public, if the public puts a value on it, must pay for it directly through the state.

Furthermore, if land is to be converted from one condition to another, in the absence of unlimited resources, it must be done under a system of constraints. These constraints can be grouped into four general classes: (1) incentives, (2) resource availability, (3) resource costs, and (4) demand.

AVAILABILITY OF INCENTIVES FOR CHANGE

Disincentives for Change

If a change is to be made in the condition of land, there must be some incentive for creating such change. Offsetting any incentives and acting as constraints will be certain disincentives for change. Inertia is a strong disincentive, both in the physical world and in human behavior, and inertia has been and continues to be a major disincentive to creating any change in the condition of used land. Coupled with inertia, lack of opportunity for gain has served as a disincentive to changing land condition either before or after any particular land use. Another major disincentive to the conversion of land from one condition to another has been the actual imposition of penalties for such change. The major penalties imposed so far have included revenue losses for the owner of the land and higher costs involved in using land that has been changed (e.g. landscaped) as compared to those involved in the use of the land prior to such change.

Positive Incentives for Change

Although positive incentives for changing the condition of used land have been in existence for centuries, their employment has been relatively slight in the past few decades. One of these major incentives is the potential for additional revenue, i.e., the profit motive. The opportunity to increase the income of the landowner after that land has been converted from one use to another is a major, but widely neglected incentive for creating that change. A second positive incentive for change that has been virtually ignored has been the potential for reducing the cost to the landowner or proprietor resulting from a change in land character.

Negative Incentives for Change

Unfortunately, most of the incentives for change that have been considered and applied in recent years have been negative incentives. Chief among these have been higher taxes. The imposition of increased taxes in order to encourage landowners or land users to convert used land to a character desired by the taxing authorities has been the most widely advocated form of incentive. It is obviously a negative incentive. An even more direct negative incentive that has been applied as well as considered has been the levying of fines and similar penalties on landowners or land users to force them to convert land used for one purpose, to a form in which it would be suitable for another purpose, even though that other purpose may give no positive benefit to the landowner or the current land user. Both of these negative techniques have been effective, however, in providing incentives for change. They will probably continue to be used as incentives for change. Unfortunately, they have been considered by many to be the only incentives available. They are not.

AVAILABILITY OF RESOURCES

Land

In addition to incentives for change, resources must be available to enable a change in the condition of land. Desire is not enough; there must also be the ability. Obviously, the most critical availability concerns the land itself. Surprisingly, in many instances this has been overlooked. If a land user is to convert land that he has used, however, he must control, or at least have access to, that land.

Knowledge

A second resource that is quite frequently overlooked with respect to its availability is knowledge. Even if an individual, firm, or government entity has land available to convert it from a previous state to a subsequent condition, that individual, firm, or government entity must have the knowledge with which to do so. People who convert land should know how to convert land. They should know what they are converting from and what they are converting to. Obvious? Yes, but often overlooked.

Natural Resources

In addition to the land that must be changed and the knowledge of how to do so, other resources must be used in the conversion of land from one form to another. In the arid parts of the world one of the most critical resources needed in the converting of land is water. If land is to be changed from barren waste dumps without vegetation to desert with vegetation, or to pecan groves, or to lawns, water must be available to provide for the growth of that new vegetation.

Another resource that must be available in using vegetation to convert land from one state to another is soil. If barren waste dumps or tailings ponds are to be changed from their previous use to a succeeding condition that involves the growth of either native or exotic vegetation, there must be soil to support such vegetation. If that soil has not been saved from previous surface disturbance, then it must be imported, but it must be available.

Of course, if vegetation is to be the means used by which land is to be converted from one form to another, then the biological resources themselves must be available. Will plants grow on the land to be disturbed? In Arizona and most of the southwest there are plants which will grow in the semiarid environment, but what of disturbed land in the Atacama Desert, or even parts of the Mohave? In such areas are there any biological resources that can be used to convert land from one condition to another? Do appropriate biological resources exist, at any cost, that can be used to vegetate an unvegetated waste dump? Or does it require planting with Russian thistle and thereby substituting tumbleweed blizzards for sand storms?

Human Resources

In the attempts to convert used land from one form to another, the availability of human resources is frequently not considered. In an economy where there are many unemployed, it may appear that human resources will be available in abundance. In a "full" employment economy, however, human resources simply may not be available at any reasonable price, and even where there is substantial unemployment, the right kind of human resources may not be available. By the right kind is meant human resources with the knowledge and ability to effectively convert land from one condition to another.

Capital

In addition to the labor and natural resources that must be applied to converting land from one form to another, capital must be employed. In a period of high interest rates, the availability of financial capital to buy the seed, or water, or labor, or soil required to convert disturbed land from its used condition to another form is of critical concern. In a time when the Treasury of the United States must go into the capital markets to borrow 20 to 100 billion dollars or more to finance its programs of unemployment compensation, social security benefits, and other so-called human projects, the availability of financial capital to convert a barren waste dump to a cactus garden could be a serious concern.

A second aspect of capital resources involves equipment. The equipment and facilities must be available to convert used land from one form to another. The equipment must be available to do the job when it is to be done and should not be taken from some other occupation for use in converting land from a prior use condition to a new condition ready for subsequent use.

COST CONSTRAINTS
Resource Costs

Even when resources are available for land conversion (or for any other use) the cost of those resources must be considered. The cost of resources forms a significant constraint on the amount, nature, and even the desirability of any used land conversion that may be contemplated.

First, there is the cost of the land itself. If the land is not under the control of the converter but is available, its purchase or lease price for conversion must be considered. The terms under which the land can be made available must also be considered. Of even greater importance, however, in most instances of mined land conversion are the costs, both incurred costs and opportunity costs, of retaining land control rather than disposing of it once its usefulness for mining has ceased.

Soil cost must also be considered if vegetation is to be the means of land conversion. Not only must the cost of soil that is available be considered, but the cost of getting it to the site can become a significant factor. This can be a major consideration in semiarid environments where the formation of soil is not a rapid natural process.

The cost of water must be considered if the creation of vegetative cover is to be a part of any intended used land conversion, particularly in the semiarid and arid environments of the western United States. The opportunity costs especially must be considered in determining if it is more advantageous economically to use water to grow pecans, feed grains, cotton, or lettuce, or if it is more appropriate to use limited water resources to grow mesquite, ocotillo, saguaro, or salt bush. The opportunity costs of diverting water from useful purposes in terms of producing food and fiber, meeting domestic needs, or satisfying industrial requirements for water resources should be estimated. To result in an efficient allocation of scarce resources, the process of used land conversion should compete for water resources in an open market situation.

A further supply restraint is the cost of biological resources available for used land conversion. As users of mined land compete for suitable vegetation the cost of cactus conceivably could rise. A eucalyptus tree could be cheaper in terms of purchase price than a saguaro for use in converting desert land from a previous use to any succeeding use. The cost of grass seed is not without relevance and should be taken into account as a resource cost in used land conversion.

The other materials used in land conversion also have costs. Those costs form a significant constraint on the process of used land conversion. These include the cost of fertilizer applied to landscapes to produce vegetation where none had existed under the prior use. They include the cost of paving if an airport, road, parking lot, or some similar use is to dictate the desired conversion characteristic. These materials will have costs, and those costs can form a significant part of the total cost of used land conversion.

Labor Costs

Although frequently unmeasured and often not even considered, the cost of labor used in converting land from one form to another forms a significant constraint on used land conversion. This is true whether the labor used in such conversion is that employed by the preceding user of the land or whether employed by someone else, a contractor or a government agency. Whether the work be done by bull gangs from the open pit, by college students, or by a landscape contractor, labor costs are involved in used land conversion and they should be considered a significant constraint upon that process.

Capital Costs

Capital costs for both equipment and the financing to obtain that equipment serve as significant factors in the total cost of used land conversion. Generally, land that has been disturbed by expensive equipment must be converted to another use by equally expensive equipment. Where labor is not cheap, the cost of equipment serves as a significant cost factor in any materials handling process. Because the amount of capital tied up in purchasing and using such equipment can be substantial, the financing costs, both in terms of interest rate and the opportunity cost of capital, can be significant.

Knowledge Costs

It is a well known but often overlooked fact that knowledge does not come cheap. Before suitable methods for converting used land to new conditions can be employed, considerable experimentation must be performed. This is particularly true in the semiarid regions of the southwest where the establishment of new vegetation on land that has been disturbed by mining operations is not simply a matter of planting and watering and watching it grow. Where previously undisturbed land has been covered by material such as mill tailings, considerable experimentation has been required in order to determine the most advantageous and economical means of converting that land to whatever subsequent state may be desired. Although it may seem obvious to many that the best way to return desert land to a desert condition is to plant natural desert vegetation, this knowledge does not come easily and in many cases has required the expensive and sometimes unsuccessful planting of many exotic species in order to find that they did not grow as well as cacti. Unfortunately, some people learn very slowly and many are not willing to accept what others have learned. This is an expensive process.

Total Costs

The total cost constraint on used land conversion is thus a sum of the costs of all of the input factors used in that process: land, soil, water, vegetation, materials, labor, capital, and knowledge. The total cost constraint, expressed mathematically, thus becomes

$$\sum_{i=1}^{i=n} C = \sum_{i=1}^{i=n} S \cdot s + \sum_{i=1}^{i=n} L \cdot l + \sum_{i=1}^{i=n} E \cdot e + \sum_{i=1}^{i=n} K$$

where S denotes the quantity of resource used and s the unit price of each type of resource; L denotes the quantity of labor used and l the unit price or wage rate of each type of labor employed; E denotes the amount of capital involved and e the unit price of each; and K represents the total cost of knowledge required for any successful conversion of used land from one form to another excluding that included directly in the cost of labor. Although opportunity costs form a very real part of the total cost picture it is considered more appropriate to treat them as negative aspects of demand constraints rather than in conjunction with incurred costs.

MARKET CONSTRAINTS

Quality of Product Demanded

If the mining process is to pay for the subsequent conversion of the land it has disturbed to a new condition or its return to a prior condition, then the demand for the product of mining forms a significant constraint on the process of used land conversion. Inherently involved in the demand for the product of mining is the quality of that product both as to purity and as to other characteristics. The purity of product derived from mining, whether it be copper, coal, gypsum, limestone, or sand and gravel, to a considerable extent affects the amount of processing required and the amount of waste product produced in achieving that end product. This can have a significant impact upon the amount of land used by the mining operation and by the nature of that use, particularly for the storage of waste products.

Other characteristics as well as purity are important in the demand for the finished product of mining. Probably the best example of these is the energy content of coal mined from western fields. The Btu content of coal also directly affects the amount of waste product that must be stored as a result of mining operations, coal preparation operations, and subsequent use as a fuel.

If, on the other hand, the cost of land conversion is to be paid by the subsequent user of the converted land, then the demand for that land, and not the product of mining, is a significant factor in that conversion. The quality of the finished product, the converted land, is usually significant in the demand for that land. One of the qualities of the land is its configuration. In some cases, level land is of much higher value than land that is not level. In some cases, just the opposite is true.

The vegetative cover upon the land after conversion can also be a significant factor in determining the quality of the finished land product. In some instances such as the use of land for housing or commercial structures, the presence of cover actually lowers the quality of the land for that particular purpose. In other instances the presence of suitable cover will increase the value of the finished converted land. Simply the appearance of the converted land can be, for some uses, a significant factor in determining the quality, and hence the value, of the converted land. In general, the overall suitability of the converted land for its intended use is what determines its value. Whether or not the land will hold water, support buildings, or grow grass, are significant characteristics that determine the quality of land for its intended uses.

Amount Demanded

Where mining is to pay for the cost of subsequent land conversion, the amount of mining product demanded is obviously a significant economic factor affecting used land conversion. The amount of mining product demanded is based on three essential components: (1) derived demand from domestic users, (2) export demand, and (3) speculative demand.

The demand for copper, for example, is a derived demand, and the final demand for products of construction, the automobile industry, and for electrical and non-electrical machinery are the major causal factors that determine the amount of copper demanded. Where coal is the product being mined, the demand for electric power, for hydrocarbon liquids, or for gases derived from coal gasification may be the factors involved in determining the amount of coal produced from a particular mining operation. Where sand and gravel or other construction minerals are the product of mining, the demand for residential, non-residential, and non-building construction, all have an impact on the amount of sand and gravel that is desired. Thus, the demand for consumer products, operating through demand for industrial products and finally exerting a demand on mineral products, is a major determinant in deciding how much land will be disturbed in order to create the product desired from mining.

Export demand, of course, is influenced by similar characteristics from non-domestic users and has had varying degrees of influence on mining in the western United States. In recent years, at least in the copper industry, speculative demand for copper metal has been a smaller, but significant item in the total demand for that metal. This may also be true of petroleum. Such speculative demand, when added to derived demand from both domestic and foreign users, can add to the amount of land that must be disturbed by mining operations as well as adding to the amount of funds available for mined land conversion. Conversely, decreases in such demand can reduce the amount of funds available for used land conversion as they reduce the demand for mineral output.

Where the conversion of land from mining use to another use is to be paid for by the product of that land conversion rather than the product of mining, demand for land of a particular quality and in a particular location forms a significant constraint on that land conversion. Private demand for land will generally be for either residential use or business use, including agriculture. Demand for residential use can affect a substantial portion of land used by the mineral industries in some areas, particularly where mountainous terrain limits the amount of level land available for construction, and inactive waste dumps and tailings dumps provide a limited amount of level surface that can be used for housing. Unfortunately, when waste and tailings storage areas reach a state of abandonment, that state is also reached by most non-mining business uses in the immediate vicinity. Of course, agriculture requires level land for all crop raising, although, for grazing purposes level land is not required and such business uses frequently replace mining.

Public demand for land can also be a significant factor in the demand for land used by mining. Demand simply for scenery has become a major component of the overall demand for land by the public sector. Public demand for scenery, unfortunately, has not been treated as an economic use of land. Nevertheless, it is, and it should be treated as such in the total demand for the amount of the product of used land conversion.

Price

The price that can and will be paid for the product of mining operations forms a major economic constraint on used land conversion where the cost of that conversion must be paid for by the mining operation. The price combined with the amount of mining product demanded obviously provides the funds that can be used, at least in part, for such used land conversion. If the cost of used land conversion rises, however, that price cannot normally be passed on (under a free competitive market system) to the users of the mining product. The prices of minerals must be, where demand is a derived demand, the prices that can be afforded by the using industries. If the price of copper gets too high for those industries that use it, then the producers of aluminum and other substitutes reap the benefits.

The price of competing materials (aluminum; plastics; steel, in the case of copper; and petroleum, hydro-electric power, and wood, in the cases of coal and construction materials) also form a significant economic constraint upon mining output. Under a free competitive market system, the cost of land conversion, if borne by mining, cannot be readily passed on to the consumers of mined products. Only under a monopoly system of resource production is this possible.

Where the cost of the used land conversion is to be borne by the conversion itself or by subsequent users of the land, the price obtainable for the product of the used land conversion forms a major constraint upon the costs that can be incurred in that conversion. Again, as with the product of mining, the price that can be charged for the product of used land conversion must be one that is affordable by the users of that land in its subsequent condition. If the potential residential users of land that is converted after mining use cannot afford to pay the price charged for that land, they will not buy it and use it for residential purposes. If agricultural users cannot afford to pay the price for land converted from a prior use, they will not use it.

Influential in determining the prices of converted lands are the prices of competing lands for residential, business, and other purposes. If competing equivalent land for residential purposes is selling at $1,000 an acre under a free market, no potential user in his right mind will pay $2,000 an acre for land simply because it has been converted from mining use. If suitable competing land is available for grazing at $50 an acre, the livestock grower would be foolish to pay more than that for the use of land that had once been disturbed by mining even after conversion to a suitable grazing surface. The price of competing land, therefore, serves as a significant constraint upon the costs that can and should be incurred in the conversion of land to those potential uses.

Expressed mathematically,

$$\sum_{i=1}^{i=n} V_1 = \sum_{i=1}^{i=n} Q\,(p_P + p_G)$$

where V, the total value of the land after conversion, is simply the product of the amount of land, Q, and its unit price. While the amount of land available for conversion is determined by the characteristics of the mining operation, the unit price that can be obtained for converted land is determined by the appropriate private land market in that particular locale. When the major demand for converted land is for public use, however, the unit price should be considered as the sum of the price in the private land market (p_P) and an increment (p_G) that taxpayers are willing to pay through government to acquire the scenic or other public values produced by land conversion.

THE EVIDENCE

The economic factors affecting both mining and the conversion of land used by mining have operated and will continue to operate as very real constraints on the use of other resources in the conversion of disturbed land. Examples are numerous throughout the west. One of the oldest is in the early gold rush area of California. There, just east and north of Sacramento along the American River, many miles of land were disturbed by gold dredging operations in the latter part of the nineteenth century. These early mining operations took gravel from the river bed and adjacent areas, washed the gold out, and then deposited the remaining sand and gravel as heaps of dredger tails. These dredger tails cover mile after mile along the river and its adjacent floodplain.

At the time the land was disturbed by mining there was no significant impetus from government to expend other resources in subsequently converting that land to any other form. Nature, therefore, was allowed to perform the conversion at her own pace.

As time went on, however, economic incentives for the conversion of that land developed. Resources became available and other inputs were made available at reasonable costs, and the demand for that land in a condition other than as piles of dredger tails increased. Consequently, early in the twentieth century many square miles of that formerly disturbed land were converted to agricultural use as citrus groves. To make this conversion, minimum input was required in terms of resources other than land, and the entire cost of the conversion was borne by the subsequent use, agriculture.

As time went on again, new demand forces arose, which increased the market value of that mined-out land, and it was converted again in the latter part of the twentieth century. This time, the conversion was from agricultural use to residential and commercial use. Again, the cost of conversion was borne entirely by the subsequent user. Many miles of American River floodplain have thus been recycled from their original natural state, to mining use, to abandoned piles of dredger tails, to natural wildlife habitat (as converted by natural processes), to agricultural use (in the form of citrus groves), to the existing state of residential and commercial use, with all costs of land conversion borne by the values created by demand for subsequent use.

Virginia City

At Virginia City, Nevada, under a more arid environment and less favorable conditions for the natural conversion of disturbed land, similar conversion has taken place. In the latter part of the nineteenth century, in an area of rugged topography precious metal mining operations created some of the only level ground available — the tops of waste dumps. Upon abandonment of the Comstock Lode mining operations, there was no incentive for change other than that provided by nature. As a result, the disturbed land was left just as it had been immediately after the cessation of mining activities. The level areas formed by waste dumps were not returned to any original contours. As time moved on into the twentieth century, the availability of level land matched an increasing demand. This increased the price of the land previously used for mining and it was converted to residential, commercial, and public facilities use with the subsequent users bearing the entire cost of land conversion.

Calico

At Calico, in San Bernardino County, California, silver mining operations in the nineteenth century created excavations, waste dumps, and tailings dumps. At the time that mining operations ceased, there was no incentive to change that used land to any form other than that which it had at that time. In the arid Mojave Desert nature did very little land conversion of her own except by erosion from wind and a few flash floods. In the 1960s, however, the demand for recreational facilities, particularly those linked to a historic subject, has provided sufficient demand for that disturbed land in the form that it was immediately after mining ended. It has been converted to an amusement enterprise operated by the County of San Bernardino. The land disturbed by mining has not been changed in form but merely applied to a use other than mining. The original disturbance has, in fact, made that land more valuable now as a recreational resource than it would have been had it been returned to the original desert condition that it held prior to mining.

Castle Dome — Pinto Valley

Near Miami, Arizona, the Castle Dome open pit copper mine ceased operations in the early 1950s. A limited part of the land, the abandoned tailings pond created by the previous milling operation, was converted to a huge catch basin to provide water for a small leaching operation continued by the owner of the property. There was no incentive, nor were there any resources available at reasonable cost, to return the existing pit and the other land that had been used by mining to the condition it had been in prior to mining. Such a change would have involved not filling up a hole but rebuilding a mountain.

For twenty years after the original abandonment of the Castle Dome Mine, the land remained in the state that it had immediately after mining operations were curtailed. As the demand for copper increased, however, and as technology provided means of utilizing lower grade ores, the remaining part of the Castle Dome deposit, now referred to as the Pinto Valley Property, became useful again as a mine. Thus, after twenty years, the land originally disturbed by mining and left in its disturbed state again became usable as a mining operation. The cost of conversion, deepening the pit, building new mill facilities, developing new waste dumps and tailings areas is being borne entirely by the subsequent user of the land, the new copper mining operation.

Bisbee

Late in 1974, the Lavender Pit operation of the Phelps Dodge Corporation, at Bisbee, Arizona, was shut down because of dwindling ore reserves. Under normal economic incentives for change, this pit would be left as it is, with erosion allowed to smooth out the sides, eliminate the rough edges of the former mining benches, oxidize some of the remaining copper minerals and provide a colorful display for tourists visiting the area.

Under the application of the negative incentives and disincentives now proposed by some, however, the prior user, the copper mining firm, would be required to return the Lavender Pit to the original contours that it held in the 1870s. This would require either taking material from the waste dumps, which are currently being used in a leach operation as a secondary source of copper, or the old tailings ponds and putting that material back into the hole that was excavated to form the Lavender Pit. Some of those waste dumps and old tailings ponds, however, are still being used and the material is not available for fill. Thus, to fill the Lavender Pit back up again would require the excavation of new material from a new hole in the ground. This new hole, of course, would have to be filled up by material from another hole in the ground. All of this reclamation would produce a smoothed-over Lavender Pit that would provide land that would be of questionable value for residential or business purposes in a community that had lost its major source of personal and business income.

Current economic forces, on the other hand, seem to indicate that the Lavender Pit would be much more valuable to the community, to the state, and even to the existing owners of that land if it were left in its present condition as a tourist attraction similar to that currently enjoyed by the disturbed land at Calico, in California. If left on their own, the natural economic forces governing the recreational resource and land markets in the Bisbee area will determine the most beneficial use for that land which has been used by mining and will prevent the misallocation of other resources. Would it not be sensible to allow those forces to operate everywhere?

10.   Political, Governmental, and Social Considerations

Michael D. Bradley

THE MINERAL INDUSTRY has affected the natural environment in the past, but more important are the coming effects of environmental concerns upon the mineral industry in the future. Between the years 1975 and 2000, environmental concerns will constitute a major, if not the major, factor in industry operations, with ultimate consequences that will force unanticipated change in traditional methods of mineral exploration and processing (Flawn 1973). The costs are as yet unclear, as is the willingness of the general public to pay them, either directly when passed on as price increases for pollution control, or indirectly, as reduced industrial capacity to find and work minerals due to public restrictions and controls. Nevertheless, public policy is crystallizing around nebulous concerns of environmental quality, and the mineral industry faces a future of rapid change, social responsibility, and responsive planning for policy-making.

A major manifestation of change is public concern over the reclamation of mining and processing wastes and devastated lands. The standards of public acceptance for mining wastes and industrial residuals have changed considerably since the early part of the twentieth century, leading to legislative proposals with far reaching consequences for industry. Reclamation planning occurs more frequently now than before, and offers insights into the nature of the social responsibility of both government and industry. Public concern over reclamation illuminates possible planning methods and systems appropriate for future environmental policy. Among the more important ideas are the technical and economic interdependencies in reclamation, the changing social expectations in a no-exit system that provides public goods, the ability of residuals management and ecological planning to increase loyalty and reduce turbulence for the mining industry, and the necessity for future-responsive long-range natural resources planning.

Portions of this research were supported under the provisions of Title II of the Water Resource Planning Act of 1965 by the Office of Water Research and Technology, U.S. Department of the Interior.

RESOURCES PLANNING AND RECLAMATION

Minerals demand comes mainly from developed economies, specifically from chemical and manufacturing industries and from agriculture. Large quantities of raw materials are essential for maintaining output, and continuing output is vital to a healthy industrial economy. Greatest demands (apart from fuels) are in iron, copper, aluminum, and the fertilizer minerals. In addition, rare metals found in the earth are essential for fabricating alloys and for sophisticated industrial purposes. The future supply of non-fuel minerals is a function of the demands of culture and population, that is, of the real growth rate of the Gross National Product (GNP) to specific populations, and of particular cultural types. Demand, either small amounts per capita for increasing population or large amounts per capita for static population of improving material standards, is the key concept for future-responsive resources planning. The disparity between demand forecasts and actual economic output or measured performance has again and again been subject to increasingly critical review. The lessons to be learned from these reviews are often elusive, but important to avoid creating conditions for further uneconomic, or naive resources planning.

It is important to define the terms derelict land, disturbed land, damaged land, and devastated land. Derelict land is so damaged by industrial processes that it is incapable of beneficial use without treatment. It includes stripmine workings, spoil heaps, other waste tips, dilapidated structures, and land damaged by subsidence. Disturbed land has some use, however small. This usually refers to land after treatment such as infilling, leveling, revegetation, or other reclamation. Derelict and disturbed land are often found together, and referred to as damaged land (Goodman 1974). Devastated land refers to land so exploited that irreplaceable constituents are removed, leading to physical and chemical destruction (Myers 1946). Some residuals from minerals processing devastate the storage site over long time periods, but as with most extreme problems, this occurs infrequently and in limited scale.

The production of industrial and minerals waste and residuals is an inevitable consequence of modern life. Land is exploited for many purposes. Nevertheless, permanent damage is avoidable, indeed it is possible to prevent further devastation and eliminate many eyesores and damaged areas that are a legacy from the past (Knabe 1965). Societal choices and trade-offs are becoming more clear, and the interaction of forward planning, economic incentives, taxes, and enlightened public policy is capable of guiding reclamation policy to valued social ends.

Public policy needs to be based upon clear understandings of the magnitude and complexity of the task. The effects of mining waste exist at two levels: the direct biophysical, and the indirect socio-economic. The direct biophysical problems of minerals residuals are often so repellent to potential reclamation or development that derelict land remains unused. The geographical location of a mineral deposit and the technology used for its extraction have consequences upon the local ecosystem; that is, the complex levels of organization between populations, communities, and the biosphere. These interactions often combine with the direct residuals of minerals waste and tailings. For instance, copper extraction produces many mine and mill residues, some important for the local ecosystem and some not. Toxic substances are found in residuals from copper mining, as are nutrient deficiencies for plant growth — such as nitrogen, phosphorus, and potassium. The waste is often acidic, highly saline, physically unfavorable for reuse — problems include crusting, leaching, and low water retention capability — and the absence of soil microorganisms retards vegetation and reclamation (Peterson and Nidson 1973). While these problems can be overcome, successful reclamation is expensive in both money and physical energy, and the benefits, such as wildlife population increase based upon newly established vegetation, tend to be second order and non-reimbursable (Holland 1973).

The indirect socio-economic consequences of minerals waste and reclamation are less readily identifiable and often more pervasive. The patterns of land-use near mines and the economic force of dominant employing industries, such as mines in rural areas, brings consequences difficult to foresee and nearly impossible to resolve. Landownership patterns often make reclamation difficult. Poor access to remote locations can work against reclamation, yet local governments are understandably reluctant to build costly roads in the hope of attracting other industry to a derelict area. Poor housing and the absence of services such as electric power, water, and sewerage are factors that add to the temporary nature of mining camps and towns, decreasing a long-run local identity with the location, and its eventual reclamation. Finally, flash flooding and unstable spoils piles are sometimes a hazard to local residents, and are seen as fearsome objects, not as possible investments in reclamation (Goodman 1970, Weisz 1970).

CHANGING SOCIAL EXPECTATIONS

Conservation, or the responsible and wise use of natural resources, is a philosophy long held by many cultures, expressed in many different ways. In developed nations increasing lifespans, material production, and leisure time have induced a greater concern for the environmental future, which appears threatened by both material shortages and declining quality. Concern for these problems can be traced to the early nineteenth century, when Malthus analyzed the conflict between increasing human population, and fixed or decreasing resources. Early American conservationists, notably Gifford Pinchot and Theodore Roosevelt, also responded to anticipated depletions and resources scarcities. Time has not proved resources in short supply. Usually technological advances enable new uses to respond to new markets; however, time has sharpened the public concern for environmental quality. Responding to the changing social expectations in environmental quality is a relatively new role for the minerals industry, without many precedents or rules to guide appropriate investments and policies.

Until recently, the accepted practice was, and to some extent is still, to mine accessible deposits as cheaply as possible, with little or no consideration of the resulting social and environmental costs or externalities. Today, however, most of the mining industry recognizes the consequences of mining operations upon the environmental quality of a locality or region. While the absolute amounts of damaged or disturbed lands are quite small, the proximity of some of these sites to either urban populations or recreation facilities — for example, Tucson, Arizona, and Red River, New Mexico — calls for special consideration of altered or lost amenity values. Serious disamenity is a value function on another scale than the amount of land involved in mineral operations. Every acre of derelict or disturbed land near a large town or scenic area is a far greater disamenity than thousands of similar acres in a remote, uninhabited place (Goodman 1970). Conflicts over disamenity in derelict land are also more than just geographical problems; the economics of disamenity also come into play. Disamenity is another term for the social cost of private mineral production or the cost of economic growth (Kapp 1967, Mishan 1967). But the disamenity problem with mining waste and derelict land contains a quirk that was unrecognized until recently. Mining disamenities are a no-exit system in a crucial industry, a situation leading to expansive use of the voice option for conflict or policy resolution.

The recently developed theory of exit and voice as social responses to decline in firms, organizations, and states has provided insights and illumination for a wide range of social, political, and even moral phenomena (Hirschman 1970). The ideas are simple. Firms (including mining firms) produce saleable goods and services. Through time, a firm's performance suffers deterioration for many reasons, and the quality of the product or service declines. When this happens, customers stop buying the firm's products or members leave the organization. This is the *exit option.* As revenues drop, management hopefully seeks ways to correct whatever faults led to exit. Another possible behavior is for customers or employees to express dissatisfaction directly to management, or to generally protest the quality decline. This is the *voice option.* As a result, management again strives to cure the source of dissatisfaction (Hirschman 1970). The interplay of these optional behaviors provides insight into many parts of the environmental quality and land reclamation puzzle, and the policy responses appropriate for industry.

It is first necessary to understand three important points: mining is a critical, no-exit industry; minerals production includes environmental public goods; and voice is a form of loyalist behavior. That the mining industry and its products are crucial to the continued economic and technical health of the nation is obvious. Modern industrial economies have few if any substitutes for the fuels, ores, and metals vital for producing a range of goods and services ranging from heavy equipment to surgical needles. In fact, the minerals industry as a whole represents industrial values even more important than those natural monopolies called public utilities. Mining is also a location specific activity. Industry must mine ore where it occurs, and the location of production is a dependent variable in minerals output. For these reasons, minerals production is a no-exit industry. Substitutes are not available for valued ores and minerals, and location is specific for each resource occurrence. The mining industry can neither disengage operations and move on whim to remote locations nor reduce on whim the output of raw materials vital to the industrial economy. By the same token, local residents can neither wish noisy, dusty, and unsightly mines away nor demand that output of a nationally vital resource be reduced or produced in quantities that maintain local-environmental purity. The theory of the exit option is here reversed, dissatisfied regional or local residents, facing silted streams, dust, and pollution cannot exit or stop buying the firm's products. Usually local residents do not constitute the market for industrial raw materials, and unless they are members of the mining organization, the exit option for both parties is closed.

On the other hand, raw materials for basic industries are not the only product of the minerals companies; ores and metals may be the only profitable product, but they are not the only physical product. Mining operations also produce waste piles, silt, dust and other forms of pollution. The residuals of industrial operations may not be saleable goods, but they are public goods in a sense. Public goods are produced by both the public and private sectors, and consumed by all members of a community, country, or geographical area in such a way that one member does not detract from consumption or use by another. Examples include crime prevention and national defense, as well as such commonly shared accomplishments as international prestige, literacy, and public health. The distinguishing feature of public goods is not that they can be consumed by all, but that there is no escape from consuming them unless one were to leave the community by which they are provided. Thus, public goods are also public evils (Hirschman 1970). Public goods for some community members, such as freedom from public regulation regardless of the environmental costs of an operation, may be public evils for others, such as deterioration of scenic amenity by industrial operations. As environmental quality becomes a more active goal for federal and other public bodies, providing a "quality" environment will become recognized and accepted as a worthwhile public good, and pollution and land disturbances from industrial operations will be increasingly seen as a public evil. In a no-exit system between industry and the local community, conflict is the likely result.

Further, in the face of declining quality in goods and services, that is, as the public good of environmental quality deteriorates with increasing pollution, the no-exit system practically guarantees the increased use of the voice option. If people cannot avoid a public evil by exit, they will attempt to change the practices, policies, and outputs of the firm by voice. Examples of voice might include individual or collective petition to the management directly in charge, appeal to a higher or public authority with intention of forcing a change, or various types of actions and protests. Two points are of interest here: first, the voice option is in the grand tradition of American politics, it is often called "interest articulation;" and second, voice is a loyalist behavior. Voice does not necessarily require outside agitators, trouble-makers, or social reformers. Instead, voice often represents local residents faced with declining environmental quality. It is an articulation by those who reject the crude admonitions of some minerals industry spokesmen, who in recent years insisted upon the acceptance of slag heaps and tailings dumps as the ransom for local payrolls and prosperity. Postponing exit in spite of dissatisfaction and qualms is loyalist behavior. Long-range problems are likely to encourage those with strong, local identities and goals, not transients.

It is not possible to state precisely the numbers of people engaging in exit, loyalty, or voice in response to environmental problems near mines, but the trend of recent affairs can offer rough indices of the rise of voice in environmental policy. Since 1970, the environmental movement has specialized in the application of voice to environmental affairs. That success, and the strong support environmental voice has generated, will encourage further applications of voice in the future. The results will have many forms, some already discernible: boycotts and disruptive social action; continued conflict over the allocation of scarce natural resources, such as groundwater from the Tucson aquifer; further dissatisfaction over the industrial use of publicly shared and enjoyed environmental amenities; zoning restrictions for drilling and exploring on private land; environmental lawsuits against industrial polluters; environmental impact assessment as required by the National Environmental Policy Act of 1969; legislation for land use planning; strip and surface mining regulations; continued critical examination of minerals tax advantages and allowances, and so forth. The future social environment for the minerals industry is unlikely to be the stable, benign, accepting and rewarding system it was in the recent past. Instead, the future is likely to be uncertain, plagued with scarcities and complexities, presenting turmoil above all else. The important lessons to learn now in preparation for the future relate to planning under conditions of social uncertainty and turmoil.

Future-responsive planning under conditions of social uncertainty and turmoil is a new field of interest to a few concerned resource professionals. As yet, no theory of long-range planning exists that offers organized and satisfying concepts and interrelationships. By experience, long-range planners are beginning to understand the organizational behaviors conducive to long-range planning. The importance of organizational feedback is being recognized, as are the substantial possibilities for industrial response to help modify turbulence, by such policies as formal residuals management, ecological decision-making, and long-range natural resources planning.

ORGANIZATIONAL FEEDBACK

Feedback is a concept more easily understood than applied, and its role in the social responsibilities of business and industry is a recent perception. Feedback can be defined as the total information system process through which primary and secondary order effects of organizational action are fed back to the organization and compared with desired performance (Rosenthal and Weiss 1966). In the free enterprise, competitive market economy, exit was traditionally considered appropriate "feedback." A merchant or manufacturer faced declining sales and revenues as an indication of consumer dissatisfaction, and hastened to change his operations in line with the requirements of a new fashion or market. In modern society, things are neither so simple nor so clear. Industry, including the minerals industry, produces not only private products but also public goods and evils. In the strongly interrelated social and biophysical system of modern society, exit is no longer an appropriate feedback; no "there" exists to exit to. Voice is rapidly becoming a dominant social feedback mechanism, and its use will likely increase in the future. The question now becomes, how do we change industrial and governmental organizations to respond effectively to voice as feedback?

In theory, organizational change through voice as feedback should come about easily; in actuality, organizational change is usually difficult, working best with economic incentives, and worst with social or philosophical incentives. Two straightforward reasons account for this behavior: illusory immunity and information distortion. In the past, public and private organizations appeared immune to many of the consequences of their actions; exit was the expected feedback, and the secondary or social consequences of a program or project were considered beyond the interest, or even the expertise of the organization. Examples include strong, locally dominant industries such as mining, fishing, or forestry; constituent-responsive public agencies such as the Corps of Engineers, the Bureau of Reclamation, and the U.S. Forest Service; and public utilities such as electrical power producers. Competitive market forces provided inadequate feedback, and immunity was taken for granted. Also, organizations reviewing feedback often developed elaborate mechanisms for buffering the information. Instead of a clear indication that revised goals were in order, feedback was often distorted to support goals accepted by the organization.

What now is clear is that feedback contains important information for planning and policy-making. Public agencies are developing programs of public participation to encourage feedback. Private organizations are developing boundary spanning roles within their structures. For example, the recent rise of environmentalists employed by minerals and energy companies is an important way of capturing information, and including it in plans and projects. The boundary spanning role balances external concerns, such as environmental quality expressed by voice, with internal standards of acceptable conduct (Michael 1973). This leads to altered appreciations of the expected and acceptable, an important source of organizational change.

TURBULENCE AND LOYALTY

The most important role of feedback, however, is as a management tool to respond to social turbulence and to develop institutional loyalty. Through feedback, three important trends for reclamation planning and natural resources policy are becoming more obvious, and more important as part of governmental and industrial responses to the future: residuals management, ecological planning and policy-making, and long-range natural resources planning.

Reclamation and Residuals Management. It is easy to mistake the recent interest in reclamation as a passing fancy and a call for short-term cosmetic responses. In fact, reclamation of mining wastes is another expression of environmental policy; more specifically, reclamation belongs to that socioeconomic endeavor termed residuals management (Kneese, Ayres and D'Arge 1971, Bower 1971). The ideas of residuals management are common to civilized cultures as directives to clean up after oneself, to show mutual consideration to each other's situation or predicament by conform-

ing to rules of conduct which appreciate the longer view, and figures consequences as well as immediate personal or pecuniary advantage. On the
other hand, a characteristic of barbarism — which, fundamentally, is the
inability to make oneself understood, or to understand what others mean —
is to grasp at momentary advantage, without regard to the consequences
or the convenience of other people (Myers 1946). Residuals management
incorporates the civilized impulses of modern, industrial man into a philosophy and technology of waste control. As public policy, residuals management deals with those common properties, environmental resources, and
values that are subject to either congestion, overuse, or both. An additional
user of such resources imposes externalities on other members of society,
for he rarely takes into account the loss of amenity or economic value
imposed on others by selfish overuse or exploitation.

Several facts of life need to be understood about residuals management
(Bower 1971). First, residuals generation is pervasive; all human activities
produce residuals. Therefore, the absolute prohibition of all residuals, or
the total reclamation of all mine wastes, is socially and technically impossible. Not only do the first two laws of thermodynamics (conservation and
entropy) guarantee no physical process, including cleansing processes, to
be absolutely efficient, but also the money and the energy cost of purity is
prohibitive. The incremental costs of residuals removal increases dramatically as one hundred percent removal is approached. It is impossible to
remove completely residuals from an industrial process. This first principle
provides an important insight. All production and consumption use the
assimilative capacities of land, air, and water systems. These receiving media
are usually underpriced or unpriced in organizational cost accounting, and
assigning acceptable dollar values to these services is difficult. What is
needed are market substitutes; that is, public environmental planning and
regulation to encourage balanced, harmonious use, and to prevent over-
loading or exploitation. Second, physical, technological, and economic relationships exist between the two major kinds of residuals — energy and
materials — and effective, equitable residuals management must deal, not
with individual residuals, but instead with a complex of relationships. For
example, the energy requirements expressed by the common denominator
of the British thermal unit (Btu) can serve as planning variables. Btus
allow a resources planner to assess more than the money cost of a production function or conservation practice; energy and its equitable use can
also serve to guide public policy. The energy requirements — the Btu equiva-
lents — of capital, labor, fuel, fertilizers, seed, and water, can be calculated
and subtracted from the net energy requirements of mineral production.
Net energy costs may be more enlightening than the economic cost of trade-
offs between producing from marginal or uneconomical mines and expending resources for land reclamation (Odum 1971, *Energy and State Government* 1973, *Energy Study* 1974).

Residuals management is more than a new method of cost accounting,
it is also a proposal for more enlightened social choices. The throw-away
philosophy of earlier decades is changing at both the individual and the
social level, and residuals management is an expression of the changed
social values supporting environmental quality, resources conservation, and

industrial responsibility. While environmental quality varies in both time and place, the past failure to consider quality considerations in both local and national arenas has left many industries, including minerals processing, with hard and expensive choices in today's business environment. Past mistakes can be considered sunk costs if the present and future industrial response to environmental quality is reasonable, rational, and carefully considered. Residuals management offers precisely such a system for the problems of industrial and mining waste.

Ecological and Landscape Planning. A second response to the problems of future social turbulence lies within the disciplines of natural resources ecology, landscape architecture, and their conjunctions. These interrelated fields are termed ecological planning or landscape planning (Hackett 1971, Hills 1974). The source of a renewed interest in land use and landscape planning is the environmental design arts and sciences. Its concern is with the most appropriate land uses based upon ecologic and aesthetic factors. Two important considerations for reclamation of mining wastes are addressed by landscape planning: first, the prevention of land disturbance, devastation, and dereliction by forward planning; and second, the correction of disturbed, devastated, and derelict land ecosystems which are unavoidable or already exist (Tandy 1973).

Ecological and landscape planning is more than a new practice, it is a philosophy of land use ecology finding expression in public policy, notably legislation. Concepts of ecological planning are features of the proposed land use planning and surface-strip mining legislative proposals that were introduced in the 93rd Congress, bills that will be resubmitted in the 94th Congress. An important expression of these ideas is the National Environmental Policy Act of 1969 (NEPA). Action-forcing provisions of the law require an environmental impact statement (EIS), analyzing the environmental consequences and costs of an action or project in advance of expenditures or commitments of resources. The provisions of NEPA and the response of federal agencies is a subject of considerable controversy. NEPA does, however, focus public policy upon the goals of providing safe, healthful, productive, and aesthetically and culturally pleasing surroundings; attaining the widest beneficial use of environmental resources without degradation, health and safety risks, or other undesirable or unanticipated consequences; and enhancing renewable resources quality while maximizing the recycling of depletable resources (NEPA 1969). An important corollary is the development of another idea in good currency, the concept of assessment. Assessment deals with the second-order consequences and the tertiary impacts of proposed actions or projects. Secondary consequences often affect human ecosystems more strongly than the direct effects of an action. Demands are induced for new, expanded facilities or to change significantly natural conditions, often by individually trivial but cumulatively significant acts. Assessment is important for two further reasons. First, it provides public policy a method to reduce the imperatives of "hypocritical geography" (Carey and Greenberg 1974). Often projects or activities provide strictly local benefits and greater than local costs. Reducing the inconsistent spread of local benefits and regional costs — internalizing the externalities, as economists call it — is a particularly important and appropriate role for resources assessment. Second, resources assessment is part of a larger paradigm shift in the theories of regional and resources planning (Kuhn 1970). Planning is changing, according to one theorist, from the models of allocative planning that stress comprehensiveness, system balance, quantitative analysis, and functional rationality to the concepts of innovative planning, stressing institutional change, experimental and action orientation, and resource mobilization (Friedmann 1973). Paradigm shifts in planning theories represent new social expectations for the minerals industry, and new standards by which conduct, responsiveness, and responsibility can be compared.

Long-Range Natural Resources Planning. A third response to a social future of increased turbulence and turmoil is long-range natural resources planning. The challenges of long-range planning under social conditions of scarcity and complexity have only recently been recognized by public officials (Train 1974). Natural resource scientists and managers in public agencies also recognize the growing importance of long-range assessments, and are expressing keen interest in training and continuing education in this area (George & Dubin 1971). Few planning or resource management scholars have responded to this growing demand, although serious thought is starting toward long-range social planning (Michael 1973). Important planning efforts in long-range analysis need the attention of more than just public planners; the perspectives, skills, and concerns of industrial policy-makers are also a necessary ingredient for successful long-range natural resources planning. Institutional adoption and change will be necessary, and the mechanisms for such change have yet to be developed.

In summary, the problems of mining residuals and land reclamation cannot be fully understood from a limited focus, or a single purpose response to seemingly irrational outside pressures. Instead, reclamation and residuals management are part of larger social attention upon resources planning and environmental quality. Reclamation is more than charity; it is a civilized duty in an interrelated human ecosystem. The standards of acceptable conduct in and toward a shared biosphere are changing, and their expression is shifting from patterns of exit to patterns of voice. The theories of resources planning necessary to deal with future scarcity, turbulence, and turmoil are also changing; from single purpose, profit maximizing, efficient engineering to greater concern with residuals management, innovative assessment, and long-range natural resources planning. The costs and consequences are as yet unclear; but the challenge is clear. Developing guidance systems for natural resources use that are equitable, responsive and capable of acceptance is a task adequate to engage the best public and private talent for as far ahead as can be seen.

REFERENCES

Bower, Blair T. 1971. Interpretation. residuals and environmental management. *J. Am. Inst. of Planners.* Vol. 37, no. 4: 218–20.

Carey, George W. and Greenberg, Michael R. 1974. Toward a geographical theory of hypocritical decision-making. *Human Ecolo: An Interdisciplinary J.* Vol. 2, no. 4: 243–57.

Council on Environmental Quality. 1973. *Federal Register.* Preparation of Environmental Impact Statements: Guidelines, vol. 38, no. 147: 20551–52.

Flawn, Peter T. 1973. Impact of environmental concerns on the mineral industry, 1975–2000. *The mineral position of the United States: 1975–2000.* Ed. Eugene N. Cameron. Pp 95–108. Madison: University of Wisconsin Press.

Friedmann, John. 1973. *Retracking America: a theory of transactive planning,* Garden City, New York: Anchor Press/Doubleday.

George, John L. and Dubin, Samuel S. 1971. *Continuing education needs of natural resource managers and scientists.* The Pennsylvania State University Press: Dept. of Planning Studies in Continuing Education.

Goodman, G. T. 1970. Ecological Aspects of Reclamation. pp. 254–55 in *Environmental Side Effects of Rising Industrial Output.* Ed. Alfred J. Van Tassel. Lexington, Massachusetts: D. C. Heath and Company.

Goodman, G. T. 1974. Ecological aspects of the reclamation of derelict land. Pp. 251–64 in *Conservation in practice.* Eds. A. Warren and F. B. Goldsmith. London: John Wiley and Sons.

Hackett, Brian. 1971. *Landscape planning: an introduction to theory and practice.* Newcastle upon Thyne, England: Oriel Press.

Hills, G. A. 1974. *Landscape planning.* Landscape Planning: An Overview. Vol. 1, no. 1, pp. 107–10.

Hirschman, Albert O. 1970. *Exit, voice and loyalty.* Cambridge, Massachusetts: Harvard University Press.

Holland, Frank R. 1973. Wildlife Benefits from Strip-Mine Reclamation. Ecology and Reclamation of Devastated Land, vol. 1, ed. R. J. Hutnik and G. Davis, pp. 377–88. New York: Gordon and Breach.

Kapp, K. William. 1950. *The social costs of private enterprise.* Cambridge, Massachusetts: Harvard University Press.

Knabe, Wilhelem. 1965. Observations on world-wide efforts to reclaim industrial waste land. In *Ecology and the industrial society.* Eds. Gordon T. Goodman, R. W. Edwards and J. M. Lambert, pp. 263–96. New York: John Wiley and Sons, Inc.

Kneese, Allen V.; Ayres, Robert U.; and D'Arge, Ralph C. 1971. *Economics and the environment: a materials balance approach.* Baltimore: The Johns Hopkins Press for Resources for the Future, Inc.

Kuhn, Thomas S. 1970. *The structure of scientific revolutions,* 2nd. Edition. Chicago: University of Chicago Press.

Michael, Donald N. 1973. *On learning to plan — and planning to learn.* San Francisco: Jossey-Bass Publishers, Inc.

Mishan, Ezra J. 1967. *The Costs of Economic Growth.* New York: Frederick A. Praeger, Publisher.

Myers, Sir John. 1946. Devastation. *Nature,* vol. 158, no. 4018: 605.

National Environmental Policy Act of 1969, Pub. L. 91–190; U.S. Statutes-at-Large, vol. 83, p. 852.

Odum, Howard T. 1971. *Environment, power, and society.* New York: Wiley-Interscience.

Oregon, State of. 1973. Energy and State Government: A Decision-Making System designed to Integrate Social, Economics, and Environmental Processes. Salem: Office of Energy Research and Planning. pp. 1–34.

Oregon, State of: 1974. *Energy Study: Interim Report.* Salem: Office of Energy Research and Planning. pp. i–iv.

Peterson, H. B. and Nidson, Rex F. 1973. Toxicities and deficiencies in mine tailings. pp. 15–25 In *Ecology and reclamation of devastated land.* Volume 1. Eds. Russell J. Hutnik and Grant Davis. New York: Gordon and Breach.

Rosenthal, Robert A. and Weiss, Robert S. 1966. Problems of Organizational Feedback Processes. In *Social indicators.* Ed. Raymond A. Bauer, pp. 302–40. Cambridge, Massachusetts: The MIT Press.

Tandy, Clifford R. V. 1973. Industrial land use and dereliction. In *Land use and landscape planning.* Ed. Derek Lovejoy, pp. 197–233. New York: Harper and Row Publishers, Inc.

Train, Russell. 1974. The challenge of scarcity. *Cry California.* vol. 9, no. 4: 2–12.

Weisz, John A. 1970. The Environmental Effects of Surface Mining and Mineral Waste Generation, pp. 291–312 in *Environmental Side Effects of Rising Industrial Output.* Alfred J. Van Tassel (ed.) Lexington, Massachusetts: D. C. Heath and Company.

PART III

MINING AND THE ENVIRONMENT

AMERICANS ARE STRONGLY CONCERNED about the quality of their environment and broadly support environmental programs. One segment of the
population tends toward legislative solution to environmental problems,
and has developed considerable political power which still may grow in
the future. A milestone in legislative action for environmental land management was the creation of the National Environmental Policy Act. Gary
L. Widman discusses a few of the highlights of NEPA and the environmental
impact statements it requires. S. Norman Keston suggests how the mining
industry can learn to live with the impact statement and use it to advantage
in environmental planning for a new mining operation.

The large-scale land modification required by a new mining operation
is capable of generating a completely new ecosystem, desirable or undesirable, which nevertheless opens a challenging area for the applied ecologist.
Mohan K. Wali explores this concept and suggests that perhaps there should
be a new subdivision of classical ecology: mining ecology.

No issue is more closely concerned with the demand for conservation
and the current energy question than mining of the vast fossil fuel deposits

in the southwest. The effects of power generation and minerals processing on the environment can be equal to or greater than the mining operations themselves. Donald Ermak, J. R. Kercher and Ronald L. Ritschard present a methodology, tailored to the environment and circumstances that exist in the southwest, for evaluating the environmental effects of a coal fired power generation station. The sources of environmental degradation (mining, coal transportation, power plant operation, and power transmission) and the possible transport pathways of their pollutants through the environment are identified. The technology of shale oil development, its environmental impacts, and reclamation plans to minimize impacts are discussed by Paul D. Kilburn. Coal and oil shale development will have an important effect on the local and to some extent the regional hydrology.

The effects of mining on the hydrologic environment of arid areas is of great concern throughout the southwest. The demands on limited water supplies are already great in much of the region. The hydrologic environment may be affected not only by the additional water demands of mining and consumptive use of water by vegetation on reclaimed lands, but also by contamination or disruption of existing supplies. The hydrologic effects of an open pit copper mine in southern Arizona are discussed by Leonard C. Halpenny, and those of a coal strip mine in northern Arizona are discussed by Tika Verma.

11. Environmental Law and Mining

Gary L. Widman

THE NATIONAL ENVIRONMENTAL POLICY ACT (NEPA) is the major land-mark environmental law designed to improve the federal government's decision processes. It says very little about what a government agency must do in terms of imposing substantive environmental restrictions. What it does provide is the information base with which an agency makes a decision, and the use of that information in its decisions. It says that a government decision (including, among other decisions, those for the issuance of per-mits, whether under future strip mining laws, or for rights of way) must proceed only with the benefit of certain information that may have been overlooked in the past.

NEPA is probably a reaction to the failure of specialized decision making in the 60s. When Congress considered agencies as diverse as the Bureau of Land Management and the Atomic Energy Commission, it probably saw similar failings in government decision making. The agencies were making decisions which made sense from the limited perspective of the predominant profession in that agency. But when these decisions were viewed with hindsight, it could be seen that many factors had been over-looked. In order to attack that problem in the future, NEPA was adapted to bring more diverse and interdisciplinary information into these "narrow" government decisions.

NEPA also addresses a need for long-range planning. There is at least the same need for long-range planning in government projects and in its decisions to award permits, as there is for a private enterprise that needs to study the engineering and tax implication of its proposed structures. The enterprise needs to know, to the best of its predictive ability, what will happen tomorrow, and next year and in the next decade, and so does the government. Government has historically been an institution with a short-range focus, and Congress saw need for more thorough study of longer-range environmental effects. NEPA, in part, satisfies that need.

NEPA is also a strong management tool that authorizes a government manager to spend money to develop information on the effects of decision options. These funds frequently would not have been available to the far-sighted government executive in the past.

[97]

The act also focuses an agency's decision processes. It encourages the manager to define the point in his operating structure when he must make a decision to choose a particular location, or to grant or deny a permit. Prior to NEPA, many decisions were not really "made," but simply evolved. The regional office would call the state office, and the state office would then call Washington. The problem would pass through two or three layers in the main office, and would be passed back down to the state, many telephone conversations later. An acceptable course of action would evolve by consensus. But there may not have been a point where the manager would have said that a decision had clearly been made. To some extent, NEPA has changed that. It is a force that has caused decisions to be structured and responsibilities to be identified, and has encouraged better management in the whole system.

A fourth important function of NEPA is to let the public in on "what is going on in the agency." These benefits run two ways. Some benefits, of course, accrue to the public, which has advance notice of what the government is going to do about a particular problem. The public has some time in which to provide information to the government agencies or take whatever other steps may be appropriate.

At the same time, the public review process avoids some of the problems of specialized decision making. The public, after all, has a tremendously wide spectrum of interest and expertise. The public includes everyone from a classic "little old lady in tennis shoes" to an entire department of a university that may choose to comment on a particular project. So the quality and value of comments which an agency receives may vary greatly. In some cases, responses from the public may represent greater expertise than the agency itself has brought to bear in drafting its initial report. The agency clearly benefits from public review.

Finally, there seems to be a need by state and local agencies for more extensive descriptions of proposed federal projects. The Environmental Impact Statement may partly satisfy that need. Twenty-two states now have their own "NEPA-type" requirements, so that even if there were no NEPA, state laws would require similar reports. And if these reports were not required for state "little NEPA" laws, they might well be required by state, regional or local land use planning processes. The EIS may satisfy the needs of local zoning bodies, who wish to understand the effect of a proposed project.

In general, the EIS process involves several steps. When a project, involving federal funding or licensing, is first proposed, an agency will be expected to make a preliminary assessment. The agency first looks at the proposal to see if there is a possibility that agency actions or decisions could have any significant environmental effects. If it determines that no such effect will be created, it will usually prepare a "negative determination," — a statement that no significant impact is foreseeable. But if it foresees a possibility of such an impact, it would then prepare its draft environmental impact statement.

Environmental Impact Statements have taken many forms ranging from the three-page documents that were submitted shortly after the law was passed in 1970, up to some 10,000-page documents that have been virtually impossible to read within the review period. There is a need to shorten the average statement. There seems to be no reason why, in the average case, a statement could not be confined to eighty pages. There has been much emphasis on detailed biological descriptions of project settings. Biological studies should be done, but unless a study has relevance for a particular decision problem, extensive biological cataloging need not be included in the statement. It might be referenced and made available, but frequently need not be included.

Sections dealing with "impacts" and "alternatives" should probably receive heavier emphasis. In the impact sections, more explicit discussions of how the decision or project will change things are needed. What will be the impacts? How will the surrounding areas accommodate the impacts? Certain types of project structures may well fit without problems in certain rural parts of the country, but the same projects may create serious problems if they are within five miles of a metropolitan area. For other projects, the reverse will be true.

The "alternative" section deserves a great deal more attention than it has received in the past because the alternative section has become the focus for the entire statement. It is really the alternatives section that is the most useful to the government decision maker. It is the section that gets the first look from politicians interested in a particular project. It is the section that gets first priority in judicial attention when attorneys are battling over the legality of a statement.

Also statements should be prepared earlier in the agencies' decision-making processes — even at the sacrifice of some completeness. To avoid an artificial exercise of assembling nice facts after a decision has already been made, the whole thrust of the process is to develop and report the information early, to report it in a form which can be used by the government decision maker, and then to use it meaningfully.

After the draft is prepared, it is sent to the Council on Environmental Quality (CEQ) and to federal and state agencies that have expertise on the problem. The CEQ has a small staff of seventy, which reviews only those statements that involve large-scale projects, important issues or important precedents that are of concern to the CEQ or the administration.

In the cases where the CEQ chooses to comment, its comments will be informal and will be made orally to the agency's representative. In the absence of special circumstances, the CEQ staff will contact the agency involved, describe CEQ's problems with the project or statement, attempt to understand the issue from the agency's perspective and suggest possibilities for resolving the issue. It is estimated that in over 90% of the cases, a mutually agreeable solution has been found at that point.

On other occasions no agreement is reached, and a final statement contains the same problems as the draft. The CEQ may still believe that the statement is not adequate. It again contacts the sponsoring agency, and if the problem still cannot be resolved, it will put its views in writing. These views then become part of the administrative record on the project. In his Warm Springs decision of the summer of 1974, Justice William O. Douglas suggested that on questions of adequacy of the statement, the CEQ's views, rather than those of the project agency, should be given preference in determining a statement's legal adequacy.

After an agency prepares a legally adequate statement, it makes a balancing decision that weighs the project's benefits against its environmental and other costs. Even in this exercise of agency expertise, the agency is not free to ignore the environmental information in a statement. The agency must also be guided by the need to refrain from actions prohibited by the Administrative Procedure Act. The agency must also act within the limits of the general policies set out in Sections 2, 101 and 102 of NEPA. But acting within those limits, it makes the balancing decision to reject, modify, condition or proceed with the project.

Some agencies are developing other mechanisms to help streamline the impact statement process, and to make it more useful. An innovation that has received recent judicial attention is the program statement. When an agency faces a class of actions that have common elements, it looks at those common elements in advance to resolve common issues as far as they can be resolved, consistent with other needs to consider site-specific problems. If there are site-specific impacts that were not considered in the generic process, there will be a need for supplemental statements and public review of those site-specific problems.

12. Planning for New Mining

S. Norman Kesten

THE ENVIRONMENTAL IMPACT STATEMENT, considered an onerous imposition by many people in industry, need not be as bad as it often is made out to be. Used with restraint it could be the means by which all those concerned keep all the aspects of a situation out front in planning any kind of an action and recognize the trade-offs being made each step of the way. This is what it is all about, is it not, this process of planning a new mining operation in the context of the environmental ethic? The EIS, by any name, is the analysis of the operation's impacts upon the environment and steps that must, or should, or can be taken to minimize them.

Planning of any kind must be preceded by the collection and collation of data, and today the planning of a mining operation often includes a new and highly specialized data-gathering process called the environmental study, or the environmental inventory, or the baseline study. There are at least six reasons for making the environmental study, and whether or not all of these reasons apply to a particular operation depends in part upon the attitude of the local or regional public, which includes government, toward development in that area, in part upon whether or not public lands will be used, and in part upon the operating company's policies.

1. A knowledge of all the environmental aspects of a proposed operation will give the planners an opportunity to foresee the possible impacts upon the area. Some of these impacts they will want to avoid. They will want to lessen the severity of some of them, and they will want to plan the repair, restoration, reclamation or rehabilitation of some of them. They might also capitalize on some impacts. In specific states and under certain circumstances what the planners *want* to do, whatever that might be, is superseded in part by what they *have* to do. In any event, whether this aspect of planning is mandatory or merely desirable, estimates of its cost will enter into the outcome calculation.

2. In these days it is necessary, in order to put a mining property into production, to obtain a much larger number of permits and licenses than even five years ago, and many, if not most, of the additional requirements for pieces of paper are related directly to environmental considerations. Some states have their own versions of the National Environmental Policy Act and because of that any major state action might be subject to the preparation of an EIS. In these states the granting of a permit to operate, to build a powerline, or even a transformer station, or for water lines might be considered a major state action, particularly if related to a large mining operation or a controversial mining operation, and the agency with that particular granting authority has to prepare an EIS. Unfortunately, most states that have their own "little NEPAs" have not seen fit to provide the funding for carrying out the new procedures so it is the applicant who has to provide the environmental data that goes into a statement if his application is not to be delayed indefinitely. The same situation might obtain if the plans include the use of federal lands for waste dumps, tailings ponds, roads, and so forth. The granting of a Special Use Permit might be considered a major federal action and an EIS might be considered necessary, prepared from data supplied by the applicant. Finally, in this category, are the new regulations being imposed by the Forest Service. These require the submission of mining plans for any operation on the national forests, which include unpatented claims, and the plans must include measures for reclamation. A prerequisite for approval of the mining plans and reclamation plans is an environmental analysis. There is some serious question as to whether or not it might be considered a major federal action requiring an EIS.

3. All the environmental policy laws that require an EIS place the responsibility for preparation of the statement upon the agency taking the action, but in very many instances the operator is expected to supply the data. Even if he were not, however, he had better, for his own protection, accumulate his own data, and make his own study. There have been instances where agency-prepared statements have been less than accurate and have lacked the objectivity required of such a statement. Without his own studies, the operator would not be in a position to enter an effective challenge.

4. Government agencies at all levels are subject to political pressures. In these days some of the strongest pressures are brought to bear by self-appointed guardians of the public interest in the environment. Unless these sentinals are satisfied that every effort is being made within a project to reach their goals and objectives, they are quite capable of influencing the course of events to the extent that a project might be expensively delayed or even destroyed. The planners and operators cannot even begin to understand the complaints of these people, let alone refute their allegations or accede to their demands, without a full environ-

mental study. Indeed, they are quite capable of jeopardizing an operation by nothing more nor less than public harassment.

5. In many states the only impact statements required are those to be prepared by the *federal* agencies, and the only environmental laws to be satisfied are the standard clean air and clean water laws. A scenario could be drawn of what could happen. An operation gets underway and is going well. Suddenly the manager is shocked by a headline in the nearest big city newspaper: "Mine Kills Fish in Creek." It develops that someone making some casual observations has found a small concentration of cadmium in the stream flowing by the mine and as a result an investigator has proposed a study to investigate the effect upon the aquatic species, stating that the probable source of the cadmium is the mine. A newspaper reporter hears about it and his paper makes a headline. Now, this in itself cannot close the mine down, but when the public becomes excited, the manager might spend a great deal of his time defending his operation. Furthermore, if the excitement grows, the politicians see a popular issue, and in the next session of the legislature they introduce a highly restrictive mining law or environmental policy law which could make the development of a new mining operation in the state very difficult, if not impossible. On the other hand, the problem might have been avoided if there had been reports available by reputable and disinterested scientists to show that the stream had those concentrations of cadmium before mining operations were even considered, to list the species and populations of fish in that stream before mining, and to project the probable future effects of the operation on water quality and the organisms present. A baseline environmental inventory is cheap insurance.

6. The requirements of an environmental baseline study are that it generate enough data to satisfy the needs of the permitting agency which might have to prepare an impact statement, to make it possible to plan the operation in such a manner as to take into account the protection and reclamation of environmental values and to provide insurance against future challenges. Indeed, the environmental study should be well along in time for the findings to be part of the input into the feasibility study.

The principal component parts of an environmental baseline study are listed below. In some states *all* of them are essential; in others, most of them. It must be emphasized that in the past many of these considerations were part of mine planning but perhaps in a far less formal manner than is sometimes required today.

Biological Survey

One of the two most important parts of the study is the biological survey. This is divided into two parts, terrestrial and aquatic.

Terrestrial Biological Survey. The study of species, numbers, habitat and feeding habits of game animals is a basic requirement of an environmental evaluation, partly because of the great interest of local people in hunting and in the possible impact of the operation upon this activity. However, possibly more important is the concern felt for this aspect of the environment by the U.S. Forest Service, if it is manager of the habitat, and by the State Department of Fish and Game, manager of the wildlife. Fur-bearing animals, which are trapped in some areas, and game birds, are of only slightly lesser importance. Vegetation and small mammals are essential parts of the food web and, therefore, must also be inventoried. It is also important to ascertain whether or not any rare and/or endangered species are likely to be affected by the project. An important point here is that everything is related to everything else and everything is affected by everything else. To study only the plants and animals of particular interest gives information on only an isolated portion of the system.

Aquatic Biological Survey. The study of all matters relating to fish in the streams and lakes of the area is important in order that adverse impacts might be avoided, if possible, and value or potential value of the fishery resource might be ascertained. Numbers and varieties of fish are related to food available in the lakes and streams; therefore, any evaluation must involve a study of other organisms, both plant and animal, present in the waters and along the banks.

Hydrology and Water Quality

Equally important with the study of the terrestrial and aquatic ecosystems is the definition of the hydrological system in the entire area to be affected by the operation. The elements of that system are surface water, subsurface water, precipitation and evaporation. It is essential to know the *quality* of both the surface and subsurface water prior to any disturbance so that the operators know what is available to them, what they have to protect and how much protection is needed. It is just as essential to know the *quantity* of surface and subsurface flow, not only in order to be able to decide upon a water supply source but also to be able to predict the effect of existing flows upon discharges whether from decant lines or from seepage into the ground, and the effects of discharges upon existing flows. The amount of precipitation, the form in which it occurs and how it is spread over time must be known not only because of its possible seasonal effect upon streams and underground pools but because arrangements might have to be made to *collect* whatever water falls upon the plant area in any given short time span and dispose of it; and to *divert* whatever rainfall or melted snow might run off neighboring hills into the plant area or the tailings area in any given short time span. All this requires periodic monitoring of quality and quantity which should start five years before construction of the plant but seldom can start much more than one year in advance. Fortunately, there is often some regional information available, which can be used with caution by projection and extrapolation. In general, the hydrological study always has been necessary to provide for long-term stability for the tailings pipelines, roads and plant sites.

Meteorology

Related to the hydrological study, but in addition, is weather monitoring. It is important for the operator to know about wind direction and velocities and about seasonal and diurnal fluctuations of temperature. Not only precipitation but also humidity and barometric pressure enter into the study of the hydrological system.

Soils Analysis

Today it is desirable and, in many instances, mandatory to stabilize disturbed areas and tailings accumulations. Often the best way of doing so is by the use of vegetation. This is usually also the best way of making piles of mine waste aesthetically acceptable to the public. Accordingly, it is now rapidly becoming standard practice to strip from a few inches to a few feet of overburden from an area that is to be disturbed or on which tailings or waste are to be accumulated, stockpile it and then use it later as a capping so that vegetation can be grown. Before the stripping is done it is necessary to sample and analyze the overburden to determine whether or not it is valuable for this purpose and the depth to which it can be stripped and stockpiled.

Air Quality

In the desert, particularly in areas that have been disturbed for agricultural purposes and then abandoned, fugitive dust is not uncommon even with relatively light winds. Continuous monitoring of the concentrations of dust prior to any mining-related disturbance will later give something of a measure of the amount of dust that the operation contributes to the air, and help determine the extent of the corrective measures that might be necessary in the mine area.

Socio-Economic Studies

Increasing concern is being expressed for the effect that the establishment of a mining operation will have upon people, the way they live and, in general, the amenities of their lives. The concern is for the adequacy of water supplies and sewage systems in the areas where people will live, for educational facilities, for impacts of increased populations upon roads and recreational areas, and so on. Of course, this is the business of local government and should bring into play the planning mechanisms in counties and towns. However, too often the county in which a new mining operation is to be established has only the most rudimentary capability for planning and might look to the mining company for assistance. If an influx of construction workers and then of miners creates a sociological disaster area, some segments of the public will blame the mining company. Again, the first step in any planning process is to discover the facts and this requires study of populations, employment, unemployment, incomes, capacities of various facilities and so on. A projection can then be made of the effects of the mining operation upon the social, economic, and cultural life of the area. With this projection in hand the company can work with local government toward softening the impacts and improving the fabric of life. One result might be that the company will attract and retain desirable employees.

People are also expressing concern about what happens when a mining operation is abandoned. They want to know, and the government agencies need to know, the number of unemployed miners for whom they are going to have to plan either jobs or welfare, what portions of the enlarged capacities of the utilities systems are going to be redundant, how many school houses will become idle, how many surplus teachers they will have to terminate, how many derelict trailer parks will be available for other uses, and so on. They must also have some knowledge of the timing. With the cooperation of the operator, the community might turn the shutdown into an advantage, for which the company should receive full credit.

Archaeological Survey

No impact statement is complete unless the area to be disturbed has been surveyed by qualified archaeologists and a report has been prepared stating whether or not there are archaeological or historic sites of importance in the operating area and what should be done about them. This kind of survey is made necessary by the American Antiquities Act of 1906, by some state statutes, and by common sense.

Probably the one aspect of operational planning that will make the most use of all of this information is the siting of the tailings accumulation areas. One reason is that a large area is involved, an area that often has other values, values of importance to the ecosystem, such as winter range. Winter range is the key to the survival of any herd and if that range is no longer usable, the herd (particularly if it is deer) might not look for an alternative area, and might die.

Although an operator might be astounded to discover that the feeding habits of a few deer might prevent him from obtaining an operating permit, such might very well be the case. However, the game biologists consulting on the project might be able to suggest some way of mitigating this impact. On the other hand, other members of the study team might assist in finding a tailings disposal site where the impact would be less severe.

Government agencies and segments of the public are genuinely and seriously concerned about the stability of accumulations of tailings, not only for the life of the mining operation, but also for the long term. Even though the history of tailings from metal mining operations gives little cause for concern, there are bound to be fears that picture a variety of dangers poised to strike. Environmental planning must demonstrate that such fears are unfounded by making the tailings accumulation proof against:

1. The ravages of rainfall, heavy runoff and flood;
2. The entry of water below a reasonable standard of quality into the hydrological system;

3. The capacity of an errant breeze or a gale-force wind to pick up fine, dry particles from the surface of the pile and deposit them downwind to damage vegetation and destroy or damage fish habitat.

Enumerated on earlier pages were some reasons for making environmental studies. One was omitted, possibly because it should be self-evident. Miners, too, want to work and live in and leave behind an environment that will add to, rather than detract from, the enjoyment of life.

13.   Ecology and Mining or Mining Ecology?

Mohan K. Wali and Alden L. Kollman

MINING FOR MINERALS has been one of the constant accompaniments of human civilization. Records of mining go back in early history and a recent report† cites that well established mining practices were in existence as far back as 1400 B.C. As population has increased, and is increasing at an enormous pace in the globe, the demand for more minerals has grown. In fact the intensity of utilization of these minerals is directly taken as a measure of the industrialization as well as the strength of nations.

Modern technology has found many uses of minerals during the last half century, and mining has consequently been intensified. The convergence of a greater demand through a simultaneous population growth and the multiple use through technological advancement has made the mineral base a gauge of the economic health of the industrialized nations. Since they have adopted this mineral-intensive life style, new sources must be discovered, known reserves must be tapped, and cheaper ways must be found both for mineral processing and for transportation of the finished marketable product.

Arguments on the development of known mineral resources may be grouped under the following categories:

1. "There ought to be zero-energy growth." This argument has the merit that it almost instantly stabilizes the entire mineral picture. Though ideologically sound, it neither accounts for the per capita rising demand in energy nor for the psychological urge of the humans to develop further. If one were to follow this tenet, then the checks and balance sheets will have to be maintained rigidly. For every mineral development that is phased in, some must be phased out. From a pragmatic standpoint, this argument appears untenable.

2. The reverse end of the spectrum would be to follow a trend of an unplanned and unprecedented development, a grab-what-you-can-get philosophy, and to rest on the laurels of the argument that since man has always found a way out in the past, he will do so now too. The proponents of this philosophy have labeled as "prophets of doom" those who have

---

For very helpful comments, our sincere thanks are due Drs. Robert R. Curry, Charles C. Boley and Tika R. Verma.

† "The oldest mine?" p. 65, *Time,* Jan. 15, 1975.

[108]

voiced concern over the finiteness of resources. Since all peer judgments are based on the current state of knowledge, one would naturally feel pessimistic amidst an unprecedented human population growth as well as the growth in, and of, mineral intensive societies.

3. Somewhere between the above two arguments lies the option of maintaining a planned and steady but restricted growth rate in mineral utilization. This, when coupled with strong conservation measures, is the most logical choice and would involve an ordered development based on a long-term plan. It falls somewhere in the middle of the development gradient and is more than likely to appeal to the majority of developers, users, policy makers, and those who have voiced a strong conviction in environmental protection. This would naturally ensure a more stable system with minimal fluxes and abate the current state of confusion.

We are more concerned in this paper with coal development and the innumerable debates on the so-called coal boom needed to sustain energy at an ever growing pace. Thus, most of us have heard of the genuine concern for air, land and water, for agriculture, for farm and family economy, for the welfare of states, nations, the human society, and so on. And these problems and concerns are not unique to the United States; the problems vary only in the degree of severity in many parts of the world today. When, for example, trillions of tons of coal reserves are cited, one turns gray in visualizing the many changes that would accompany such a resource development. But there is a big gulf of difference between what may exist and what can be readily obtained. For example, of the 226,317 km^2 (141,448 sq. miles) of northern Great Plains considered in the Northern Great Plains Resources Program study (NGPRP, 1975), only 2.8 percent are considered as surface minable even under a "high coal development profile." This constitutes only 0.86 percent of the total area of Montana, Nebraska, North Dakota, South Dakota and Wyoming considered in this study.

Two other important variables have to be considered in conjunction: the economic feasibility, and the problem of trade-offs. Should all of us desire the ideal rehabilitation of the environment so disturbed, the price tag of energy is likely to be high; perhaps that is as it should be. Trade-offs, on the other hand, may partially be circular arguments. For example, how much current and potential agricultural land will be out of such production because of strip mining? But then, how much energy does current, and will future agriculture need? Lerza (1974) stated: ". . . the technological miracle that U.S. farming supposedly represents is completely dependent on a never-ending flow of cheap fossil fuels and uses more energy than it produces." It is here that the balancing of the equations will have to be thorough.

We refer to ecology in its scientific usage. It was pointed out recently (Wali 1975) that like any other ecological process, the ecology of mining is complex with all its attendant intricacies, ramifications and unknowns, characterized by many cause-effect pathways (Fig. 13.1). In the past many years, many terms like the ecology of "derelict," "drastically degraded," "devastated," "orphaned," and "disturbed" lands have been used to focus attention on many of the environmental problems connected with strip mining. As a natural curiosity, the immediate reaction is, who caused these lands to become derelict, drastically degraded, devastated, orphaned and

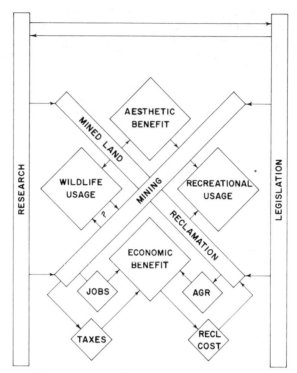

Fig. 13.1.   Cyclic and interlocking relationships of strip mining and mined land reclamation with aesthetic, ecological, economic, and social considerations. (From Wali 1975)

disturbed? Superficially, the answer is simple: not one individual, not one company nor one government. On deeper thought, the words of Weins (1972) apply:

> The roots of environmental crisis are, of course, more than just technological, and involve more than ecosystem balance alone. Religious beliefs, ethics, sociological patterns, governmental philosophies, demography, economics, various behavioral and morphological vestiges of our primate ancestry, and a host of other culturally fostered attitudes are also implicated and must be included in the search for solutions.

Problems attendant on strip mining are sufficiently distinct and interconnected (Wali 1975) to warrant their inclusion in what we propose here as a subdivision of ecology — Mining Ecology. We shall discuss only one aspect of mining ecology: the rehabilitation of land. Coaldrake (1973) proposed the development of a full capability designed and devoted specifically to the rehabilitation of mined areas. Should the proposing of this branch of ecology help to bring together the operators and specialists, and help thwart what might presently (and erroneously) be viewed as adversary roles, so much the better. In support of the creation of mining ecology, we present the following arguments.

Surface mining of land for any mineral, even sand and gravel, scars the land surface. The greater the overburden that is to be removed, the greater will be the disturbance to that as well as to the immediately adjoining area. Since modern equipment has the capability of removing anywhere from 45 to 90 cubic yards of soil in one "scoopful," the integrity of the native soils is permanently lost, together with all propagules, disseminules, seeds, and so forth of the native vegetation. Deeply buried parent materials are exposed to the surface, and their chemical character may bear very little resemblance to the original disturbed soils. As a natural course, migration of plant species from the neighboring areas to these freshly exposed areas will occur; only the pioneer and the hardy species will survive. Since many pioneer species of the semiarid and arid regions are poor soil binders, substantial erosional losses are likely in high wind and intense rainfall areas, particularly if the parent materials are fine. It is also likely that salt concentrations may become high if the parent materials are rich in salts, and if the rainfall is typical of the arid or the semiarid type of distribution. Whether these processes act singly or in conjunction, soil stability and the enrichment of organic matter will require substantially long periods of time.

It is believed that it took from the dawn of history until 1965 to disturb 3.2 million acres of land in the United States (Department of the Interior 1968) and that it will now take less than 20 years to disturb an equivalent area (Morgan 1973). By present estimates, the United States has one-third of the world's recoverable coal reserves followed by Russia (about 25 percent), China (about 20 percent) and the rest of the world (about 20 percent). In the northern Great Plains alone, over 80 billion tons are considered surface minable in a 63-county area, representing 37 percent of the total minable coal reserve by weight and 60 percent of the nation's surface minable coal (NGPRP 1975). Although it is now considered unrealistic, *Coal Age* (1973) projected growth by a factor of 10 for coal production in the western United States from 1970 to 1990; tripling the production is considered most realistic (Atwood 1975). Besides, there are many minerals other than coal that will be surface mined throughout the world. Thus, the problem is not unique in the United States; disturbance from intensified mining will occur globally. This emphasizes the need to find effective methods of ensuring the return of these disturbed areas to both stability and biological productivity.

From an ecological standpoint, most approaches to the rehabilitation of land, until recently, have been fragmentary, often looking only at some problems in isolation and not in a holistic or ecosystematic sense. Thus, agriculturists have been concentrating their efforts on the growth of cash crops, foresters on the growth of tree species, range managers on fodder species, etc. Consequently, if the growth potential of the spoil materials was low for the limited number of species selected, the land areas were pronounced as completely unsuitable for growth. For better or for worse, the bias of these early growth experiments has greatly dictated the course of legislation in many states. Problems of land rehabilitation and reclamation, however, are many and varied and may not be solved by a simple extension of a few agricultural, silvicultural, and other applied science practices that have been proved successful in drastically different situations. It remains to be seen, for example, if the standard fertilizers applied to the parent materials excavated during mining will bear the same response as when applied to naturally developed soil profiles. Lithological discontinuities, additionally, may be such that these, and the parent material on which the native soils developed in the past, may be very different.

In an appraisal of the problem, Packer (1974) has listed some of the most pressing research needs in the area of land reclamation which bear out the dimensions of the problem. A perusal of his list shows that work is needed in diverse fields from studying the physiological and ecological tolerances of plant ecotypes to spoil segregations and configurations, from the application of tissue culture techniques and hormonal stimulation of plant roots to soil amendments, fertilizers and farming practices. Clearly, such a broadly based research effort has to be based on sound ecological principles and a coalescence of the concepts of classical ecology as well as the practices of many applied sciences. Bringing the two together would logically result in mining ecology. By no means are we suggesting that the boundaries of mining ecology stop there; the inputs of mining engineers, geologists, economic analysts and others are essential.

We perceive mining ecology to fall somewhere between pure and applied science (Fig. 13.2) for each of these has particular and important contributions to make. Whatever the reasons for the deceptive insularities of the applied and pure sciences in the past, our suggestion is simply a plea for a return to the unity of purpose in science so that one can draw from the whole. Rehabilitation of strip mined lands badly needs the inputs from both pure and applied sciences; comprehensive studies (e.g. Schramm 1966) are completely lacking in the west.

Imperative before the initiation of any mining activity, is the understanding of the species diversity, the community structure, the microenvironmental features and the edaphic considerations of a given ecogeographic region. The availability of such information should allow for a reasonable forecasting of the potential as well as the expected productivity of mined areas (Wali and Freeman 1973, Fisser and Ries 1975). The often repeated statement that "these lands will be, or should be returned to original or better than original productivity" should be familiar to many who have been in this field for some time. We believe that the first part of the statement is substantially possible in many regions; the second part is a myth because we shall never be able to change all the environmental variables, particularly those of the climate. Differences in microclimate in mining areas within a distance of less than 48 km have been shown to be sufficiently distinct (Wali and Sandoval 1975) to be of ecological significance in revegetation.

Successional studies in areas supporting mining activity are of paramount importance. Primary succession is thought to be a directional and ordered march of communities, one replacing another, on habitats that have not supported vegetation in the past. Spoil materials whereon growth is not aided by human activity and that take a natural course of revegetation fall within the realm of primary succession. Recently however, many state laws

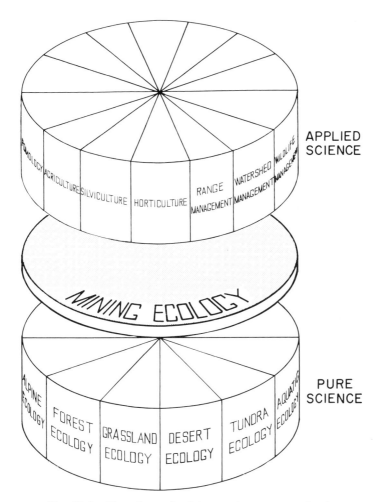

APPLIED
SCIENCE

MINING ECOLOGY

PURE
SCIENCE

Fig. 13.2. The place of mining ecology as a natural blend of pure and applied sciences; pedologic, edaphic, biogeochemical considerations, etc., are integral components of the pure science.

require the replacement of "topsoil" that had earlier been scraped off, and then is spread over the recontoured and reconfigured spoil banks. Respreading of this topsoil, with its existing plant propagules, disseminules, and so forth might create the impression that these areas are analogous to those wherein secondary succession is taking place. However, in secondary succession, soils usually retain their structural integrity while there may be a drastic change in the vegetative cover (e.g. in abandoned fields, after logging, fire). In most cases materials replaced on recontoured spoil banks are stockpiled for at least one, but in some cases, two to three years. These stockpiles of topsoils lack original profile sequencing, and there are indications that they undergo tremendous changes in their edaphic considerations from the microorganismal and nutrient standpoint. True, in their natural state, these topsoils have had a proven potentiality for revegetation. The question is not whether topsoils should be replaced, but whether one can expect revegetation to proceed in a manner similar to an old field succession, or one that bears more resemblance to a primary "soil forming" succession.

Not entirely distinct from above, but extremely useful in revegetation, would be the study of vegetation establishment as litho- and chronosequences in areas that have been mined for some time. Only a few such studies are available (Leisman 1957, Wali and Freeman 1973). In these studies, a time gradient exceeding 50 years was used. While Leisman's study was conducted in a forested region and Wali and Freeman's in grassland, both demonstrated that growth rates of the native species, increase in organic matter, and soil profile development were low. Such studies aid in the selection of treatments that can be used to enhance organic matter enrichment and vegetation establishment.

Autecological investigation of species to be used for revegetation are lacking. Of particular concern should be the study of native plant populations responses to water imbalances, nutrient conditions, trace metal toxicities, individual species tolerances, etc. Professor A. D. Bradshaw and his co-workers, for example, are directing their particular attention to rehabilitation of land through the use of metal tolerant species. Smith and Bradshaw (1972) pointed out the dangers of reclaiming land with edible species without an adequate consideration of the geochemistry of the area and the subsequent cycling of trace elements by plants.

These are but a few of the problems from the standpoint of which reclamation of land should be achieved. We believe that both basic and applied data generated as a result of the intensification of mining activities will not only prove useful in better understanding the structure and the functioning of the ecosystems, but may also result in a better selection of species for revegetation, ensuring a better stability and resiliency of these ecosystems. Their productivity on a long-term cycle may be assured. As a spin-off, better range management may be carried out in areas where land is seriously overgrazed. Multipurpose land use plans, and an educational program bearing upon it, may prove to be one of the boons of this accelerated mining activity.

Despite its elementary nature, we have offered the above discussion along the lines of Cantlon (1970) who viewed the potential role of ecology as a bridge between the sciences and the technological professions. We realize that the mere institutionalizing of a term will not solve any problems, but work generated along such lines can.

REFERENCES

Atwood, G. 1975. The strip-mining of western coal. *Sci. Amer.* 233(6):23–29.

Cantlon, J. E. 1970. Ecological bridges. *Bull. Ecol. Soc. Amer.* 51(4):5–10.

Coal Age. 1973. The challenges and opportunities in mining western coal. *Coal Age* 78(5):41–48.

Coaldrake, J. E. 1973. Conservation problems of coastal sand and open-cast mining, pp. 299–313 in *Nature Conservation in the Pacific*. Canberra: Australian National University.

Department of the Interior. 1968. Effects of surface mining on fish and wildlife resources of the United States, A Special Report. Fish and Wildlife Serv.

Fisser, H. G. and R. E. Ries. 1975. Pre-disturbance ecological studies improve and define potential for surface mine reclamation, pp. 128–134 in *Third Symp. on Surface Mining and Reclamation*, Vol. 1, NCA/BCR Coal Conf. & Expo II, Oct. 21–23, 1975, Louisville, Ky.

Leisman, G. A. 1957. A vegetation and soil chronosequence of the Mesabi Iron Range spoil banks, Minnesota. *Ecol. Monogr.* 27:221–245.

Lerza, C. 1974. The new food chain. *Environ. Action,* March 2:7–10.

Morgan, R. L. 1973. Environmental impact of surface mining: The biologist's viewpoint, pp. 61–71 in *Some Environmental Aspects of Strip Mining in North Dakota,* M. K. Wali (ed.). Educ. Ser. 5. Grand Forks: North Dakota Geol. Surv.

NGPRP. 1975. Effects of Coal Development in the Northern Great Plains. Denver, Colorado: Northern Great Plains Resources Program.

Packer, P. E. 1974. Rehabilitation Potentials and Limitations of Surface-mined Land in the Northern Great Plains. USDA For. Serv., Gen. Tech. Rpt. INT-14.

Schramm, J. R. 1966. Plant colonization studies on black wastes from anthracite mining in Pennsylvania. *Trans. Amer. Phil. Soc.* n.s. 56(1):1–194.

Smith, R. A. H. and A. D. Bradshaw. 1972. Stabilization of toxic mine wastes by the use of tolerant plant populations. *Trans. Inst. Mining Metall.,* Sec. A, 81:A230–A237.

Wali, M. K. 1975. The problem of land reclamation viewed in a systems context, pp. 1–17 in *Practices and Problems of Land Reclamation in Western North America*. M. K. Wali (ed.). Grand Forks: The University of North Dakota Press.

Wali, M. K. and P. G. Freeman. 1973. Ecology of some mined areas in North Dakota, pp. 24–47 in *Some Environmental Aspects of Strip Mining in North Dakota*. M. K. Wali (ed.). Educ. Ser. 5. Grand Forks: North Dakota Geol. Surv.

Wali, M. K. and F. M. Sandoval. 1975. Regional site factors and revegetation studies in western North Dakota, pp. 133–153 in *Practices and Problems of Land Reclamation in Western North America*. M. K. Wali (ed.). Grand Forks: The University of North Dakota Press.

Weins, J. A., ed. 1972. *Ecosystem Structure and Function*. Corvallis: Oregon State Univ. Press.

14. The Effects of Coal-fired Electricity

E. L. Ermak, J. R. Kercher, and R. L. Ritschard

THE NATION'S RISING ENERGY DEMANDS coupled with the desire for environmentally clean energy production have focused attention on the vast deposits of low sulfur coal in the southwestern United States. Coal strip mining operations are currently underway on Black Mesa in northern Arizona and in the San Juan Basin in northwestern New Mexico. These mines supply coal for electric power generation at the Mohave power plant located at the southern tip of Nevada, the Navajo power plant at Page, Arizona, and the Four Corners and San Juan power plants both located in the northwestern corner of New Mexico. Plans already exist for the expansion of these operations and the development of the coal reserves in the Kaiparowits Plateau in southern Utah. As energy needs increase, this geographical region can be expected to be called upon for expansion of coal extraction and electric power generation.

All phases of the electric power production process can result in significant impacts upon the environment. Assessment of these impacts is necessary to insure that reasonable action is taken to avoid unnecessary environmental damage and to evaluate the unavoidable costs associated with energy production. This need is heightened in the southwest by the fact that most of the benefits of energy production in terms of available electric power are received outside the energy producing region which bears the burden of environmental degradation.

To assess the environmental effects upon a large and complicated region such as the southwest requires one must evaluate and integrate the results of many research programs which are directed at various components of the ecosystem (air, soil, water, vegetation, animals). Our objective is to provide a comprehensive approach to the evaluation of the environmental and human health consequences of coal-fired electricity production in the southwestern United States. The approach considers the entire power production process (Fig. 14.1) and attempts to identify:

This work was performed under the auspices of the U.S. Energy Research and Development Administration, Contract #W 7405-ENG-48.

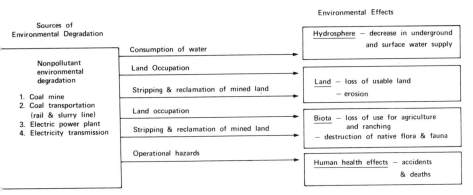

Fig. 14.1.   The environmental effects of coal extraction and utilization in the southwest. Non-pollutant degradations produce a direct effect on the environment.

1. The pollutant emissions released into the environment and the sources of non-pollutant environmental degradation.
2. The data and models necessary to describe the transport of pollutants through the environment.
3. The anticipated environmental effects and the methods for assessing the magnitude of these effects from monitoring data and model calculations.

The approach has been tailored to the circumstances and environment that exist in the southwest and includes specific examples taken from the northern Arizona system of the Black Mesa mines and the Mohave and Navajo power plants.

ELECTRIC POWER PRODUCTION: SOURCE OF ENVIRONMENTAL DEGRADATION

The electric power producing process is divided into four components: coal strip mining; coal transportation; electric power generation; and electricity transmission. This section identifies the possible sources of environmental degradation associated with each one.

Coal Strip Mining

The coal extraction process consists of stripping off the overburden covering the coal and spoiling it onto the previous cut; mining the coal; trucking the coal to a processing center for stockpiling, crushing and cleaning; and transporting the coal by conveyor belt to either a slurry pipeline or a train for transportation to the power plants. The spoils piles are recontoured into rolling hills and reseeded. Coal extraction processes are potential sources of environmental degradation because:

1. Strip mining operations are a source of local air pollution in the form of dust.
2. The mine, roads, and support facilities occupy land making it unavailable for alternate uses.
3. The stripping process alters the physical, chemical, and biological properties of the soil. The possibility exists for the introduction near the surface of either toxic trace materials or large quantities of inorganic salts.
4. The mining and recontouring operations alter topography and remove surface vegetation. Re-use of the land will depend upon the success of reclamation efforts.
5. Groundwater is used for dust control and preparation of coal slurry.
6. The unleached mine spoils are a source of fresh minerals and could potentially contaminate rivers and lakes with dissolved solids and silt in runoff.
7. Occupational hazards exist in operating the coal mine.

Coal Transportation

Environmental degradation results from the use of land by the slurry pipeline and railroad for the right of ways, access roads, and work areas.

Electric Power Generation

The electric power generation component includes: receiving, storage, preparation, and burning of the coal; steam generation and the subsequent liquefaction; control and disposal of the exhaust gases, ash, and cooling water; and electricity production and initial transmission. Adverse effects upon the environment and human health may result from the following activities:

1. Emissions to the atmosphere from the furnace stacks and the cooling towers.
2. Disposal of bottom and fly ash.
3. Disposal of waste cooling water (generally in evaporation ponds).
4. The use of large quantities of water to run the cooling system.
5. The occupation of the land by the power plant site.
6. Occupational human hazards due to construction and operation of the power plant.

Transmission of Electric Power

Environmental degradation results from the use of land for the electric powerlines and access roads.

POLLUTANT TRANSPORT IN THE ENVIRONMENT

Pollutant Pathways

Pollutants emitted into the environment often travel large distances from their source. Transfers from one part of the environment to another, such as from air to soil and soil to water, also occur, as summarized in Fig. 14.2. While emissions tend to move away from the initial source, transport

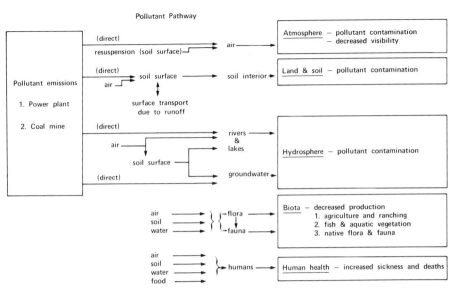

Fig. 14.2.   Pollutant effects of coal extraction and utilization in the southwest.

over the long run may be cyclic rather than in one direction. For example, trace elements taken up from the soil by plants will return to the soil when the plant dies. While a complete study of pollutant transport must consider all possible pathways, not all pathways are significant. The pathways which are deemed most important for consideration are shown in Fig. 14.3. These pathways were chosen because they were judged to have either a high

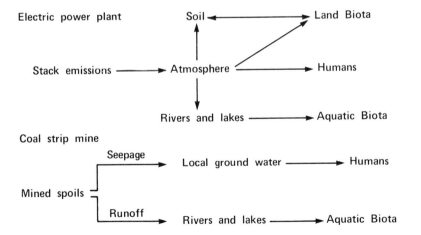

Fig. 14.3.   Major pollutant sources and pollutant pathways for environmental effects of power production which have either a high probability of occurrence or a high potential for environmental damage should they occur.

probability of occurring or a high potential for environmental damage should they occur.

The major pollutant source from power plants is stack emissions. Due to the comparatively rapid transport which occurs within the atmosphere, stack emissions have the potential for contaminating large areas. Pollutants can reach living organisms via several pathways. The main pathway for humans is by direct inhalation. The food and water pathways to humans are possible secondary sources. The pathways to plants are direct atmospheric disposition onto the plants and root-uptake of pollutants deposited on the soil. The pathways to animals are direct inhalation and the consumption of contaminated plants or animals. Rivers and lakes can be contaminated by direct atmospheric deposition and soil runoff resulting in the exposure of aquatic life.

The major pollution concern associated with coal mining is the contamination of the hydrosphere by the mined spoils. Toxic elements in the spoils could possibly enter the local groundwater by seepage. This process is not considered very probable (see Effects section), although if this did occur, the effect could be quite serious. Increased particulate, salt, and trace element contamination of rivers and lakes within the coal mine watershed due to soil runoff from the mined spoils can also occur. Recontouring of the spoils piles to minimize runoff or to trap it in ponds will greatly reduce this problem.

Pollutant Transport Evaluation

Assessment of pollutant transport and effects requires an integrated approach of field measurements, laboratory experiments, and model calculations. Field measurements and laboratory experiments provide the needed data for building and applying models from which predictions of environmental impacts can be made. The need for a predictive capability in the area of pollutant effects is due to the high potential for environmental degradation, the long time scales (on the order of years) required for the application of emission controls, and the long time scales over which contamination occurs. In addition, a monitoring program, initiated prior to and continuing after construction of future power operations, should be conducted to ensure the accuracy of model calculations.

Atmospheric Transport

Atmospheric models and atmospheric monitoring are needed to determine the resultant air concentration of pollutants and the surface concentration of ground-deposited pollutants as a function of time and position for known pollutant sources. The concentration measurements sought are average levels over a specified time period. The time periods over which averages are to be taken are determined by the applicable air quality standard. Model predictions are sought which predict the maximum concentration at each position within the study region when averaged over time intervals on the order of an hour, a day, and a year. The spatial size of the study region must be sufficiently large to ensure that pollutant concentration levels outside the region are essentially unchanged by the sources of concern.

In order to calculate pollution levels using atmospheric transport models, a considerable amount of input information is often needed. These data include: (1) characterization of the pollutant emissions as to kind, chemical and physical state, and emission rate; (2) the deposition, chemical kinetic, and physical transformation rates for the various pollutants; (3) topographical information; and (4) meteorological and climatic data. Accurate deposition and transformation data are the most difficult to obtain and the associated physical processes are also the most poorly represented in simulation models. The complexity of the models varies from Gaussian plume type models to diffusion equation models with complicated boundary conditions requiring numerical techniques and large computer codes for solution.

Numerous atmospheric pollution transport models are available. The theoretical basis and detailed technical descriptions of atmospheric models can be found throughout the open literature and in numerous government reports. (For example: Sutton 1953; Gifford 1960; Pasquill 1962; Slade 1968; Briggs 1969; Lang 1973; Frankenberg *et al.,* 1973; Ragland 1973; Hameed 1974; Scriven and Fisher 1975.) Two bibliographies of atmospheric transport models which include a brief description of each model and its availability are given by Milane (1973) and Knox and Sherman (1974).

Hydrologic Transport

The consumption of surface water by the power plants will obviously reduce the quantity of available water downstream. It will also increase the downstream concentration of dissolved solids since the evaporative losses are essentially unaffected by the change in water flow. Coal mining can increase river concentrations of suspended solids within the watershed due to erosion of the mined spoils. Consequently, surface water models are needed to predict changes in flow rate and pollutant concentration due to direct water consumption, direct emissions of wastes into surface waters, and runoff of solid wastes on the soil surface.

The pollutants of concern for groundwater contamination are the dissolved solids. Models are needed to predict the possible: seepage of pollutants through the soil into the groundwater; head declines of near surface aquifers due to penetration of the aquifer during the strip mining operation; head declines in aquifer layers due to pumping; groundwater flow; and the transport of pollutants within the groundwater.

Several hydrologic models have been developed and applied to the southwest. Model predictions have been made of the changes in water quantity and quality of the Colorado River due to water consumption by the power plants and head declines of the Black Mesa groundwater due to pumping (Department of the Interior 1972; Dove *et al.* 1974). Thames and Verma (1975) have studied runoff on Black Mesa and conducted model calculations of water movement into the mined spoils. Aquifer models have been proposed including those by Freeze (1969), Prickett and Lannquist (1971), and Gates (1972). A bibliography of aquatic transport models is included in Knox and Sherman (1974).

Ecologic Transport

In order to predict ecological impacts one must characterize all pollutant inputs, ecosystem transfers, compartment concentrations, and physicochemical form in each ecosystem compartment. Field monitoring can provide a detailed account of where physico-chemical changes take place. Laboratory experiments delineate uptake, turnover, excretion rates, the variables that determine them, the factors important in physico-chemical form transformations, and the parameters of soil transport.

Terrestrial Systems. Inputs to terrestrial systems are from airborne pollution, onto either soil surfaces or plant surfaces, and from direct input into soil by stripping. These inputs must be determined by field measurements. Transfer rates within the system must be known. Soil movement is determined by soil characteristics, rainfall, infiltration, and physical and chemical characteristics of the trace contaminants. Plant uptake and retention; consumer feeding rates, uptake, retention; and population dynamics must be known. The nature of the food web should be determined. Decomposition rates and movement from litter to soil must be characterized. These variables determine movement and concentration of trace elements in the ecosystems. If they are known, then it is possible to project ranges for future dynamic behavior of the trace element in question.

A preliminary field study has been completed by F. B. Turner *et al.* (1975) for trace elements in soils and vegetation in the vicinity of the Mohave Generating Station. Research is currently being funded by the Southern California Edison Company on trace element cycling near the Mohave Generating Station (Dr. Ian Straughan, personal communication). The Environmental Impact Studies Group at Northern Arizona University (Zaleski and Gaud 1974) has conducted a baseline investigation of the trace element content in soils, vegetation, and water surrounding the Navajo Generation Station.

Preliminary work near the Mohave and Navajo stations reports extremely high variability in samples. This means that very extensive sampling procedures must be used for trace element assessments and model development. Given the high variability, it might prove difficult to parameterize and validate a complex model as developed by the Desert Biome of the International Biological Program (Goodall *et al.* 1973). Simple linear cycling models of very few compartments in which many compartments are lumped together are more feasible. Then data collection can center on acquiring good means and variance values of a few compartments.

Aquatic Systems. The inputs to aquatic systems are from airborne pollution, stream runoff, and groundwater flow. Each of these inputs must be characterized. There are dissolved solids and suspended solids in stream runoff. Both are potential pollutants of aquatic systems. Stream transport of sediments must be determined. Transfers in and out of the water compartment with respect to all other compartments (i.e., sediments, primary producers, and consumers) must be determined. Transfers in the system are determined by the physico-chemical form of the contaminant in each compartment, uptake rates, elimination rates, turnover time of populations, and chemical properties of the substance in natural aquatic systems. These properties depend on metabolism of substance in the individual, differences between species, sexes, season, and age and growth of the individual.

Work has been done on Lake Powell (Standiford et al. 1973) in which mercury content of the water, sediment, and food web compartments have been determined. The data from this report have not been incorporated into a dynamic model of trace element transport. The Desert Biome (Goodall et al. 1973) has a very complete aquatic model which would require extensive sampling to parameterize for a particular aquatic system in the southwest.

ENVIRONMENTAL EFFECTS

The environmental effects are divided into five categories: atmospheric effects; land and soil effects; hydrospheric effects; biotic effects; and human health effects. These are summarized in Table 14.1.

Atmospheric Effects

Power plant emissions are the most significant source of air pollution and can ultimately impact the atmosphere, land, water, and biota. The pollutants of concern are particulates, SO_2, NO_2, converted SO_2 and NO_2 to particulate sulfates and nitrates, trace elements, and H_2O. Decreases in atmospheric visibility result mainly from emissions of particulates and water vapor. Additional reductions in visibility will result from the conversion of SO_2 and NO_2 to particulate sulfates and nitrates; however, the chemical reaction and conversion rates which control these processes are not well known. Increased dust level due to strip mining is a local effect in the mining area and resuspension of ground deposited atmospheric pollutants is expected to be insignificant in comparison to the direct emissions.

In order to determine the fate and concentration of pollutants in the environment, the emitted pollutants must be characterized at the source as to: the pollutants emitted; the rate of emission; and the chemical and mechanical (gas, particle and size) state. The emission rates of these pollutants can be predicted from direct calculations using estimates of the pollutant concentrations in the pre-burned coal and the ash waste, the rate of coal consumption, and the efficiency of any pollutant control equipment (such as electrostatic precipitators which reduce particulate emissions).

Predictive evaluations of the atmospheric effects can be made using atmospheric transport models (see previous section). To ensure that pollutant levels are not above designed specifications after construction of the power plant, monitoring of pollutant concentrations should be conducted both at the stack and downwind from the power plant where the concentrations are expected to be largest. Model predictions have been made for the Navajo and Mohave power plants (Department of the Interior 1972; Dove et al. 1974) and these results should be compared with currently available monitoring data.

TABLE 14.1

Summary of Environmental Effects

Source	Method of Evaluation	Effect
I. Atmospheric		
1. Pollutant emissions from the coal mines and power plants	1. Pollutant transport models — atmospheric dispersion & deposition — pollutant resuspension	1. Increased pollutant concentrations in atmosphere
2. Resuspension of soil deposited pollutants	2. Monitoring of pollutant concentrations in air	2. Increased pollutant deposition rates on soil, water & biotic surfaces
		3. Decreased visibility
II. Hydrospheric		
1. Water use in coal mining and power plant operations	1. Planned water usage 2. Monitoring water usage	1. Decrease in water resource
2. Pollutant sources — direct emissions from coal mine and power plant — atmospheric deposition — soil surface runoff — soil seepage	1. Pollutant transport models — atmospheric deposition — soil surface runoff — soil seepage 2. Monitoring of pollutant concentrations in surface and ground water	2. Increase in pollutant concentrations in rivers, lakes, and groundwater
III. Land and Soil		
1. Land use for coal mining, coal transportation, power generation, and power transmission	1. Planned land usage 2. Monitoring land usage	1. Loss of alternative land use
2. Strip mining and land reclamation	1. Erosion models 2. Monitoring soil quality and erosion	2. Altered quality of mined land to support plant growth
		3. Loss of nutrients and salts through increased erosion

TABLE 14.1 — cont.

Summary of Environmental Effects

Source	Method of Evaluation	Effect
III. Land and Soil — cont.		
3. Pollutant sources — ash disposal — atmospheric deposition — deposition from overland flow — transport in soil column	1. Pollutant transport models — atmospheric deposition — overland flow — soil column transport 2. Monitoring pollutant concentrations in soil	4. Increased pollutant concentration in soils
IV. Biotic		
1. Land usage — habitat removal — community disruption	1. Ecosystem response (succession) models to land disturbance and pollutant stress	1. Loss of land for agriculture/pasture land
2. Strip mining (alteration of physical, chemical and biological soil properties)	2. Laboratory experiments on critical biological processes (growth, fecundity, etc.)	2. Disruption of native community 3. Decreased productivity or reproductive potential of plants and animals
3. Pollutant sources — atmospheric deposition — soil uptake — rivers and lakes	3. Monitoring condition of biota	4. Increased mortality 5. Alterations of behavior 6. Alterations of population, trophic, and community structure 7. Alteration of ecosystem nutrient cycling
V. Human Health		
1. Occupational hazards	1. Models based on historical data	1. Increased number of human injuries, sickness and death
2. Pollutant sources — air — water — food — soil	2. Models which predict: — pollutant dose to man — human effects	

Hydrospheric Effects

Power plants consume large quantities of water each year for cooling purposes. When all units are in operation the Mohave power plant will use 30,000 acre-ft. per year and the Navajo plant will use 34,000 acre-ft. per year of water from the Colorado River (Department of the Interior, 1972). The development of additional power plants requires coordination with existing users and other projected uses of the river water. The mining operation on Black Mesa is expected to consume an average of 2,400 acre-ft. of water per year for use in the slurry line. This water is obtained from five wells which are drilled to a depth of about 3,000 ft. Model predictions by the Southwest Energy Study Group and the Arizona Study Group indicate that there will be no significant reduction in head levels at surrounding Indian wells. The U.S. Geological Survey operates a monitoring program to ensure that the local supply of water is not endangered. A combined modeling and monitoring program such as this should be used to evaluate the effects of groundwater pumping on the water supply.

Increased pollutant concentrations in the surface waters can result from: (1) the consumption of surface water for cooling at the power plant resulting in a reduction in dilution capacity; (2) direct deposition of airborne emissions from the power plant stacks; (3) soil surface runoff of ground-deposited power plant stack emissions; and (4) increased soil surface runoff from the coal mine spoils. Models are available for predicting the extent of these effects (see Transport section); however, an assessment should also include monitoring of surface water quality before and after installation of the operation.

Acid mine drainage is not expected in the southwest. The coal is low in sulfur compared to eastern coals, rainfall is low, evaporation rates are high, infiltration rates are slow, and the soil is strongly buffered by calcium carbonate. Hence, acids would not be formed or transported in large quantities. Disruption or pollution of the groundwater by strip mining is also unexpected. Although coal seams are aquifers in the west, disruption of the water supply for local wells is unlikely since wells are generally about 300 ft. deep while strip mining extends to a depth of only 120 ft. Low infiltration rates and the existence of several impervious layers between the mine and well depths restrict the possibility of contamination.

Groundwater contamination from power plant liquid and solid wastes appears remote due to the elaborate control measures presently being taken to prevent seepage to the groundwater. Monitoring efforts are being conducted at the Navajo and Mohave plants as an additional safeguard.

Land and Soil Effects

Occupation. The electric power production process requires the use of land. Such use effectively makes land unavailable for alternate purposes at least for a length of time, which depends on the projected life of the power plant, roads, transmission corridors, and transportation corridors, or on revegetation in the case of forestry or grazing as the alternate use.

Erosion. The strip mine is the largest single component of the power production process that is subject to erosion.

For a regional evaluation, the total area disturbed and the acreage within various slope classes should be known. Empirical relationships on erosion as a function of ground cover, slope, and climate can be used to estimate expected impact. On Black Mesa, the fraction of the total watershed to be stripped is only 0.7 percent, the land is recontoured to contain any runoff, and the effects of erosion will be limited to the mined area. Erosion on Black Mesa mine sites is being investigated by Thames and coworkers (Thames and Verma 1975).

Contamination. Contaminants can be classified into two groups: (1) trace elements, and (2) large concentrations of salts and acids. Land contamination can be the result of (1) the direct replacement of soil by overburden in strip mining, (2) air-borne pollutants emitted by power plants, and (3) irrigating with water of reduced quality downstream from the mine or power plant. To determine contamination caused by soil replacement requires the comparison of the chemical analysis of recontoured spoils with that of undisturbed sites. Evaluating contamination of land from power plant emissions requires an integrated program of the kind discussed above in the pollutant transport section. Soil sampling must be extensive because of the high variability in the samples (Turner et al. 1975). Evaluation of soil contamination from irrigation downstream from the electric power generation system must be done using on-site monitoring coupled with hydrologic transport and soil transport models.

Alteration of physical, chemical, and biological properties of soils. The properties which can be altered are pH, texture structure, water holding capacity, cation exchange capacity, concentrations of chemical constituents, organic matter, microorganisms, and bulk density. These properties can affect plant growth. Runoff characteristics (infiltration and runoff rates, suspended and dissolved solids) determine water quality downstream from mined areas. All the above properties must be compared by measurements of disturbed and undisturbed sites. Thames and Verma (1975) have investigated moisture retention properties, nutrient levels, textural composition, and salt levels on mined and unmined land.

Ecologic Effects (Biotic)

Effects on the biota in a stressed ecosystem occur at several levels. They occur to the individual in the form of both lethal and sublethal effects. Sublethal effects can alter the individual's productivity, life cycle behavior, reproductivity, ability to compete, or genetic composition. At the community level these changes can alter age-class structure, population dynamics, population size, biomass, or genetic distribution. At the trophic level of organization this might alter the competitive balance for the same resource. In food webs, a change in the species composition of any one level can change the species composition in the trophic level it feeds upon or for which it is the prey. Changes can occur in the composition of the entire community including diversity. Effects can also occur in the functioning of the entire ecosystem, for example the disruption of the cycling of a nutrient (Lefohn et al. 1974).

Non-Pollutant Degradation. The non-pollutant effects can be divided into two main types: (1) removal of vegetation and the above-ground community with minimal disturbance to the soil and (2) disruption due to removing the biotic community and significantly altering the properties of the soil. The magnitude of the effect of above-ground community disruption depends on the success rate of revegetation and succession. Hevly (1974) reports on success rate of revegetation on lands disturbed by construction of the railroad and Navajo Generating Station. For an assessment of land disturbance of biota it would be useful to have a succession model for predicting revegetation under various environmental conditions.

The most useful approach in assessing revegetation potential of disturbed land is to set up controlled plots of different treatments (fertilizers, mulches, and so forth). This has been done for other environments (Farmer *et al.* 1974; Hevly 1974) and should be done on Black Mesa mines. Important factors include type of community disturbed, grazing pressure, rainfall and climate, seed source, and quality of reclaimed land (Packer 1974; Hevly 1974).

The Peabody Coal Company reseeding program using a mixture of native grasses expects to reseed with no additional irrigation. Productivity, yield, and erosion control are the primary indicators of a successful reseeding program. It is important that the re-established system have high diversity and high reproduction success.

Pollutant Degradation. Three types of pollutants can affect the biota: (1) gaseous pollutants; (2) surface deposited particulates from the power plants; and (3) pollutants introduced directly into the soil by the mining operation.

1. Four gaseous effluents from power plants are of possible concern: sulfur dioxide, oxides of nitrogen, ozone, and fluorine. SO_2 and NO_3 have been identified as primary effluents associated with nearly all coal-fired plants. Ozone has been suggested as a possible secondary effluent (Davis et al. 1974). SO_2 and fluorine have been identified as a possible problem of the Navajo power plant (Dove et al. 1974). These gaseous pollutants all inhibit photosynthesis, reduce growth at lower levels, and cause leaf necrosis at higher levels. The effect is a function of exposure level, exposure frequency, exposure duration, plant age, plant species, and environmental factors such as available soil moisture, temperature, and relative humidity. In addition, these pollutants, when combined together, are suspected of producing synergistic effects. SO_2 data may be found in Johnson and Hevly (1973), Hill et al. (1974), and Davis et al. (1966). Work on NO_3 has been reported by Tingey et al. (1971), Hill et al. (1974), Bennett and Hill (1973), Taylor and Eaton (1965), and Hill and Bennett (1970). Ozone effects are discussed by Davis and Coppolino (1974), Davis and Wood (1973), and Hill and Littlefield (1969). Fluorine data is presented by McCune and Weinstein (1971) and Hindawi (1968).

The estimation of the regional effect of these gases requires the distribution of the vegetation types to be known throughout the region of impact. Each vegetation type has to be characterized by typical patterns of growth, productivity, distribution, and age class structure on a seasonal basis. A validated, predictive model of the behavior of the various species as functions of age, environmental conditions, doses, and synergisms should be used to simulate the changes in productivity and growth produced by the range of pollutant levels expected over the region of impact. To estimate the total vegetation impact, model results should be combined with the known distributions of vegetation types. Carbon and nutrient dynamics models could then be used to determine the impact of this change on the consumer and decomposer portions of the ecosystem.

Gaud (1974) has developed a preliminary model which has been used to simulate the effects of SO_2 on grassland and conifer vegetation types. Growth, reproduction, death, and respiration were simulated as functions of SO_2, insolation, and available water in a soil storage compartment. Since the predicted impact was too small to be detected by conventional field sampling, Gaud suggests that validation should be done by controlled fumigation experiments which measure SO_2 uptake.

The Desert Biome model (Goodall et al. 1973) could be modified to predict alterations in ecosystem structure due to gaseous pollutant effects. Such effects will require a long time to manifest themselves in the field if the pollutant levels are as low as expected.

2. The trace elements *Ag,* Al, As, B, Ba, Be, Bi, Cd, *Co,* Cr, Cu, F, *Hg, Ni, Pb,* Sb, Se, *Sn,* Te, Tl, U, V, W, Zn, and Zr are toxic to plants (Bowen 1966). The italicized elements are considered to be very toxic to plants. The trace elements *Ag, As, Be,* Bi, Cd, *Cr, Cu,* F, *Hg,* Mn, Mo, Ni, Pb, Po, Sb, *Se,* Sn, Te, *Tl,* U, V, are toxic to animals. Italicized elements are considered to be very toxic. The low level of elements released from power plants is expected to show few effects short of a long-term contamination (Turner et al. 1975).

Because long-term contamination is involved, it will be necessary to model the effects of these elements in order to make an assessment. Baseline data, necessary for these models, are concentrations of these elements in the various ecosystem components; the productivities, population, and standing crops of the various plant and animal populations; food web structure; community structure and diversity; and the normal behavior and transport of these elements in the system. The transport models discussed above should be integrated with ecologic effect models. Such models require data on the effect of the various trace elements on reproduction, feeding behavior, competitive behavior, mortality, and morbidity. For a regional evaluation, these models should be run for each of the communities in the region. Long-term monitoring should continue for continual model updating and validation. These considerations apply to both terrestrial and aquatic systems.

The Environmental Impact Studies Group of Northern Arizona University (NAU) has been collecting terrestrial and aquatic baseline data for the area around the Navajo Generating Station (NAU 1973; NAU 1974; Zaleski and Gaud 1974). The only trace element effects work done in the case study area has been on effects of mercury on Lake Powell diatoms (Tompkins and Blinn 1947). Since the Desert Biome models (Goodall et al. 1973) simulate primary production, consumer population dynamics, decomposition, and nutrient cycling they would be appropriate for adaptation to the case study region.

3. High levels of trace elements toxic to plants can inhibit plant growth or could differentially allow only the most resistant species to grow. To predict any effect, the overburden material would have to be analyzed and then compared to materials with known toxic levels. To estimate toxic effects for animals, concentration factors for plants, animal intake, and body burdens must be determined. Thus, a knowledge of the food chain portion of the ecologic transport model is necessary.

Soil samples should be taken throughout the strip mined area to establish the distribution of salt in the new surface material. This should be evaluated to determine which, if any, species will be affected. Thames and Verma (1975) report on one such sample.

Human Health Effects

Human health effects are discussed under the following categories:

1. Increases in human injury, sickness, and death resulting from occupational hazards.
2. Increases in human sickness and death as a result of air pollution.

The first category is restricted to the power production operation. Activities which pose significant hazards are coal mining, coal transportation by rail, and electric power generation. Using historical data from the Bureau of Mines, Hamilton and Morris (1974) have predicted the annual average number of injuries, hearing loss, and deaths for a typical 1,000 MWe (megawatt electric) operation. Chronic lung disease is generally not associated with surface mining. There appears to be no acid mine drainage problem in southwestern mines. A more accurate evaluation for the southwest might be made from current human health data of the mines and power plants in this region.

Human health effects of air pollution are a function of the pollutant types, the exposure levels, and the population being exposed. The relationship between air pollution and increased morbidity and mortality from respiratory diseases has been documented by several studies (Winkelstein et al. 1967; Lave and Freeburg, 1974; Schwing and McDonald 1974). There appears to be a measurable difference in mortality as it relates to air quality even if socioeconomic distinctions are controlled (Winkelstein et al. 1967, 1968). Epidemiological studies have been conducted using a single pollutant as an index of exposure with generally good correlation between health effects and total suspended particulate matter, and fair correlation with SO_2 concentration (McJilton, et al., 1973; EPA 1973).

A survey of morbidity and mortality studies related to environmental pollution illustrates the following points, which may limit evaluation of human health effects:

1. There are, in most cases, insufficient data available to make accurate assessments of health effects. For example, a complete understanding does not exist of the mechanisms by which each of the various pollutants affects human health.
2. In small populations, like that of the case study area, mortality studies are unlikely to produce useful results because of the vari-

ability in the number of deaths from year to year and the differences in diagnosis and reporting.

Consequently, evaluation of the potential human hazards from air pollution can be based only upon the pollutant concentrations, the population density, and the above mentioned mortality relationships.

CONCLUSION

A comprehensive approach has been presented for evaluating the environmental and human health effects of coal-fired electricity production in the southwestern region of the United States. The possible effects upon the atmosphere, land and soil, hydrosphere, biota, and humans have been identified. Methods for evaluating the extent of the effects have been presented. In general, evaluations will require the combined results of laboratory experiments, field experiments and monitoring, and predictive model simulations. Of all the effects considered, the major concerns in the southwest are: (1) pollutant effects on all parts of the environment due to atmospheric emissions of power plants; (2) increased salinity of the Colorado River due to water consumption by present and future power plants; (3) possible pollutant contamination of the hydrosphere due to mining operations; and (4) reclamation of mined land.

REFERENCES

Bennett, J. H., and A. C. Hill. 1973. Inhibition of apparent photosynthesis by air pollutants. *Journal of Environmental Quality* 2: 526–530.

Briggs, G. A. 1969. *Plume Rise*. TID-25075, USAEC, Division of Technical Information.

Davis, C. R., G. W. Morgan, and D. R. Howell. 1966. SO_2 fumigations of range grasses native to southeastern Arizona. *Journal of Range Management* 19: 60–64.

Davis, D. D. and J. B. Coppolino. 1974. Relative ozone susceptibility of selected woody ornamentals. *Hort. Science* 9: 537–539.

Davis, D. D.; G. Smith, and G. Klanber. 1974. Trace gas analysis of power plant plumes via aircraft measurement: O_3, NO_2, and SO_2 chemistry. *Science* 186: 733–736.

Davis, D. D. and F. A. Wood. 1973. The influence of plant age on the sensitivity of Virginia pine to ozone. *Phytopathology* 63: 381–388.

Department of the Interior. 1972. *Southwest Energy Study — An Evaluation of Coal-Fired Electric Power Generation in the Southwest*. Federal Task Force, Study Management Team.

Dove, F. H., D. W. Layton, E. B. Oswald, and P. G. Thorne. 1974. Physical Section, pp. 19–120. *In* Coal-Fired Energy Development on Colorado Plateau: Economic, Environmental, and Social Imports. T. G. Roefs and R. L. Gum (eds.). Report on Natural Resources Systems No. 23, Department of Hydrology and Water Resources, University of Arizona, Tucson, Arizona.

Environmental Protection Agency. 1973. *Monitoring and Air Quality Trends Report. 1972.* Research Triangle Park, North Carolina.

Farmer, E. E., R. W. Brown, B. Z. Richardson, and P. E. Packer. 1974. *Revegetation research on the Decker Coal Mine in southeastern Montana.* USDA Forest Service Research Paper INT-162, Ogden, Utah.

Frankenberg, T. T., et al. 1973. *Recommended Guide for the Prediction of the Dispersion of Airborne Effluents.* Second Edition. New York: The American Society of Mechanical Engineers.

Freeze, R. A. 1969. *Theoretical Analysis of Regional Groundwater Flow.* Inland Waters Branch, Department of Energy, Mines, and Resources, Scientific Series No. 3, Ottawa: Queen's Printer for Canada.

Gates, J. S. 1972. *Worth of Data Used in Digital-Computer Models of Ground Water Basins.* Unpublished Ph.D. Dissertation, University of Arizona.

Gaud, W. S. 1974. Modeling, pp. 343–352 in *Environmental Impact Studies of the Navajo and Kaiparowits Power Plants.* Third Annual Report. Vol. 1. Flagstaff: Northern Arizona University.

Gifford, F. A. 1960. Peak to Average Concentration Ratios According to a Fluctuating Plume Dispersion Model. *International Journal on Air Pollution* 3(4): 253–260.

Goodall, D. W., C. Gist, W. Valentine, S. Payne, J. Radford, J. Wlosinski, G. W. Minshall, J. Deacon, A. Holman, R. Kramer, F. Post, C. Stalnaker, D. Porcella, D. Koob, and I. Noz-Meir. 1973. Multi-Purpose Models in *Report of 1972 Progress.* Volume 1. Desert Biome. U.S. International Biological Program, Ecosystem Analysis Studies. Logan, Utah: Utah State University.

Hameed, S. 1974. A Modified Multi-Cell Method for Simulation of Atmospheric Transport. *Atmospheric Environment* 8: 1003–1008.

Hamilton, L. D. and S. C. Morris. 1974. *Health Effects of Fossil Fuel Power Plants.* Proceeding of Eighth Midyear Topical Symposium, Health Physics Society, Knoxville, Tenn.

Hevly, R. H. 1974. Land restoration, revegetation and plant succession on lands disturbed by construction of the Black Mesa-Lake Powell Railroad and Navajo Generating Station, Northern Arizona, pp. 139–164 in *Environmental Impact Studies of the Navajo and Kaiparowits Power Plants.* Third Annual Report. Vol. 1. Flagstaff: Northern Arizona University.

Hill, A. C. and J. H. Bennett. 1970. Inhibition of apparent photosynthesis by nitrogen oxides. *Atmospheric Environment* 4: 341–348.

Hill, A. C., S. Hill, C. Lamb, and T. W. Barrett. 1974. Sensitivity of native desert vegetation to SO_2 and to SO_2 and NO_2 combined. *Journal of the Air Pollution Control Assoc.* 24: 153–157.

Hill, A. C. and N. Littlefield. 1969. Ozone. Effect on apparent photosynthesis, rate of transpiration, and stomatal closure in plants. *Environmental Science and Technology* 3: 52–56.

Hindawi, J. I. 1968. Injury by SO_2, HF, and chlorine as observed and reflected on vegetation in the field. *Journal of the Air Pollution Control Assoc.* 18: 307–312.

Johnson, Z. and R. H. Hevly. 1973. A preliminary report of the effects of SO_2 on growth of three native grass species of the Kaiparowits Basin, pp. 139–152 in *Environmental Impact Studies of the Navajo and Kaiparowits Power Plants.* Second Annual Report. Flagstaff: Northern Arizona University.

Knox, J. B., and C. A. Sherman. 1974. *Catalog of Regional Modeling Capabilities; Energy Related Studies Program.* USAG 74-9. California: Lawrence Livermore Laboratory.

Lang, R. 1973. *ADPIC, A Three-Dimensional Computer Code for the Study of Pollutant Dispersal and Deposition Under Complex Conditions.* UCRL-51462. California: Lawrence Livermore Laboratory.

Lave, L. B. and L. C. Freeburg. 1974. Health Cost to the Consumer Per Megawatt-Hour of Electricity in *Energy, The Environment, and Human Health.* A. J. Finkel (ed.). Acton, Massachusetts: Publishing Sciences Group, Inc.

Lefohn, A. S., R. A. Lewis, and N. R. Glass. 1974. An approach to the investigation of the bioenvironmental impact of air pollution from fossil fuel power plants. EPA-NERL 74–206. Presented at Air Pollution Control Association Meeting held in Denver, Colorado, June 9–13, 1974.

McCune, D. C. and L. H. Weinstein. 1971. Metabolic effects of atmospheric fluorides on plants. *Environmental Pollution* 1: 169–174.

McJilton, C., R. Frank and R. Charlson. 1973. Role of Relative Humidity in the Synergistic Effect of a Sulfur Dioxide-Aerosol Mixture on the Lung. *Science* 182: 503–504.

Milane, M. P. 1973. *Bibliography on Air Pollution Forecasting By Computer and Diffusion Models.* PB-233484, National Technical Information Service, U.S. Department of Commerce.

NAU. 1973. *Environmental Impact Studies of the Navaho and Kaiparowits Power Plants.* Second Annual Report, 1 June 1972–31 May 1973. Flagstaff: Northern Arizona University.

NAU. 1974. *Environmental Impact Studies of the Navaho and Kaiparowits Power Plants.* Third Annual Report, 1 June 1973–31 May 1974. Flagstaff: Northern Arizona University.

Packer, P. E. 1974. *Rehabilitation Potentials and Limitations of Surface-mined Land in the Northern Great Plains.* USDA Forest Service General Technical Report INT-14. Ogden, Utah.

Pasquill, F. 1962. *Atmospheric Diffusion.* London: D. Van Nostrand Co.

Prickett, T. A. and C. G. Lannquist. 1971. *Selected Digital Computer Techniques for Ground Water Resource Evaluation.* Illinois State Water Survey, Bulletin 55, Urbana: The State of Illinois.

Ragland, K. W. 1973. Multiple Box Model for Dispersion of Air Pollutants from Area Sources. *Atmospheric Environment* 7: 1017–1032.

Schwing, R. C. and G. C. McDonald. 1974. Measure of Association of Some Air Pollutants, Natural Ionizing Radiation and Cigarette Smoking With Mortality Rates. Research Publication GMR-1573. Warren, Michigan: General Motors Research Laboratory.

Scriven, R. A. and B. E. A. Fisher. 1975. The Long Range Transport of Airborne Material and Its Removal By Deposition and Washout — II. The Effect of Turbulent Diffusion. *Atmospheric Environment* 9: 59–68.

Slade, D. H., ed. 1968. *Meteorology and Atomic Energy, 1968.* TID-24190, USAEC, Division of Technical Information.

Standiford, D. R., L. D. Potter, and D. E. Kidd. 1973. Mercury in the Lake Powell Ecosystem. Lake Powell Research Project Bulletin No. 1. University of California, Los Angeles: Institute of Geophysics and Planetary Physics.

Sutton, O. G. 1953. *Micrometeorology.* New York: McGraw-Hill Book Co.

Taylor, O. C. and F. M. Eaton. 1965. Suppression of plant growth by nitrogen oxide. *Plant Physiology* 41: 132–135.

Thames, J. L. and T. R. Verma. 1975. Coal Mine Reclamation on the Black Mesa and the Four Corners Areas of Northeastern Arizona. Presented at Practices and Problems of Land Reclamation in western North America, U.S. Dept. of the Interior Symposium held at Grand Forks, North Dakota, January 1975.

Tingey, D. T., R. A. Reinert, J. A. Dunning, and W. N. Heck. 1971. Vegetation injury from the interaction of NO_2 and SO_2. *Phytopathology* 61: 1506–1511.

Tompkins, T. and D. Blinn. 1974. The effect of mercury on Lake Powell diatoms, *Fragillaria crotonensis* and *Asterionella formosa*, pp. 353–376 in *Environmental Impact Studies of the Navajo and Kaiparowits Power Plants,* Third Annual Report. Vol. 1. Flagstaff: Northern Arizona University.

Turner, F. B., E. M. Romney, R. F. Logan, V. D. Leavitt, T. L. Ackerman, G. V. Alexander, B. G. Maza, P. A. Medica, and A. T. Vollmer. 1975. Preliminary analyses of soils and vegetation in the vicinity of the Mohave Generating Station in southern Nevada. UCLA #12-990. Los Angeles: Laboratory of Nuclear Medicine and Radiation Biology, University of California.

Winkelstein, W., S. Kantor, E. W. Davis, C. S. Maneri, and W. E. Mosher. 1967. The Relationship of Air Pollution and Economic Status to Total Mortality and Selected Respiratory System Mortality in Men. I. Suspended Particulates. *Arch. Environmental Health* 14: 162–171.

Winkelstein, W., S. Kantor, E. W. Davis, C. S. Maneri, and W. E. Mosher. 1968. The Relationship of Air Pollution and Economic Status to Total Mortality and Selected Respiratory System Mortality in Men. II. Oxides of Sulfur. *Arch. Environmental Health* 16: 401–405.

Zaleski, L. and W. S. Gaud. 1974. Trace element analysis of vegetation, soils, and water from the region around Page, Arizona, pp. 189–206 in *Environmental Impact Studies of the Navajo and Kaiparowits Power Plants*. Third Annual Report. Vol. 1. Flagstaff: Northern Arizona University.

15. The Implications of Oil Shale Development

Paul D. Kilburn

ORGANIC MATERIALS are commonly associated with inorganic substances in sedimentary rock. Such materials are sometimes converted to petroleum and natural gas. In some cases, when the conversion process is perhaps less complete, the organic material is converted into a waxy solid material known as kerogen.

Kerogen is present in a great many shale-like rocks, which, when the kerogen content is high, are usually referred to as "oil shale," even though they may contain carbonate minerals and in many cases are really marlstones. Oil shales occur in all continents, have been recognized in many countries, and have been used intermittently and in a very minor fashion for centuries.

All shales contain some organic materials. Even rich deposits, however, have always lacked sufficient hydrocarbon content to make the effort expended in mining, crushing, and retorting economically feasible. As petroleum and gas supplies get more critical, however, oil shales are attracting increasing interest.

The huge deposits of oil shales in the western United States are among the world's largest, and the total hydrocarbon in the marlstone, often referred to as oil shale, probably exceeds the total that will be found in the Middle East. This enormous potential source of energy is the subject of intensive technical, economic and environmental study; is the location for several pilot and semi-works extraction programs; and may well be on the verge of being tapped as one more input to help ease the world energy crisis.

Commercial extraction of crude petroleum from shale — which forms as a liquid condensate from the retorting and distillation of the waxy kerogen permeating the raw shale — began in 1838 in France and in 1850 in Scotland, and was subsequently carried on in Australia, Estonia, Sweden, Spain, Manchuria, South Africa, Germany and Brazil (Schramm 1970; East and Gardner 1964). As of the mid 1970s, only the Estonian and Manchurian operations are proceeding commercially. Deposits are only partially investigated, but the largest deposits to date are in the U.S. and Brazil (Schramm, 1970).

In the U.S.A., Savage (1974, p. 17) notes that "In pioneer days, the shale had been distilled to make axle grease for covered wagons." He also states that prior to the first retort in 1917, "these shales already had been used to heat peach orchards at Palisade and as domestic fuels." He goes on to note that, "The first oil shale retort in Colorado was built in 1917 on Dry Fork, northwest of Debeque. The next few years saw hundreds of retorts invented, dozens built, and several which produced shale oil."

The Piceance Creek basin deposits have been well summarized by Murray and Haun (1974, pp. 33 and 37) as follows:

> "Approximately 80 percent of the oil shale resources of the Eocene Green River Formation are in the Piceance Creek basin; this amounts to an estimated 1,200 billion barrels of oil-equivalent. . . . includes the thickest and richest deposits, and is one of the largest deposits of petroleum known anywhere in the world. Of this total resource, approximately 600 billion barrels of oil are contained in oil shale that averages 25 or more gallons of oil/ton. The higher grade sections of oil shale vary from zero to approximately 1,600 ft. The thickest interval of oil shale, which in part corresponds areally with near-maximum over-burden, is located in the north-central part of the basin. Approximately 75 percent of Colorado's oil shale resources are situated on public lands administered by the Bureau of Land Management. . . ."

It is perhaps fortunate that no large-scale mining in the area has yet taken place, for the nation's recent environmental impetus will require application of the stringent environmental constraints to assure that commercial operations proceed in an environmentally acceptable manner. This will occur for several reasons. First, because the federal government is the major owner of surface land and mineral rights in this region; second, because federal, state and local governments are devoting a great deal of effort to assure sound environmental management; third, because the public at large is both informed and vociferous in these matters; and finally, because industry is concerned that development minimize environmental damage.

Predicting and describing the environmental implications of oil shale development in the region has, as its major problem, the uncertainty of the various operations, each of which varies a great deal in planning and development schedules. One processor has developed a semi-works operation, and has published extensive environmental studies and predictions. This is Colony Development Operation which plans on utilizing the The Oil Shale Corporation (TOSCO) II retorting system. For this reason, discussion will be centered on the environmental implications of technology; wherever possible, discussions will be expanded to include basin-wide impacts.

TECHNOLOGY

Several technologies exist for converting raw oil shale to crude petroleum. Of these processes, only TOSCO II and Union have been tested at the semi-works level in this country. Several other companies have conducted pilot testing and some are gearing up for semi-works levels. Others will no doubt evolve and develop in response to feedback from these procedures.

The technologies most likely to be used in the first generation shale oil plants in the U.S. include, in addition to the TOSCO II and Union, the

NTU and Paraho process. All are above-ground retort systems and are more fully described elsewhere (Kilburn, et al. 1974). One additional, the Garrett in-situ process, has not had semi-works testing by 1975, but is briefly discussed owing to its differing environmental impacts and its capability to utilize the lower grade shales which would not be commercially feasible with the above-ground technologies. The in-situ process would (according to Cameron Engineers, 1974) develop "chimneys" of crushed shale that would be retorted in place using the deposit itself as the container. Controlled combustion within the chimneys would convert the kerogen from the rock to liquid petroleum which would be drawn off below. The spent shale would remain in place filling the chimney. Environmentally, the process is particularly attractive in that it does not need an above-ground spent shale (tailings) embankment; it is less attractive with regard to its impact on groundwater which, at present, is unknown, but which could be a major factor. Its attractiveness to the nation is furthermore reduced by the percent Btus extracted in place. In-situ processing has always given low yields — on the order of 20%–30% in published accounts, although Occidental is now claiming far higher yields. The controversy is far from settled and additional data are needed to approach an answer.

THE ENVIRONMENTAL SETTING

The present-day setting of the Piceance Creek basin is described in detail in Colony's EIA (Colony Development Operation, 1974a, Part One, pp. 31–33), and the following discussion is digested from that introduction.

The diverse environmental character of the Piceance Creek basin is attributable to the broad range of climatological characteristics and topographical features found within this relatively small area. These varied conditions have produced a remarkable number of distinct ecosystems. The most spectacular features are the sheer cliffs formed by the resistant marlstones of the Parachute Creek member of the Green River formation which rise over 4,000 feet above the Colorado River at the southern edge of the basin. The escarpment is more or less continuous from the Grand Hogback near Rifle to the massive Book Cliffs and Cathedral Bluffs near Grand Junction, broken only where tributaries of the Colorado River have cut deeply into the sedimentary deposits. Parachute Creek Valley and Roan Creek Valley, the largest of these tributary canyons, vary from 2,000 to 3,000 feet in depth and the Roan Creek Valley is more than 30 miles long. The upland areas surrounding the canyons form a gently rolling surface (Roan Plateau) that slopes northward to the White River.

The regional climate of the Piceance Creek basin is characterized by abundant sunshine, warm temperatures throughout the growing season and limited amounts (10–20 inches) of precipitation. The elevation ranges from 5,000 feet to over 9,000 feet and produces climatic variations ranging from semiarid conditions on the valley floors and lower elevations to cooler, moister conditions on the upland portions of the Roan Plateau. Winters are harsh on the plateau with snow accumulations exceeding five feet in some places. In the valley, individual storms may produce substantial snow accumulations, but these usually do not persist for very long, and the valley floors are snow free much of the time.

Most of the basin is covered with similar amounts of shrubland and woodland. These types are intermixed in the basin and their location is controlled primarily by available moisture and soil depth. Ward et al. (1974) described and defined these types, while Terwilliger (1973) mapped the major vegetation types.

Forests dominate only a minor portion of the basin and consist primarily of gallery cottonwood-box elder forests along streams at lower elevations and conifer and aspen forests, slightly intermixed, on the northern and eastern slopes at middle and upper elevations. Such forests are often isolated patches on those slopes that have more favorable moisture conditions.

Shrublands and woodlands dominate most of the basin. Shrublands dominate middle elevations and consist primarily of two types. The first is mixed mountain shrub at the higher elevations, particularly on northerly and easterly slopes and more favorable situations; while the second, the big sagebrush type, dominates on drier situations at all elevations.

The woodland is dominated by pinyon pine and juniper, a common vegetation type in the western United States, and is widespread over thinner soil areas at lower elevations. Deeper soils within these elevations often support grassland and a disclimax sagebrush community resulting from past grazing practices.

Sedgley (1974) reports how man has decidedly altered the pre-settlement vegetation in the area through heavy grazing and deer population build-up. He describes it as follows (p. 111):

"The deep-soiled valleys now covered by coarse sagebrush, greasewood, and rabbitbrush, were once dominated by perennial grasses. The upland slopes and benches now dominated by big sagebrush and invaded by pinyon-juniper were also once productive grasslands intermingled with a variety of shrubs and forbs."

Information on mammals, cold-blooded vertebrates, birds, invertebrates, and aquatic systems is available for the southern area in Colony's Environmental Impact Analysis (Colony Development Operation, 1947a) and is expanded in a report by Thorne Ecological Institute. Small animals and bird species also show a close relationship to vegetation distribution.

Larger mammals range throughout the area and include the largest migratory mule deer herd in the world, some elk, black bear and even a few wild horses. Others, like the mountain lions and ringtail cats, are more common in rocky habitats in the southern canyon country. The region is rich in bird life and is an area where golden eagles live year-round and a few bald eagles winter. Peregrine falcons have been reported, but may only be migrants.

With settlement in the mid-to-late nineteenth century man has become the most important mammal of the area and has drastically affected all ecosystems through his habitations, grazing regimes, hunting camps and other activities. His interest in the area was based originally on a source of grazing for his domestic cattle and sheep, and no area of the basin has been devoid of this impact. Recently, the U.S. Soil Conservation Service removed large areas of pinyon and juniper trees in order to convert certain areas of limited grazing potential to areas much more productive for livestock.

Deer hunting has made the area famous throughout the country. The Colorado Division of Wildlife has compiled statistics for this region for many years, and it notes that the average number of hunters to use this area over the past 20 years has been 28,632 (L. Roper, personal communication). The average deer harvest has been 25,779. If this harvest is assumed to remove about 25% of the herd, then the total deer population would exceed 100,000. While the area covered by this figure includes more than the Piceance Creek basin (Units 22, 23, 34, 31, 32), one still gets a good view of the magnitude of the deer resource. Hunter interest in this resource is great and will undoubtedly *intensify;* of added interest is the fact that large numbers of the hunters now come from California and Texas and the out-of-state use approximates the local use. Good game and hunter management of the area is imperative.

Apart from hunting, little recreational use is made of the Piceance Creek basin, as it is considered one of Colorado's less scenic and least desirable areas to visit. Roads are unpaved and usually unimproved, scenery is only average and water is lacking over large areas. Few people, other than those on hunting or shale-oriented visits, actually enter the area. The impact by man on shale development should be viewed in this light.

The first settlers were ranchers who utilized the surrounding rangelands for raising cattle and in later years for raising sheep. Irrigation ditches constructed early in the settlement period allowed for production of fodder for feeding livestock during the winter months. Numerous orchards and garden plots were planted in the valley and the people were more or less self-sufficient. The most active agricultural period lasted until the mid-twentieth century, and residents describe Grand Valley as a "nice farming community" as late as 1940. Since World War II, there has been a decrease in the number of people in the area and currently many of the ranches are leased from owners who live outside the area.

The Piceance Creek basin lies within two sparsely populated counties in western Colorado. These counties had in 1970 an estimated population of 19,663 in an area of 6,620 square miles, or only 3.2 people per square mile (Wengert 1973). This small population and low density would be even less if it did not include Glenwood Springs, the county seat of Garfield County and the only city with 4,000 people.

Mesa County, to the immediate south, contains Grand Junction, a city with over 50,000 people, the regional center for the three-county area, and the largest city in western Colorado.

The Colorado and White River valleys serve as the location for urban development in the three-county region. Communities are located in these valleys at periodic intervals and provide goods and services to the adjacent population. This urban settlement pattern has followed the western pattern of the search for water and level land.

PROPOSED OPERATIONS

The only project to publish its proposed action in detail has been Colony Development Operation, termed Colony in this paper (Colony Development Operation, 1974a, Parts One-Four), and for this reason the following summary is presented.

Colony is a joint venture which is comprised of the following four active members: Atlantic-Richfield Company (A.R. Co.), operator; Ashland Oil Company; Shell Oil; and TOSCO. It proposes to construct and operate a shale oil complex designed to process, with the TOSCO II process, approximately 66,000 tons per day of oil shale and to produce and upgrade some 46,000 barrels per day of petroleum product. Construction was to have begun in 1975 but it was delayed both for economic reasons and because of the absence of a national energy policy.

Development of the retorting technology known as the TOSCO II process began in 1956 (Kilburn, et al. 1974). Based on research conducted by the Denver Research Institute of Denver, Colorado for TOSCO, the first pilot plant was built near Littleton, Colorado, in 1957. In 1964 a 1,000-ton-per-day semi-works retort was constructed and a full-scale pilot mine was started on the Dow property in Middle Fork Canyon on Parachute Creek approximately 17 miles north of the town of Grand Valley, Colorado. The semi-works plant and mine were operated from 1965 through the fall of 1967. As a result of these operations, the design of a commercial oil shale complex was initiated. Further testing of the semi-works plant was required to complete this design. The additional testing was undertaken in 1969 under the direction of Atlantic Richfield Company as operator for the joint venture.

The four major components of shale oil production are mining, crushing, retorting, and upgrading (product recovery).

The mining operation would be conducted from a mine bench located in the upper portion of Middle Fork Canyon of Parachute Creek. Initial crushing would take place at this bench, from where the shale would be conveyed to the plant site for final crushing to ½"-minus size. The retorting system and the upgrading operation would be located together on the Roan Plateau. The disposal operation would be conducted in adjacent Davis Gulch which is also on the Roan Plateau. The primary commercial product, low sulfur fuel oil, would then be transported from the complex by an underground pipeline to an existing oil gathering station in Lisbon Valley, Utah, a distance of approximately 180 miles. The other products, ammonia and liquefied petroleum gas, would probably be pipelined to Grand Valley and transported further by rail or by truck.

At this stage the full-scale development of a shale industry is, of course, impossible to predict. A useful reference point that was utilized in the leasing impact statement by the U.S. Department of the Interior (1973) predicted development of a million-barrel-a-day industry by 1985. This date is clearly too early, but such a scale of development prior to the year 2000 is feasible. At the rate of extraction required for a million-barrel-a-day industry, utilizing only the better shales (e.g. 30 gal./ton in deposits 30' thick), and at complete extraction, such an industry would have more than a 300-year life. The use of leaner shales, the normal pattern in the development of a resource, could greatly extend this time span to more than 3,000 years (based on complete extraction of 15 gal./ton shales)*. While such figures are highly optimistic within today's economic and technical conditions, they nevertheless serve to illustrate the magnitude of the potential resource.

The picture that emerges, then, is one of a long-lived shale industry, and one that consists of several large plants operating simultaneously. Each operation would continue for 25 to 30 years on a site, prior to moving to a new location. Revegetation and rehabilitation should be completed on each site within that time, and although woody plants would not have reached full size, they should be well established on disturbed areas and spent shale embankments.

ENVIRONMENTAL STUDIES

The information base from which to make environmental predictions is becoming more voluminous with each passing year. Colony alone has concluded over one hundred environmental studies, which have been described in a series of papers (Hutchins, et al. 1971; Kilburn, 1973; Kilburn and Legatski 1974) and are published in a twenty-volume Environmental Impact Analysis (Colony Development Operation, 1974a). Extensive studies have been done in each of eight areas: air, water, processed shale, revegetation, ecology, pipeline, plant and mine, and socio-economics. Completion of the studies cost more than $3 million. The multi-volume Impact Analysis includes three basic Colony-written volumes on mine and plant, pipeline, and socio-economic impacts. The other seventeen volumes include only the most recent and comprehensive environmental studies made since 1969. All are available from the Colony offices in Denver. These materials form the basis for the material pertaining to Colony utilized in this paper. The interested reader is referred to this set for more complete information.

Any assessment of impacts on the entire basin must at present be far less precise than that for Colony alone, owing to the less intensive study of both the environmental setting and the uncertainty of proposed actions. The latter cannot be detailed until each operator has had time to study mining plans, evaluate processing information, make detailed site engineering studies, or even select the processing technology to be utilized.

A better understanding of the environmental impact will be available after publication and analysis of the four "COSEP" studies established by the Committee on Oil Shale Environmental Problems, hence the term COSEP. They are funded equally by the State of Colorado, the Department of the Interior and various industries, primarily oil companies. They are being administered by the Department of Natural Resources of Colorado.

The studies include one on revegetation already published (Cook 1974), another on ecology by Thorne Ecological Institute, a third on hydrology by the U.S. Geological Survey, and a fourth on socio-economic mat-

---

*Miscellaneous Energy Conversion Tables and Data in Murray, D. K. (ed.) 1974.

ters by three local consulting firms (published by THK Associates, et al. 1974). These reports provide considerable additional data on the environmental setting, as well as an assessment of impact on all the interrelated ecosystems and components in the Piceance Creek basin. After these reports have been analyzed and discussed, a more accurate appraisal of impact within the basin can be established.

The six-volume oil shale leasing impact statement (U.S. Department of Interior, 1973), extensive though it is, was really a first attempt at integrating all published and unpublished information that could be focused on possible shale development in the entire Green River formation areas in Colorado, Wyoming and Utah. As such, it provided some highly useful baseline information and some target estimates against which to compare recent and future refinements. While the statement was as complete as could be reasonably expected, the actual field data from the Piceance basin on which it was based were far less than will be provided by the COSEP studies.

ENVIRONMENTAL IMPACTS

Hydrology, Water Quality and Salinity

Discussion of the water-related features subject to environmental impact in shale development includes effects on water quality, surface flow and groundwater. Prediction of impact on these features is further complicated by the dual drainage basin character of the Piceance Basin. The southern edge drains downstream to the south and the Colorado River, an area with ample available water owing to federal impoundments. Most of the remainder of the basin drains northward to the White River; it is here that water supply problems could be encountered owing to the lack of available reservoirs.

The proposed plant would use from three to four barrels of water per barrel of oil produced. For a 50,000 bbl./day plant the total daily water use would be approximately 170,000 barrels, about 11 c.f.s., or about 7,950 acre-feet per year. This is the amount of water needed to irrigate about 700 acres of peach orchards in this area per year. Although these requirements are significant, they are less than one percent of the so-called "Cameo Demand" diversion of 1,200 c.f.s. near Grand Junction, primarily used as irrigation water for agricultural purposes. This water would be totally consumed in the industrial operations, with no planned discharges to surface water bodies or groundwater aquifers.

Urban water use requirements resulting from the population growth accompanying a commercial plant would amount to an additional ten to twenty percent.

During normal flow of the Colorado River, this total quantity of water will be available for diversion. During periods of low flow, water will be available from other sources such as the nearby Green Mountain Reservoir.

Inasmuch as the plant will be totally consumptive of water, there would normally be no discharges of waters containing contaminants from the plant or processed shale embankment into adjacent surface or groundwaters.

While it is clear that a single commercial operation would use a rather small amount of water, these amounts could become significant with the initiation of other operations and the development of an industry. In the latter case, water consumption could even become limiting. While undoubtedly water conservation could result in less water consumption per barrel of oil, it is likely that large water consumption would be a necessary trade-off for an oil shale industry. The environmental impacts, their size, and magnitude are still subject to argument and questions which can only be resolved by additional study and actual operations.

It is important to keep in mind that there is sufficient water available in Colorado from its historic allotment for an industry larger than currently contemplated. Some 852,000 c.f.s. are now available within Colorado for future development, a figure which is 100 times the amount required by Colony alone. The problems will occur in delivering this amount to the right place at the right time in an economic manner. Thus, it is the allocation problem that is of more concern than total water usage.

Water quality is the second part of the water question and is increasing in importance. Colony has demonstrated that its operations would cause minimal impacts to water quality, owing to the fact that the operation would be totally water consumptive, and that all solid and liquid wastes would be contained in or within their tailings (processed shale) disposal embankment.

Two features are fundamental to the prevention of accidental discharge of polluted water and the leaking of toxic materials to surface or groundwater. The first is the processed shale embankment itself, which, when compacted, is extremely impermeable. Process liquids used to moisten the processed shale would be "locked" into the embankment, thus preventing their effluence to streams.

Second, a catchment dam constructed below the processed shale embankment would confine any runoff from the embankment or the plant complex, both of which are located in the Davis Gulch watershed. Any water containing dissolved solids leached from the top few feet of the processed shale embankment or process waters from the plant would be contained and circulated back to the plant area for the moisturizing of processed shale.

In short, only following unusual precipitation could contaminated water flow over the spillway and go downstream. Should this ever occur, the water flowing over the catchment dam would be highly diluted, and any resultant impact should be minor.

Colony's water use will have only a minor impact on downstream salinity. Diversions from the Colorado River at Grand Valley would be about three-tenths of one percent of the average flow at that point. Based upon one set of reasonable assumptions, Skogerboe (1974) has estimated that withdrawals from the river would cause a very minute increase in the salinity of the Colorado River at Hoover Dam of about 0.12 mg./l., or one-sixtieth of one percent of the present 730 mg./l.

As previously mentioned, water utilized by the complex during low flow periods could be obtained from the Green Mountain Reservoir. The reservoir has for years been operated as a working reservoir, releases being for the general benefit of the western slope and the Colorado River. Colony expects that its requests for delivery of water will only slightly alter historic operation of the reservoir and would provide only minor new impacts.

To monitor accurately and document the non-discharge status of its operations, Colony has for several years been monitoring the quality of Parachute Creek and its tributaries at several points. The program is dynamic, and the number of sample points and their locations is modified from time to time. Some forty-four analyses have been performed on samples. The list of water quality parameters has been modified and has increased over the course of the program, as knowledge of the process, new information on the toxic character of certain elements, and an understanding of stream features have been obtained and evaluated.

With the information obtained during this program and continued water monitoring programs being evaluated during construction and operation, Colony anticipates no water pollution and minor or no impact on water use patterns on the western slope of Colorado.

In addition to the quality of surface water, groundwater is of similar concern. Colony studies of its experimental mine (Colony Development Operation, 1974a) have revealed very little groundwater, much less than could be utilized for dust control in mining. The groundwater problem may be of little importance in the southern part of the basin.

Toward the center of the basin, to the north of the Colony tract, the area drains north to the White River, and groundwater aquifers are extensive (Tait 1972; U.S. Geological Survey 1974; and Coffin et al. 1971) and appear to be items of concern in the mining operation. At the lease tracts, C-a, and C-b, for example, there is a fresh water aquifer above the mahogany zone and a lower quality aquifer below this zone. Dewatering required for mining in this zone could cause a mixture of the two waters that might degrade the fresh water with the lower quality water. This is possible, though by no means certain, in the initial operation. The appropriate strategy to prevent discharge of degraded waters into Piceance Creek or the White River could be developed if necessary.

An additional matter that is discussed in an unpublished report by the USGS on hydrology, as part of the COSEP studies, is the impact of mine dewatering on surface streams in the area. Preliminary computer modeling by the USGS, Water Resources Division, has indicated that stream flows may be affected by dewatering. Results from the model depend on input values from various water movement parameters which were in 1975 incomplete. Additional study is needed to clarify this point.

Downstream salinity impacts from basin developments are more complex than that described for Colony, owing to differences in processes, geography and groundwater. While the one and one-half percent Colorado River salinity increase predicted for extensive commercial development (U.S. Department of Interior 1973) is undoubtedly too high, the problem requires continual attention and surveillance. Weaver (1974) has pointed out that use and consumption of groundwater, high in dissolved solids, in mining could actually reduce contribution of these materials to the surface water and result in an actual reduction downstream. Skogerboe (1974) has similarly pointed out several mitigatory measures that are available for preventing increased salinity detriments.

Surface Disposal of Processed Shale

The production of shale oil by surface retorting will generate large volumes of waste material consisting chiefly of spent shale. The spent shale, termed "processed shale" in the TOSCO II process, is a fine, black, powder-like material. The dark color is attributable to a small amount of residual carbon which coats the dust particles. It is powder-like because retorting is conducted under processing temperatures that are not high enough to produce the clinker-type chunks characteristic of certain other pyrolysis process waste materials.

Environmental studies in this area have included analyses of the permeability of the processed shale embankment, leaching characteristics of processed shale, the structural integrity of processed shale compacted to various densities (Dames and Moore 1971b), the appropriate compaction methods for processed shale (Heley and Terrell 1971), and the liquefaction potential of the embankment (Dames and Moore 1971a). The design of the embankment has been influenced substantially by the findings and recommendations set forth in these studies. Heley (1974) previously described these matters in considerable detail.

Four hundred million tons of this processed shale will be disposed of in a plateau valley during twenty years of commercial operation by the Colony operation. This particular small valley has a very limited watershed and low stream flow, and the flooding hazard will be minimal. The moistened processed shale will be deposited initially in the side draws of Davis Gulch to avoid contact with natural aquifers. Eventually, about eight hundred acres will be covered with the processed shale.

After compaction, the processed shale is nearly impermeable and does not permit leaching or percolation of water into surrounding aquifers. Engineering studies indicate that a 1:3 slope (vertical:horizontal) on the face is adequate from a safety point of view, and Colony plans to use a slope of less than 1:4 as an additional safety factor. Due to the possibility of surface erosion from the action of rain and melting snow, the surface will be benched at regular intervals. This will reduce surface water flow velocities and allow any sediment being carried to settle out. These benches will be drained to prevent low quality water from the higher benches affecting any established vegetation on the lower benches.

A catchment dam will be placed downstream of the pile to collect any surface runoff from the pile. The surface runoff will be returned to the processed shale moisturizer and thus eventually returned to the processed shale pile.

A study of processed shale ash (Culbertson and Nevens 1972) has analyzed the potential use of processed shale for a variety of commercial purposes, including the manufacture of building blocks, concrete, road substrate, bricks, and paneling material. As of 1975, not one of these alternatives was economically feasible, and more testing was needed to make them so.

Furthermore, even after improvement and additional development provides an economic use of these materials, the study suggests that only three percent (ca. 500,000 tons per year) of the processed shale produced by a Colony-sized plant could conceivably be used in the near future on *all* possible products. Any disposal of a sizeable fraction of processed shale for use materials is clearly unattainable.

Spent shales produced in other processes vary considerably in physical and chemical characteristics, particularly those relating to strength, compressibility and permeability. These features must be carefully investigated in order to assess the best environmental means of placement, particularly to diminish erosion and adverse leaching impacts. Such investigations await production of the sizeable amounts of this material required for such tests; only after the completion of such studies will it be possible to assess the probable impacts.

As permanent surfaces are created in the spent shale embankments, revegetation would begin at the first available planting time to control wind and water erosion. Before starting the revegetation program, water spraying would be utilized to control fugitive dust.

The growth of plants on the processed shale embankment is important to surface stability, reduced erosion potential, proper water balance, preservation of aesthetic values, and restoration of the area to a balanced ecosystem (Duffield 1974). For these reasons a considerable number of vegetation studies have been conducted at the Colony site since 1965 (Block and Kilburn 1973). These programs include seven separate plot studies on processed shale which continue to provide useful information on the effects of varying slope, soil cover, mulch, fertilization, and irrigation; the migration of salt through root zones; and the suitability of various native, naturalized and exotic plant species for use in the ultimate revegetation process.

Regular watering during the first year is important to rapid establishment of vegetation. Results indicate that only infrequent supplements to natural rainfall are required during the second growing season. After two seasons the vegetation may require no further watering. Similarly, adequate fertilization the first year, and regular nitrogen additions in subsequent years, should provide suitable nutrient materials.

Successful growth of native woody plants, such as juniper, skunkbush, and four-winged salt bush in more recently established plots, indicates that these species may maintain continued growth and vigor in the same manner as several species of grass and forbs have done in the past. The satisfactory growth of native and exotic grasses, forbs and woody plant species on processed shale should not be surprising in view of the fact that the material is actually a substrate similar to the native soils derived from the weathered shale. The natural soils have had many years to weather and leach; consequently, they support a wide variety of plants and vegetation types. To establish vegetation quickly on processed shale, the vegetation and leaching process must be accelerated. This can be done by supplemental watering for a limited period of time to create a satisfactory soil medium.

The results from these test plots show clearly that revegetation of processed shale not only is feasible, but also could yield stands of vegetation that are more productive than existing vegetation. Present plots show that several grass species, both exotic and native, are capable of producing a dense cover in a single growing season with, of course, suitable leaching, watering, mulching and fertilization.

The question of permanence of re-established vegetation has been frequently raised by those concerned with the use of exotic species rather than natural species, the lack of nitrogen in the soil, and the possibility of salt migration upward following cessation of initial leaching and irrigation. None of these questions appears difficult. First, native species have been investigated extensively and will be used whenever suitable. Second, nitrogen additions will be made as long as necessary to get levels equivalent to the native soils and to establish a permanent nitrogen cycle. Finally, salt movement has not presented problems to established plants in our various plots, although continual monitoring will be intensified on this matter. In addition, a new plot to test various soil covers was established in 1975 and substrate treatments that would avoid salt migration in the event it should prove troublesome.

In summary, Colony plans to replace existing plant communities in areas covered with processed shale through prompt revegetation with seed mixtures of native and exotic grasses. After leaching has reduced the salinity of the substrate, perhaps a year after initial planting, native shrubs would be planted throughout the area. The combinations of grasses and shrubs should produce an ecosystem that would be scenically and ecologically compatible with the existing vegetation.

Information from the Colony revegetation plots should be applicable to all parts of the basin where the TOSCO II process is used, although allowances for climatic variation will be necessary in the different parts of the basin.

Revegetation work on other spent shales, all of which are coarser at the present time, has been confined to the Union Oil plots planted in 1966, and the CSU plots planted in 1973.

The Union Oil plots are located in Parachute Creek Valley, some four miles south of Colony's semi-works plant, at an elevation of about 6,000 feet. One large plot was established with seed, fertilizer and irrigation. Although no published data exist on this work, several species of grass planted in 1966 persisted for at least 10 years in a permanent appearing stand and indicate that permanent revegetation of this coarse material is a reasonable goal.

The CSU work includes one plot at Anvil Points at 5,590 feet and a second plot near the center of the Piceance Basin at 7,200 feet. Assessment of two growing seasons' data should soon be available. These data will be a most valuable background for revegetation techniques, but several seasons of growth are essential to accurate predictions.

It would appear that revegetation of spent shale embankments on coarser spent shale from various processes will be feasible, although a great deal more work is needed on the various spent shales to determine potential productivity and the appropriate cultural methods required for suitable establishment.

Ecology

The impact of the natural systems of the Piceance Basin has been the object of several major Colony studies, including one primarily focused on the Parachute Creek property by Thorne Ecological Institute (1973); another on the access corridor from Grand Valley by Geoecology Associates (1974); another on the potential pipeline corridors by Thorne Ecological Institute and the Laboratory of Mountain Ecology for Man (Marr et al. 1973); and another on the impact assessment of the products pipeline route from Parachute Creek to Lisbon Field, Utah (Utah Environmental and Agricultural Consultants, 1973); as well as other less comprehensive reports.

These reports emphasize that the ecosystem impact would stem particularly from three features: (1) physical coverage and modification of various ecosystems and habitats by complex structures, spent shale disposal embankments, roads, powerlines, etc.; (2) interruption of such natural processes as erosion, stream flow and animal migration by dams, reservoirs, roads or other diversion systems; (3) increased human use of presently little visited or disturbed areas, magnified by visual and noise impacts.

In addition to the published work for Colony, a great deal of unpublished work is available from the Soil Conservation Service and the Bureau of Land Management, the two federal agencies of primary responsibility in the basin, and the Colorado Division of Wildlife. In addition, a comprehensive final draft of a major ecological impact study was completed by Thorne Ecological Institute in 1974, and the final publication was expected soon thereafter.

The Colony studies have been keyed to actual operation plans and are discussed for this reason. The studies show that conventional underground mining would disturb about thirty percent of the surface above the area to be mined (Colony Development Operation, 1974a). Two-thirds of this figure results from the above-ground emplacement of spent shale. If the aforementioned seven oil shale operations are going on simultaneously in the Piceance Basin, this could result in an area of about twenty-one square miles of disturbed area during operations. For the 1,200 square mile Piceance Basin, this amounts to less than two percent of the area. Following completion of operations, this area would be completely revegetated and rehabilitated.

While such a percentage is clearly a minor ecosystem impact, when coupled with the "beyond boundary" impacts of process interruption and human disturbance, it is certainly worthy of strong environmental control. Additional development would increase the percentage.

Natural process interruption is more difficult to pinpoint. Colony (Colony Development Operation, 1974a) has detailed its overview of such impacts regarding effects on geology and stream flow. These would be of moderate impact on its own property, but should have only minor impacts outside of this area.

With regard to wildlife movement, some impact would impinge on the deer population owing to their mobility. Perhaps most serious would be increased deer accidents from road traffic, particularly in winter when they are concentrated near roads. The wide ranging raptors would be adversely affected owing to loss of habitat and disturbance. Other impacts on animals should more nearly approximate habitat loss, and this would be related to surface disturbance.

The impact on the aquatic habitat should be minor owing to the strict planning for maintenance of present water quality.

The third feature, increased human use, has been discussed in considerable detail (Colony Development Operation, 1974a) and is felt to be the major ecological impact by Stoecker who notes (1974, p. 85), "The most serious impacts will occur from miscellaneous human pressures: noise, visual disturbances — the indirect disturbances caused by more people using the area." Stoecker's feeling is that this feature is the most serious one of all for wildlife, and he discusses this for many of these species in his earlier report in Geoecology Associates (1974). Still, however, the uncertainty that prevails for most wildlife impacts is apparent owing to the uncertainty of operator actions and the irregularity of animal behavior.

Delineation of ecological impact beyond the Colony operation is particularly difficult to define owing to the uncertain size and timing of future shale development. Certainly first generation development will have minor impacts when viewed in the basin-wide context; full scale development may be too far in the future to see either the technological or the ecological impact.

Air Quality

The difficulty in predicting air quality impacts is at least redoubled when the baseline air quality and meteorology are only partially understood (as in the Piceance), and when the actual shale plant details are unknown. This difficulty is further compounded by the uncertain nature of Colorado air standards (now undergoing extensive review). Furthermore, the emission level achieved depends on the investment made and equipment utilized.

Colony's approach to the air quality situation is therefore instructive and is, for this reason, summarized in the following paragraphs in considerable detail. This material follows closely the discussion by Kilburn et al. (1974).

Most of the Colony air quality studies can be classified into three categories: (1) determination of existing conditions, (2) estimation of air contaminant emission rates, and (3) evaluation of the effects of the proposed plant operations upon existing air quality. These categories correspond to the terms of "inventory," "proposed action," and "impact," each an integral part of impact analysis.

The "inventory" of the Parachute Creek region includes measurements of existing air quality, climatology, and meteorology. There have been three major programs within this area: monitoring background air quality, area meteorology measurements, and addition of new ambient stations. These are discussed in more detail by Kilburn and Legatski (1974).

In the second or "proposed action" category of studies, Colony has estimated, and will continually update as design becomes more definitive, the air contaminant emissions that would result from commercial shale oil operations. This category includes both experimental and design studies.

Experimental programs have included many studies, such as the determination of appropriate stack sampling procedures, as well as evaluations of scrubbing and incineration efficiencies which are necessary for the design of the retorting systems and the evaluation of combustion characteristics of various shale oil fractions.

Two principal conclusions emerge from these experimental programs: (1) the plant can be designed to conform with all applicable Colorado emission regulations, although some rather substantial costs are required; (2) a final oil product can be produced that will satisfy EPA "new source performance standards" if the product is consumed as a utility fuel, and will be far superior to most conventional fuel oils from the standpoint of sulfur oxide emissions.

In addition to experimental evaluations, a very substantial effort has been devoted to air emissions by the engineering staff and managing contractor. A commercial plant could require over forty stacks; one for the primary crusher; several for the secondary crushers; six for each of the preheat systems, the retorting systems, and the processed shale moisturizing systems; and a dozen or so for furnaces and heaters; as well as some for miscellaneous storage and transfer operations. The design of all of these systems is currently being completed by the engineering contractor and subcontractors, and many of the details concerning emission rates, stack heights, and so forth are not yet precisely defined. One of the prime decision factors in final selection of stack heights, exit temperatures, and emission rates is the result from diffusion modeling techniques, which were developed from the extensive studies made on the property over a period of several years.

Emission rates from stacks are controlled by state regulations and vary with the type of pollutant and source. These regulations are, of course, one of the prime inputs and limitations to the design of any of the equipment which will result in emissions to the atmosphere. In some cases, this causes a substantial increase in the cost of the basic processing equipment.

A third category of studies involves the evaluation of "impacts" from the proposed operation. This is the most difficult of all the areas studied in that prediction methods give only approximate answers owing to the many meteorological factors which can vary simultaneously.

Industrial plant siting has become increasingly more complex as a result of recent concern for the environment. Gaseous and particulate effluents cannot be released in unlimited quantities to the atmosphere, which formerly was considered a boundless sink. This concept of the atmosphere is no longer accepted, and industry is expending vast resources to limit effluent output. There remains, even with limited outputs, a potential for adverse pollution episodes if inadequate attention is given the local topography and meteorological conditions during site selection.

Historically, in the Meuse Valley of Belgium and at Donora, Pennsylvania, the interaction of topography and meteorology resulted in severe pollution episodes with the loss of life. It has been noted by investigators that the valleys in which these episodes occurred were only several hundred feet deep but pollution was confined during a four-to-five-day period of air stagnation. It was noted further that the situation becomes much more critical if the plant is located on the floor of a narrow, steep-sided valley. Even under non-stagnant conditions, the lateral confinement of a plume by the valley walls will increase pollutant concentrations above those that would be encountered in an unconfined location. It was necessary, therefore, to examine carefully the proposed siting of an oil shale processing plant at the confluence of East Middle Fork with the Middle Fork of Parachute Creek. The complexities of the meteorology and the terrain at Parachute Creek necessitated a trial approach to the problem and a series of atmospheric diffusion experiments was the result.

The importance of the studies is illustrated by the change in plant site location. Colony originally intended to locate a commercial plant in Parachute Creek Valley. The rather elaborate series of diffusion modeling calculations and tracer experiments that were conducted provided data which led to the conclusion that the danger of encountering "episode conditions," i.e., temperature inversions, and/or stagnant air masses, was significant. In other words, the probability of the "trapping" of air contaminants by the canyon walls and a prolonged inversion "lid" is such that short-term ambient air quality standards would probably be exceeded by a plant in the valley.

As a result of these and other environmental studies, the decision was made to locate the plant on the plateau, some 2,000 ft. above the valley, at substantial cost, and to conduct further meteorological and tracer studies to assure that this relocation would result in compliance with ambient air quality standards. The analysis of the data at the plateau site is not yet complete, but it is clear that this site is far superior from an air pollution stand-point.

It should be added that several other environmental features also exerted strong pressure for a change in plant site and processed shale disposal embankment location including a much lessened possibility of flood damage to the embankment location, the maintenance of valley areas as winter deer range, and lessened geological, aesthetic and aquatic biological damage.

In addition to extensive studies relating to the plant's air quality impacts, Colony has not ignored the impacts on the air quality due to population increases. Much work has gone into studying the impact of the home-to-work travel of employees and the impact of new communities on the air quality, in addition to the socio-economic aspects of these items.

An evaluation of the probable air quality impacts in the basin is particularly difficult due to the meteorological and topographical complexities (Thorne Ecological Institute, 1974, p. 27). As this report states:

"It appears that the local wind circulation which would control transport and diffusion characteristics of the Basin is strongly influenced by local topographic features and the coupling and uncoupling of surface flow and the synoptic gradient flow. At night, as heat is lost from the higher elevations, cold air tends to drain down into the valleys and lower depressions. As the cold air pools in the valleys and lowlands, inversions are formed in these areas which restrict the dispersion of smoke or other pollution held within the boundary layer. Height of inversion is dependent upon the depth and duration of drainage air, the topography of the area, and the surface thermal features."

The report concludes (p. 27), "From this study it appears that only two or three plants would exhaust the capacity of the Basin to dispose of stack emissions." This statement is based on consideration of present Colorado State standards which are extremely stringent; the report goes on to indicate several unanswered questions. First, it is not clear how shale activities (private) in the southern part of the basin will affect the northern areas (mainly federal lease tracts). Second, the effects of industrial development in adjacent areas on basin air quality is not known (and neither is the effect of basin air impacts outside the basin, and even extending 200 miles beyond to the Front Range). Third, what additional air degradation would occur from other developments accompanying shale development. Finally, just how great would be the impacts from a million-barrel-per-day industry?

It is doubtful that reliable quantitative answers to these questions can be provided without the initiation of some actual operations and some adequate basin-wide monitoring.

Socio-Economic Factors

Planning for new development began in 1970 (Rold 1974) when a joint federal, state, and industry effort was made, a committee called the "SCEEP" Committee was formed, and a report was issued (Special Committee . . . 1971). As a result, a three-county Oil Shale Regional Planning Commission (OSRPC) was then organized with representation from Garfield, Rio Blanco, and Mesa counties. This commission had as its goal, orderly and planned development for the region. Subsequently, a Council of Governments consisting of elected officials from Garfield, Rio Blanco, Mesa and Moffat Counties was formed and has taken over the activities of the OSRPC. At the same time, the Garfield County Board of Commissioners in Glenwood Springs completed a land use zoning effort for the county. Activity since that time snowballed quickly and studies by Colony (1974) and THK Associates (1974) have better delineated the specific problems of housing, education, utilities, transportation, government, recreation, and other community services and facilities which accompany all new urbanized development.

The Colorado West Area Council of Governments (1974) summarized much of this information and shows that shale development would compound the presently predicted growth. The Colorado West study indicates that even without shale development the population in the three-county region containing the Piceance (Garfield, Rio Blanco, and Mesa counties) could increase from its present 82,000 to 163,000 in 15 years (high projection). With the effect of moderate shale development added to this, the population could increase another 60,000 to 70,000 and result in almost a tripling of the population of the region in only fifteen years. While the growth increase would not be due to shale development alone, it could be a major contributor, and it is apparent that strong planning and execution efforts must precede and accompany the development of oil shale, if orderly growth and development are to occur. The real challenge will be to maintain the planning base and to make sure of an orderly execution of community development.

CONCLUSION

Certain areas of the Piceance Creek basin are important both for their unusual ecosystems and for other amenities. Stoecker (1974) has suggested maintaining one of the larger box canyons in the southern portion as a nature reserve. Knutson and Boardman (1973) have suggested preserving a "green belt" of the magnificent bluffs adjacent to the Colorado River. Preservation of the Cathedral Bluffs running north and south in the western portion has also been suggested. It is clear that a movement is growing to provide a system of natural areas throughout the basin. Such a system would maintain important examples of representative vegetation types, wildlife habitat, geological and archeological features, and other areas of particular ecological interest. This preserve system would maintain natural (or semi-natural) areas both as ecological baselines and as nuclei for obtaining biological "seed" for rehabilitation of disturbed areas. A study to delineate suitable areas that could be included in such a system is in order.

Man is viewing in the Piceance, the world's largest known hydrocarbon deposit, and his use of that resource is inevitable. The magnitude of the deposit certainly demands a major planning effort. One can even speculate into the future, as Knutson and Boardman (1973) have done, to visualize a concerted or systems approach to obtaining and processing this resource.

Nevertheless, until commercial activity begins and monitoring procures actual data, predictions will remain speculative. Operation of a few commercial plants, mines and rehabilitation activities is needed in order that precise predictions can be made to better delineate and quantify the environmental impacts of a major shale oil industry.

REFERENCES

Block, M. B. and P. D. Kilburn, eds. 1973. *Processed Shale Revegetation Studies, 1965–1973*. Denver, Colorado: Colony Development Operation.

Cameron Engineers. 1974. Occidental gives more details on modified in situ scheme and field test results. *Synthetic Fuels Quarterly,* 11(2): 2–30 to 2–37.

Coffin, D. L., F. A. Welder, and R. K. Glanzman. 1971. *Geohydrology in the Piceance Creek Structural Basin Between the White and Colorado Rivers, Northwestern Colorado.* U.S. Geological Survey, Hydrologic Atlas HA-370.

Colony Development Operation. 1974a. *An Environmental Impact Analysis for a Shale Oil Complex at Parachute Creek, Colorado.* Denver, Colorado.

————. 1974b. *Oil Shale. A Symposium for Environmental Leaders: The Colony Case Study.* Denver, Colorado.

Colorado West Area Council of Governments. 1974. *Oil Shale and the Future of a Region. A Summary Report.* Denver, Colorado: Council of Governments.

Cook, C. Wayne, ed. 1974. Surface Rehabilitation of Land Disturbances Resulting from Oil Shale Development. Phases I and II. Fort Collins, Colorado: Colorado State University.

Culbertson, W. J., Jr. and T. D. Nevens. 1972. *Uses of Spent Oil Shale Ash.* Denver, Colorado: Colony Development Operation.

Dames and Moore. 1971a. *Liquefaction Studies, Proposed Processed Shale Disposal Pile, Parachute Creek, Colorado.* Denver, Colorado: Colony Development Operation.

————. 1971b. *Slope Stability Studies, Proposed Processed Shale Embankment, Parachute Creek, Colorado.* Denver, Colorado: Colony Development Operation.

Duffield, W. J. 1974. Revegetation: a ten-year study. *The Green Thumb* (Journal of the Denver Botanic Gardens), 31(4): 114–117.

East, J. H., Jr. and E. D. Gardner. 1964. *Oil Shale Mining, Rifle, Colorado, 1944–56.* U.S. Department of the Interior, Bureau of Mines. Bulletin 611: 4–59.

Geoecology Associates. 1974. *Environmental Inventory and Impact Analysis of a Proposed Utilities Corridor in Parachute Creek Valley.* Denver, Colorado: Colony Development Operation.

Heley, W. 1974. Processed shale disposal for a commercial oil shale operation. *Mining Congress Journal* 60: 26–29.

Heley, W. and L. R. Terrell. 1971. *Processed Shale Embankment Study.* Denver, Colorado: Colony Development Operation.

Hutchins, J. W., W. W. Krech, and M. W. Legatski. 1971. *The Environmental Aspects of a Commercial Oil Shale Operation.* New York, N.Y.: American Institute of Mining, Metallurgical, and Petroleum Engineers, Inc.

Kilburn, P. D. 1973. *Environmental Analyses by a New Energy-producing Industry.* 74th National Meeting of AIChE, New Orleans, La. Denver, Colorado: Colony Development Operation.

Kilburn, P. D., and M. W. Legatski. 1974. *An Environmental Program for a Commercial Oil Shale Complex.* Casper, Wyoming: American Institute of Mining, Metallurgical and Petroleum Engineers.

Kilburn, P. D., M. T. Atwood, and W. M. Broman. 1974. Oil shale development in Colorado: processing technology and environmental impact, pp. 151–164 in *Guidebook to the Energy Resources of the Piceance Creek Basin, Colorado.* Denver, Colorado: Rocky Mountain Association of Geologists.

Knutson, C. F. and C. R. Boardman. 1973. Hydrology of the Piceance Basin and its Impact on Oil Shale Development. Paper no. SPE 4409. Dallas, Texas: Society of Petroleum Engineers of AIME.

Marr, J. W., D. Buckner, and C. Mutel. 1973. *Ecological Analyses of Potential Shale Oil Products Pipeline Corridors in Colorado and Utah.* Denver, Colorado: Colony Development Operation.

Murray, D. K. and J. D. Haun. 1974. Introduction to the geology of the Piceance Creek Basin and vicinity, northwestern Colorado, pp. 29–39 in *Guidebook to the Energy Resources of the Piceance Creek Basin, Colorado.* Denver, Colorado: Rocky Mountain Association of Geologists.

Rold, J. W. 1974. Research on environmental problems of oil shale development: an example of federal, state and industry cooperation, pp. 165–170 in *Guidebook to the Energy Resources of the Piceance Creek Basin, Colorado.* Denver, Colorado: Rocky Mountain Association of Geologists.

Savage, J. W. 1974. Oil shale and western Colorado, pp. 17–19 in *Guidebook to the Energy Resources of the Piceance Creek Basin, Colorado*. Denver, Colorado: Rocky Mountain Association of Geologists.

Schramm, L. W. 1970. Shale oil, pp. 183–202 in *Mineral Facts and Problems*. U.S. Department of the Interior, Bureau of Mines, Bulletin 650.

Sedgley, E. F. 1974. Is oil shale rehabilitation possible? *Green Thumb Magazine* (Journal of The Denver Botanic Gardens), 31(4): 110–113.

Skogerboe, G. V. 1974. *Colorado River Salinity: Impact of Parachute Creek Oil Shale Plant and Alternatives for Mitigation*. Denver, Colorado: Colony Development Operation.

Special Committee of the Governor's Oil Shale Advisory Committee. 1971. *Report on Economics of Environmental Protection for a Federal Oil Shale Leasing Program*. J. B. Tweedy, Chairman, Department of Natural Resources, State of Colorado.

Stoecker, R. E. 1974. Wildlife workshop, pp. 79–91 in *Oil Shale. A Symposium for Environmental Leaders. The Colony Case Study*. Denver, Colorado: Colony Development Operation.

Tait, D. B. 1972. *Geohydrology of the Piceance Creek Basin, Colorado*. Denver, Colorado: Atlantic Richfield Company.

Terwilliger, C. V. and P. G. Threlkeld. 1973. *Vegetation Map. Piceance Basin and Roan Plateau, Colorado*. Ft. Collins, Colorado: Colorado State University.

THK Associates, Inc.; Bickert, Browne, Coddington and Associates, Inc.; and the University of Denver Research Institute. 1974. *Impact Analysis and Development Patterns Related to an Oil Shale Industry: Regional Development and Land Use Study*. Colorado West Area Council of Governments and the Oil Shale Regional Planning Commission.

Thorne Ecological Institute. 1973. *Environmental Setting of the Parachute Creek Valley: An Ecological Inventory*. Denver, Colorado: Colony Development Operation.

————. 1974. *Regional Oil Shale Studies. Environmental Inventory, Analysis and Impact Study*. Final Draft. Colorado Department of Natural Resources.

U.S. Department of the Interior. 1973. *Final Environmental Statement for the Proposed Prototype Oil Shale Leasing Program*. Washington, D.C.

U.S. Geological Survey. 1974. *Hydrologic Data from the Piceance Creek Basin, Colorado*. Colorado Water Resources Basic-Data Release.

Utah Environmental and Agricultural Consultants. 1973. *Environmental Setting, Impact, Mitigation and Recommendations for a Proposed Oil Products Pipeline Between Lisbon Valley, Utah, and Parachute Creek, Colorado*. Denver, Colorado: Colony Development Operation.

Ward, R. T., W. Slauson and R. L. Dix. 1974. The natural vegetation in the landscape of the Colorado oil shale region, pp. 30–66 in *Surface Rehabilitation of Land Disturbances Resulting from Oil Shale Development*. Fort Collins, Colorado: Colorado State University.

Weaver, G. D. 1974. Possible impacts of oil shale development on land resources. *Journal of Soil and Water Conservation* 29(2): 73–76.

Wengert, N. 1973. *Oil Shale Country Fact Book: Garfield, Mesa, Moffat, and Rio Blanco Counties, Colorado*. Denver, Colorado: Colony Development Operation.

16. Environmental Aspects of Hydrology

Leonard C. Halpenny

WATER REQUIREMENTS

THE LARGEST MINING OPERATIONS IN THE SOUTHWEST are related to oxide and sulfide deposits of non-ferrous metals, mainly copper, frequently associated with molybdenum. These deposits are generally mined by open-pit methods but some are operated as underground mines. The primary water use in a sulfide operation is for grinding and concentrating the ore and for transport of tailings to the tailing ponds. Depending upon the amount of recirculation within the mill and recovery methods for reclaiming water from the tailing ponds, the amount of fresh makeup water required ranges from a low of 240 gals. per ton of ore milled to a high in the range of 1,000 gals. per ton and averages about 260 gals. per ton. The makeup water is essentially equivalent to the net amount of water used to transport tailings to the tailing ponds. By "net," I refer to total water sent to a pond less the amount of water reclaimed from the pond. The water required for operation of a lead-zinc-silver mining operation is about the same as for copper sulfides.

For vat-leach operations for recovery of copper from oxide ores, the amount of makeup water required is about 125 gals. per ton of ore. An equivalent amount is bled off from the circulating volume of leaching fluid. This bleed-off fluid is sent to evaporating ponds or, at some operations, is used for heap-leaching.

Heap-leaching of oxide ores is much more difficult to evaluate in terms of water use. Each heap-leach operation differs, in ratio of area to thickness of the leach pile and in method of application of fluid, all depending on specific character of the ore. Because economic viability of the process depends on maximum recovery of leaching fluids, special care is taken in preparing and locating the leaching pads. A lower limit for the amount of water required for heap-leaching would likely be about 125 gals. per ton.

The required amount of water for operation of uranium mining and milling has a wide range. For one underground mine, 380 gals. of makeup water are required per ton of ore. Most of this is used for transport of tailings and some is recovered from the tailing pond area. For a uranium mill only, the amount of makeup water required at one new mill is 300 gals. per ton of ore milled.

At one dry underground coal mine in the southwest, the water needed is only for dust control and for washing the coal. The amount required is in the range of 75 to 100 gals. per ton.

SOURCES OF WATER SUPPLY

Most underground mines produce some water, and many open-pit mines also produce some water. As long as the quality is reasonably acceptable, the water produced unavoidably as a part of the mining operation is generally utilized as a part of the required water supply for the entire operation. A few mines produce far more water than is needed and the excess mine drainage water must be disposed of in some manner.

In general, the amount of water required for the milling operation is more than is produced from the mine. Therefore, an outside source of water must be developed and brought to the mill by pipeline. In the southwest the water available for diversion from the surface streams generally was long ago appropriated by others and is not available for mining and milling. An exception to this is an operation in east-central Arizona which has rights to divert water from the Gila River. In most cases, however, the water supply must be obtained from wells. The hardrock, or pediment, areas in which mines are developed generally are not sufficiently prolific in yielding water to support a well field, and the water supply usually is developed in alluvial valleys several miles from the point of use.

CONSUMPTIVE USE OF WATER

For most open-pit or underground sulfide copper mining and milling operations, the consumptive use of water at the mine and mill amounts to less than 10 percent of makeup water from the fresh water source (well field and/or mine drainage water). At some operations water is recovered from the tailing pond area and recirculated to the mill, thereby providing part of the required makeup water. At others, this is not feasible. Consumptive use of water from the tailing pond area consists of evaporation and of specific retention of moisture within the pore spaces of the tailing material. At some tailing ponds, water sinks downward into the underlying material. If the pond area lies on unfractured rock, water may appear as seeps at various points around the perimeter of the pond area. If on fractured rock or on alluvial material, seepage from the tailing pond area may percolate downward to the water table and become a minor component of groundwater recharge. At one mine in South America and another in British Columbia the tailings are sluiced down a wash or pipeline and deposited in the ocean. For this operation consumptive use of makeup water is effectively 100 percent. From the conventional tailing pond areas, consumptive use resulting from evaporation and specific retention may range from about 25 to about 75 percent of makeup water from the fresh water source. Thus, for this type of operation and considering an average amount of makeup water of the order of 260 gals. per ton of ore, consumptive use could range between about 35 and 100 percent, or 90 to 260 gals. per ton.

For vat-leach operations, consumptive use of water amounts to about 125 gals. per ton, which is the amount of bleed-off fluid sent to evaporation ponds. Consumptive use by evaporation from the leach tanks is effectively negligible.

For heap-leach operations it has been stated that the amount of makeup water required is estimated at a minimum of 125 gals. per ton. Evaporation from a heap-leach operation could be a factor in the consumptive use determination, but it would be offset in that there would be less bleed-off fluid to be sent to evaporation ponds. Effectively, consumptive use is the amount of makeup water required, 125 gals. or more per ton.

Consumptive use of water in a uranium mine and mill operation is considered to be low, for only small areas are subjected to evaporation. As in sulfide copper operations, the biggest factor in consumptive use is the tailing pond area. Because of the possibility of contamination from radioactivity in uranium mining and milling operations, the Atomic Energy Commission requires that tailing pond areas be sealed to prevent leakage. All of the pond water not recovered by recirculation or held by specific retention must be evaporated. Therefore, consumptive use of water is 100 percent of new makeup water, and most volumes of plant water are recycled.

For the one example cited of water use at an underground coal mine in the southwest, consumptive use of water was 100 percent. Water not evaporated as a result of dust control and washing of coal would be transported from the area as retained moisture in the coal exported from the mine.

DISPOSAL OF WASTE WATER

Disposal of waste water for sulfide copper operations, lead-zinc-silver operations, uranium operations, and the one example of a coal-mining operation have been discussed. This leaves us with consideration of disposal of fluid from leaching operations and disposal of excess water produced from mine drainage.

The bleed-off fluid from leaching operations generally has a low pH and a high content of iron in the ferrous form. When discharged to an evaporation pond area this fluid initially tends to percolate downward into whatever materials may underlie the pond, at a rate depending upon the extent to which the pond may have been sealed. The slightly acidic water acts with the alkaline rocks in the underlying materials and the pH gradually rises to the range of neutrality. The ferrous iron becomes oxidized and precipitates as limonite ($Fe(OH)_2$), which clogs the pore spaces and reduces the permeability of the underlying material. Not too long after the evaporation pond begins operation, downward percolation is no longer possible owing to clogging of pore spaces with precipitated ferric hydroxide, and the pond thus seals itself against further percolation losses. Henceforth, discharge from the pond can occur only by evaporation.

For the mining operations which produce more water from mine drainage than is needed for benefication of the ore, the experience of the southwest is limited to two prospects which are not yet in production. Of these, one is a prospective open-pit operation and the other an underground operation. At both, the estimated rate of pumping required to permit successful mining is in excess of 25,000 gals. per minute. The quality of water that would be produced is suitable for most uses including irrigated agriculture. In order to prevent waste of water, negotiations are being conducted to deliver the un-needed water to nearby farms for irrigation of crops. Under this type of plan, the farmers can reduce their water demands from other sources (surface water and/or production from wells) and the mines can be operated without substantial net depletion of the water resources of the semiarid southwest.

ENVIRONMENTAL EFFECTS OF MINING
IN RELATION TO HYDROLOGY

It has been shown that mining and milling operations cannot be conducted in the southwest without some consumptive use of water. Not yet discussed is the relationship between consumptive use of water for production of urgently needed minerals in terms of national needs, economic returns to the economy, nor the number of jobs created in comparison with consumptive uses for other purposes such as agriculture. These topics are in the field of economists, not hydrologists. Nevertheless, needed minerals cannot always be found in water-rich areas, and the nation cannot afford to leave needed minable ore deposits in the ground in the semiarid southwest simply because of the limited water supply, or to give unlimited priority for alternative potential water uses.

It has also been shown that for most mining operations in the southwest the primary water supply is obtained from wells in alluvial materials several miles from the site of the mining and milling operation. In some cases the withdrawal of groundwater for mines is in competition with withdrawal for irrigated agriculture or perhaps for domestic consumption. One of the problems of the semiarid regions of the United States for the past hundred years or so has been to view competition for scarce water politically or emotionally rather than objectively, and to decry the other fellow who competes for water. When the millennium arrives, there will be equitable water management under carefully balanced programs for allocation of water in accordance with local, regional, and national needs.

Another environmental aspect of the hydrology of mining and milling operations in the southwest is the spectre of permitting no returns to the hydrologic cycle to avoid any taint from pollution. Nevertheless, research in the 1960s and '70s has indicated that the threat of pollution of groundwater from trace metals in most mine-mill wastes in the southwest is so low as to be effectively negligible. There are two reasons for making this statement. First, because of current trends toward monitoring quality of water from wastes of any type, operational procedures have been greatly improved to the extent that trace metals in industrial wastes are being reduced nearly to subliminal levels. Second, "sorption" of most trace metals by soil particles in the unsaturated zone above the water table further reduces their concentration before the percolating water reaches the water table. Firm knowledge as to the extent to which "sorption" of trace metals does occur in the unsaturated zone above the water table is of vital importance to the mining industry. Further research should be done to evaluate these effects.

During the early 1970s many well meaning people have stated that tailing pond areas must be sealed against downward percolation no matter what the quality of the pond water might be. Perhaps this is a wise precaution, but how can it be done within the limits of economic feasibility? Tailing ponds require some drainage in order to maintain stable conditions. There are alternatives, however. A plan has been developed by the U.S. Bureau of Mines under which, in a new operation, the finest slimes would be deposited in a layer several feet thick at or near the base of the tailing pond area. This would substantially reduce seepage but could not eliminate it entirely. The second possibility is to construct evaporation ponds conjunctively with tailing ponds, to line them with slimes from the ponds, and to decant into them for evaporation all of the water from the ponds that can be removed and that cannot feasibly be returned to the mill for makeup water. Of course, measures such as these would increase consumptive use of water in our semiarid southwest, but they may become required at some time in the future.

17. Strip Mining and Hydrologic Environment on Black Mesa

Tika R. Verma

ARIZONA HAS ABOUT 3,000 SQ. MILES OF COAL DEPOSITS, mostly in the Black Mesa area. Large-scale strip mining of coal in the state has been going on since 1971. The largest operation is by Peabody Coal Company with a projected life of 35 years. These mines are located on the Navajo and Hopi lands of the Black Mesa in the northeast corner of Arizona. Production goals of 400 acres to be mined annually were not expected in 1976 to be reached for at least another year.

The semiarid region of the coal mine does not provide enough surface water to meet the requirements for strip mining and transportation of coal and subsequent revegetation of the disturbed land. Water is a key factor, however, in land use planning and successful reclamation of the disturbed land.

The objective of this paper is to evaluate the results of studies being carried out on the Black Mesa by the School of Renewable Natural Resources, University of Arizona, in cooperation with the Peabody Coal Company. The area has been inventoried for its biological, geological and hydrological characteristics and sparse historical climatic data have been augmented by more detailed meteorological measurements. A 5.5 acre watershed on the regraded mined land and a similar one on a nearby unmined site have been instrumented to study surface runoff, infiltration and water quality.

WATER RESOURCES

Surface and groundwater resources will be tapped to meet the additional water requirements for strip mining of coal on the mesa. Two interrelated aspects of the regional hydrology are of major concern: surface water, and groundwater.

Surface water resources are limited because the average annual precipitation is about 12 inches and the potential evapotranspiration rate of

Financial support and cooperation from the Peabody Coal Company and Cooperative State Research Service (U.S. Department of Agriculture) are gratefully acknowledged.

about 63 inches per year is high. Intense storms during the summer months produce high runoff and accelerate erosional and sedimentation processes. Winter precipitation is low and usually in the form of snow.

Groundwater resources provide the bulk of water supplies for strip mining and transportation of coal. These resources in the mesa hydrologic basin are contained in three separate multiple-aquifer systems. All three of the aquifer units — Navajo Sandstone, Coconino Sandstone, and de Chelly Sandstone — are characterized by low permeability and thick separations of sandy siltstone, siltstone and mudstone with no hydraulic interconnections. The Navajo Sandstone aquifer system is extensive and deep and is currently the source of water for strip mine operations and transportation of coal. The formation is recharged through an area to the north where it surfaces in the narrow Klathla Valley. Indian wells draw their supplies from the aquifers closer to the surface.

The quality of water resources in the Black Mesa area is of major concern since the potential uses of water are affected by it. All the major washes on the mesa carry an unusually high load of total solids. Water quality tests of the water from the Indian wells indicate high amounts of total soluble salts, sodium and bicarbonates.

WATER REQUIREMENTS

Water requirements for the strip mining operation of coal on Black Mesa consist of dust control during the operation, slurry formation to transport coal and revegetation of regraded disturbed land. The energy production processes also require substantial amounts of water. These plants are not located on Black Mesa, however, and therefore, water requirements for them are not discussed in this paper. Dust control is essential during the strip mining operation, especially during the summer months. Water requirements for dust control on the mesa range from 75 to 100 gals. per ton of coal.

Slurried coal is transported to the Mojave power plant through a pipeline. The water requirements for the operation can be up to 2,700 gals. per minute. It is estimated that the operation will utilize about 35 to 40 billion gals. of groundwater during a 35-year period.

Water requirements for revegetating the regraded disturbed land are difficult to estimate. The needs would depend upon the post reclamation use of the land. For example, if the reclaimed area is to be used for livestock grazing, the water requirements would be considerably lower compared to those if the area is used for horticultural crop production. Initial water application of 6 to 8 inches per acre per year would facilitate a good vegetation cover during the first and second year of post reclamation. Once an effective vegetation cover is established, control grazing should be practiced to ensure successful and long lasting rehabilitation of the disturbed lands.

HYDROLOGIC IMPACT OF STRIP MINING

The impact of strip mining on water resources of Black Mesa is evaluated on the basis of data collected by the School of Renewable Natural Resources at the University of Arizona. Disturbed land is regraded into an aesthetically pleasing landscape which blends in with the surrounding undisturbed area. The regarded area is practically devoid of vegetation while the undisturbed area consists of pinyon-juniper type cover intermixed with grasses and brush. Results from the two watersheds (5.5 acres each on regraded and undisturbed areas of similar slope and aspect) show that the response varies seasonally.

Snowpack that built up on the undisturbed watershed during the colder periods melted rapidly during the warm periods in February and March. There was no snow accumulation on the regraded watershed as it melted continuously during and after the storms. The continuous snowmelt is caused by warmer temperatures in the regraded spoils, probably due to subsurface oxidation of coal left in the regraded spoil material.

For the summer storms watershed response is reversed. Total volume and intensity of runoff are much greater (Table 17.1). The regraded water-

TABLE 17.1
Rainfall and Runoff on Black Mesa Sites (1974)

Date	Mined Watershed		Unmined Watershed	
	Precipitation (mm.)	Runoff (m³)	Precipitation (mm.)	Runoff (m³)
7/15	15.2	21.89	20.3	14.81
7/16	1.3	0	0	0
7/17	3.8	25.32	1.3	0
7/18	43.2	—	6.4	9.91
7/19	10.2	—	2.5	0
7/22	7.6	60.06	8.9	5.07
8/1	6.4	38.65	8.9	11.67
8/9	6.4	45.05	6.1	13.59

shed responds with higher and sharper peak flows and produces more runoff than the undisturbed watershed (Fig. 17.1). Runoff characteristics play an important role in erosion and sedimentation processes and therefore, intense runoff coupled with the lack of vegetation on the regraded watershed intensifies erosion and sediment rates. The Soil Conservation Service "A-hydrograph" (Fig. 17.2) indicates the effectiveness of vegetation cover in runoff and erosion control. Two sets of assumptions were made:

I. Present site conditions
    a. Poor or no vegetation cover
    b. Soil has a very high runoff potential
    c. Antecedent soil moisture was 1.3 cm (0.5 inch)

II. Site with a good vegetation cover
    a. Native grassland (good cover)
    b. Below average vegetation
    c. Antecedent soil moisture was 1.3 cm (0.5 inch)

Good vegetation cover is effective in reducing the runoff by 65 percent with a reduction of peakflow and increased duration. In order to rehabilitate the regraded land and to increase the on-site conservation of precipitation, a good vegetative cover is mandatory.

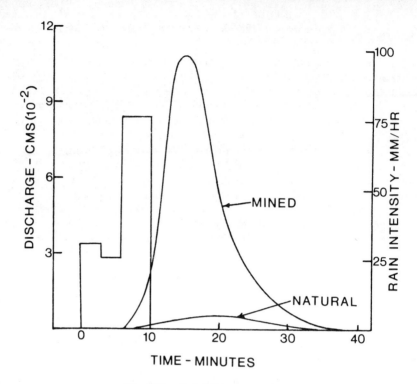

Fig. 17.1.   Rainfall hydrographs.

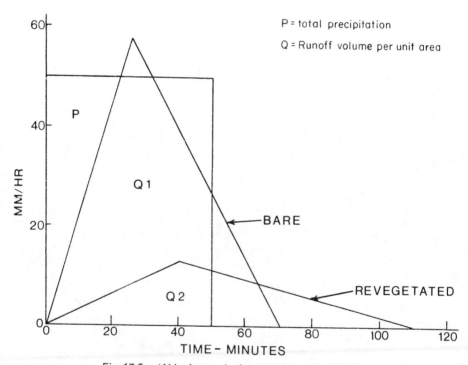

Fig. 17.2.   'A' hydrographs for recontoured watershed.

The impact of strip mining on water quality is important in the hydrologic environment analysis of any disturbed area. Reclamation practices should be aimed not only at the on-site conservation of precipitation but also at keeping the impact of strip mining on water quality to a minimum. Present and potential uses of water resources and the water quality requirements are being evaluated for use by livestock, irrigation or fishery habitat and for possible contamination of groundwater from seepage.

The quality of runoff water from the regraded and undisturbed watersheds is analyzed and compared with the water from other local sources as shown in Table 17.2. Total soluble salts are higher in the runoff water from the regraded disturbed land than in that from the undisturbed catchment. Water quality of the runoff from the regraded catchment compares well with the proposed environmental protection agency criteria.

TABLE 17.2
Water Quality Analyses of Black Mesa Samples in PPM

	Natural		Mined		Indian Well	E.P.A. Standards
	Snowmelt	Rainfall	Snowmelt	Rainfall		
TOT. SOL. SALT	177	306	310	335	782	500
Ca	24	61	41	57	79	—
Mg	3.6	7	13	18	35	—
Na	21	5	22	16	108	N.L.*
CHLORIDE	3.0	5	3	6	7	250
SO₄	65	19	100	110	318	250
CO₃	0	0	0	0	0	—
HCO₃	80	189	129	117	232	N.L.*
FLUORIDE	0.2	0.09	0.3	0.61	0.29	1.4
NO₃	2.4	20	1.6	12	0.9	45
pH	7.6	7.5	7.4	7.6	8.1	5–9

*N.L. = No Limits.

Lack of vegetation on the regraded strip mine spoils makes it vulnerable to wind and water erosion processes and consequently the suspended solids in the runoff from the strip mined catchments are 300 to 1,300 percent greater as compared with those in the runoff from the undisturbed areas (Table 17.3). It is estimated that a rainfall event such as that shown in Fig. 17.1 has the potential of producing 1.3×10^5 kilograms (60 tons) of sediment per hour.

Concerning the contamination of groundwater resources through the seepage or recharge from the impounded water, mathematical simulations based on the preliminary data indicate that moisture will move 1.8 meters

TABLE 17.3
Percent of Suspended Solids per Liquid Weight in Runoff From Black Mesa (1974)

Site	Snowmelt	Rainfall
Natural	0.1	1.4
Mined	1.4	5.6

based on the preliminary data indicate that moisture will move 1.8 meters (6 feet) into the spoil material in 17 days and the moisture content above the wetting front would be only 7 percent greater than the wilting point. The amount of water available to infiltrate into the regraded mine spoils is limited by high evaporation rates. It is estimated that about 610 mm. (2 feet) of impounded water would evaporate during summer (June through mid-September) before water could penetrate 21 meters (70 feet) of initially dry spoil material. It seems highly unlikely that surface water from the impoundments on the regraded spoil material could contribute to any significant extent to a free water buildup on the fire clay underlying the spoils, and consequently, contribute to the local groundwater resources. More detailed experimentation is underway to obtain additional data for simulating soil moisture profile under ponded conditions.

CONCLUSION

Impacts of strip mining of coal on the hydrologic environment of a semiarid region are different from those in the humid east. Reclamation practices should be aimed at erosion control and on-site conservation of precipitation. There seems to be no permanent impact of strip mining on water quality if the disturbed lands are rehabilitated and effective vegetation cover is established.

PART IV

MINING AND THE AMENITIES

AMENITY RESOURCES such as natural geological phenomena, prehistoric and historic cultural remains, and the vastness of a region are assuming important positions among our country's concerns with quality of the environment and social life. In the southwest, the greatest demand on the amenity resource is that for outdoor recreation. There is no common definition of outdoor recreation. The term is often extended to include human experience over a wide range of meanings, satisfaction and social settings. In fact, "amenity" and "recreation" often appear inseparable, at least to the extent that they both imply the aesthetics, the pleasure derived from, and the non-monetary value of real estate. The relationship of amenity resources to recreation resources and to landscape resources is discussed by Lawrence Royer, who also explores the inevitable conflicts that have and will continue to arise between the supporters of the amenity resources and those supporting the use of natural resources for the production of monetary wealth.

The traditional view that the pursuit of recreation or amenities is an indulgence having at best a marginal status among the concerns of society is now being supplanted by a philosophy that considers recreation a primary public need. Indeed, the institutions of recreation and the action programs that support them are considered by some equally as important as the institutions involved in production and consumption. Accordingly, the economics of recreation has been given a great deal of attention over the past decade, which, of course, is a result of increasing recreation demand and the concomitant increase in conflicts with other land uses such as mining.

The mining industry competes for resources that are suited for outdoor recreation and, along with others such as the transportation industry and the manufacturers of recreation equipment, increases the demand for recreation experiences associated with irreplaceable resources. It should also be pointed out that criteria for unique, rare or irreplaceable resources have not been developed. David A. King discusses opportunity costs of recreation as influenced by mining. He presents information on the outdoor recreation opportunities provided in the southwest, and estimates the economic value of recreational resources in Arizona.

An important example presented by R. Gwinn Vivian is that of a rare, unique, and irreplaceable natural resource, archaeological sites, and their relationship to mining. He presents the premise that efforts to preserve a portion of archaeological data, like measures taken to reclaim disturbed lands, may be viewed as a practice that can provide benefits to the public and enhance its attitude toward the mining industry.

Conflicts between miners and preservationists will undoubtedly continue. However, it is unlikely that the public will permit the miners to stop mining the raw materials of our technological society, but rather, demand that it be done under conditions that lower its destructive effects to levels that minimize the loss of the amenities. Aesthetics is perhaps one of the most important, and the least understood, of the amenities. Because it is also highly subjective, the values and preferences of individuals, their goals, basic orientations, attitudes, and inclinations must be considered.

There is a need for objective evaluations of public perception regarding goals for future use of mined lands. What do people expect? What do they want? An important problem is to define the aesthetic dimension as a means of identifying visual standards for rehabilitation efforts. Terry C. Daniel presents one method for quantitative assessment and prediction of public perceptions of landscape disturbances.

Fred S. Matter suggests the interesting argument that few landscapes exist which have not been modified by man and that our image of naturalness is only a regression to some previous man-made state. Given this, and the necessity of mining, he presents an idealized plan that modifies an existing mining complex which embodies considerations of the amenities, economics, and aesthetics in the design of a major mining operation.

18. Recreational Opportunity Costs

David A. King

AMENITY RESOURCES FORM THE BASIS for a large portion of the demand for recreation in the southwest. Amenity resources are defined as the natural phenomena, prehistoric and historic cultures, and vastnesses of the region. The purpose of this chapter is to demonstrate the importance of recreational opportunity costs in making land disturbance decisions.

Recreational opportunity costs are the recreational benefits foregone as a result of land disturbance or development. These (as well as other) opportunity costs are important considerations if a socially optimal pattern of land use is to be obtained. Furthermore, in disturbed land reclamation research little, if any, attention has been given to the question of what kinds of opportunity costs can be recovered through reclamation efforts. In the main, reclamation efforts appear to be aimed at stopping soil erosion, halting degradation of water quality, and clothing the "disturbances" in greenery. Are these the most important opportunity costs or are they simply the ones most easily attacked? It must also be realized that certain opportunity costs cannot be recovered through physical and biological manipulation of the disturbed site and, therefore, should be considered in the decision analysis prior to initiation of land disturbance or development. While the recreation opportunity costs have not been completely quantified, it is possible to show that they are large enough to warrant consideration in making land use and management decisions, whether or not land disturbance is involved.

The topic here is outdoor recreation that takes place away from urban areas and for which natural environments are ingredients of primary importance. As is usual when working with secondary data from diverse sources, common geographical definition is a problem. In most instances, the Four Corners Development Region will be the geographical area discussed. In a few instances, the data apply to the "Four Corner" states of Arizona, Colorado, New Mexico, and Utah.

IMPORTANCE OF OPPORTUNITY COSTS

Amenity resources, the basic recreational resources, are non-renewable in that they are gifts of God and, as Fisher, Krutilla, and Cicchetti (1972) have pointed out, "it is precisely because the losses of certain natural environments would be losses virtually in perpetuity that they are significant." They go on to point out that the non-renewable nature of amenity resources means that there is an upper bound on their absolute supply (supply cannot be expanded) and that any land development that destroys them is an irreversible investment, i.e., once an amenity resource is lost it cannot be recovered. On the other hand, a decision to preserve land is reversible. Fisher and his colleagues show how these two characteristics of amenity resources — we can never have more than we now have and those we lose cannot be recovered — lead to certain conclusions about development and preservation patterns. A summary of the results of their analysis is useful to demonstrate the importance of opportunity costs.

The first conclusion they come to is, "It will in general be optimal to refrain from development even when indicated by a comparison of current benefits and costs if, in the relatively near future, 'undevelopment' or disinvestment, which are impossible, would be indicated" (Fisher, et al. 1972). What does this mean? It means that if the benefit/cost picture over the first five years favors development and after that time it favors preservation, we should choose preservation now since development is irreversible. Such situations are not unlikely. For example, it has been demonstrated that demand for recreational uses of amenity resources is growing as a result of increasing incomes, population growth, and changing preferences. The benefits of amenity resources, therefore, will increase over time even if there is no decline in their absolute supply. Since the benefits of amenity resource preservation are the opportunity costs of development, the costs of developing natural environments will increase over time, changing the benefit/cost picture so as to favor preservation.

Extending the analysis further, as the amount of developed land increases and the supply of amenity resources decreases, the values of the remaining amenity resources will increase due to increasing scarcity. Therefore, the marginal opportunity costs of development increase with increased development. This means that the net benefits of development, taking into account the preservation opportunity costs, decrease with increased development. The conclusion is that if these opportunity costs are taken into account, as they should be if a socially optimal pattern of land development and preservation is to be achieved, less land will be disturbed than if the opportunity costs are ignored (Fisher, et al. 1972). And, since these opportunity costs are mainly of extra-market nature, the socially optimal pattern will not be achieved through the private market.

The opportunity costs under discussion up to this point arise from the actual physical obliteration of the amenity resources. These resources are non-renewable. Hence, reclamation efforts cannot recover opportunity costs of this type and we shouldn't fool ourselves, or the public, into thinking they can.

Land development or disturbance can have impacts on amenity resources short of physical obliteration. These impacts are too often ignored, but they may be of great importance and more amenable to reclamation than the direct impacts. Development can change the environmental and social context within which amenity resources exist in such a way as to decrease or cut off their flow of services, even though specific resources are not physically altered. These are the external diseconomies or negative spillover effects of land development or disturbance. For example, it is now impossible to climb any mountain peak on the Coronado National Forest without seeing disturbed land; Shiprock on the Navajo Reservation is now seen through a maze of utility lines; Grand Canyon is viewed through a haze of pollution; and the view of the Sierrita Mountains from Green Valley, south of Tucson, is blocked by mine dumps. Benefits have been lost and these, too, are opportunity costs of development or disturbance.

Reclamation may have a slightly better chance to recover the opportunity costs resulting from the indirect impacts of disturbance. The sort of reclamation efforts that will most completely do the job are those that will produce a result most closely resembling the original, undisturbed state of the land. But such efforts are also most likely to have the highest direct costs and we must consider the benefits to be gained through reclamation. The recovery of recreational opportunity costs resulting from the indirect impacts of development are one kind of benefit and we need to know what those opportunity costs are (Dials and Moore 1974).

RECREATIONAL RESOURCES AND THEIR UTILIZATION

The natural environments of the arid and semiarid southwest are powerful recreational attractants. The abundance, regionally, and the uniqueness, nationally and internationally, of geological phenomena of great scenic beauty are obvious. The visual, aesthetic appreciation of these phenomena is dependent, in turn, on the visual clarity of the air provided by the arid climate. The juxtaposition of natural features and manmade features strongly affects the aesthetic qualities of the natural features. And, until the recent past, manmade features have not existed to any great extent in the region's areas of greatest scenic beauty. Recent and proposed developments, however, indicate an increasing rate of growth in conflict situations.

That the value of these aesthetic resources has been recognized by our society is demonstrated by the many national parks and monuments that have been established in the region. An incomplete listing of these includes such national treasures as Grand Canyon, Bryce, Zion, Sunset Crater, Canyon de Chelly, Petrified Forest, Canyonlands, Dinosaur, and Arches.

Other recreational opportunities unique to the region are based on its historic and prehistoric cultures and their interpretation to the public. The value to society of areas from which we may gain knowledge about our predecessors has also been recognized through the establishment of national parks and monuments. Again, a partial listing serves to remind us of the national significance of these resources: Mesa Verde, Bandelier, Chaco Canyon, Aztec, Walnut Canyon, Montezuma's Castle, Navajo, and Casa Grande.

Of more recent occurrence, but no less important, are the resources that provide an opportunity to understand our cultural heritage and to develop an identity as a regional society. Resources to supply these opportunities have also been set aside: Tumacacori, Lee's Ferry, Hubbell Trading Post, and Coronado.

All of the areas mentioned are under the jurisdiction of the National Park Service. In the Four Corners region, the acreage under the administration of the Park Service is slightly over 5.9 million acres, 20 percent of the total area of land administered by the National Park Service (Everhart 1972). In 1973, these areas received slightly over 17.6 million visits, 8 percent of the total visitation to National Park Service units nationwide (National Park Service 1973).

In addition to these aesthetic, educational, and cultural opportunities, there are abundant, but perhaps less unique, opportunities for more hedonistic or physical recreational activities. In this region the majority of these activity opportunities are provided in national recreation areas, national forests, and on lands administered by the Bureau of Land Management (BLM).

Visits to the two major water oriented recreation areas, Glen Canyon and Lake Mead national recreation areas, amounted to more than 6.7 million in 1973 (National Park Service 1973).

Developed recreation areas on BLM administered lands are few in number. These lands, however, provide scenic backdrops for the traveler and dispersed recreational activities such as hunting, off-road vehicle travel, camping, picnicking, and fishing. In the Four Corner states the Bureau of Land Management is responsible for the management of approximately 5.7 million acres of land, 12.5 percent of the total area under their jurisdiction (Bureau of Land Management 1973). These lands received about 23 million visits in 1973, approximately 45 percent of the total visitation to BLM lands nationally (Bureau of Land Management 1973).

National forests in the region play a very important role in providing opportunities for outdoor recreation. Sixteen percent of the total area of the national forests in the U.S., or 29.4 million acres, is located in the Four Corners region (The Encyclopedia Americana 1974). These National Forest lands attracted approximately 25 million visitor days* of use in 1972, approximately 14 percent of the national total for the national forests (U.S.D.A. 1973).

I could find no data regarding the specific origins of recreational visitors to the Four Corners region. A study of camping parties at the six developed campgrounds in the Arizona Strip, however, provides origin information that is suggestive, if not completely representative, of the regional situation (Gibson and Reeves 1972). They found that 81 percent of the camping parties came from states other than those in the Four Corners region. Except for those in the two campgrounds on Lake Powell, the campers were generally engaging in sightseeing and exploring. These two findings support the proposition that amenity resources are the primary basis for recreational demand in the southwest.

---

*A visitor day is defined as a combination of visitors and time aggregating to 12 visitor hours. A visit is the entry of an individual into an area for recreational purposes.

VALUE OF RECREATIONAL RESOURCES

We are also faced with the fact that no regional studies of the economic value of recreational resources have been completed. A study of the value of recreational resources in Arizona, however, is applicable to our purposes (Martin, Gum, and Smith 1974). Using a modification of the Clawson-Hotelling Model for recreation resource valuation, they estimated consumers' surplus values and non-discriminating monopolist values* of the state's recreational resources to Arizona residents.

To provide for crude comparisons with other land uses, the non-discriminating monopolist values were also expressed as dollars per square mile (section) of land. Caution must be exercised in using these values because they are averages; any specific area or resource will have greater or lesser value depending on its attractiveness to potential recreational visitors. The values per section were estimated for each of the seven Wildlife Management regions in Arizona and for the state as a whole. For general outdoor recreation and hunting, the values ranged from $132 to $1,968 per section per year. The weighted average for the state was $417 per section per year. The region including Phoenix and Tucson has the highest value per section followed by the two regions including the Navajo Reservation and the Grand Canyon, Regions 1 and 2 (Martin, et al. 1974). Because of the distance and population factors, it would be expected that the recreational resources in the most densely populated region of the state would have the highest value. That the recreational resources of the two regions at the greatest distance from the metropolitan areas of the state, Regions 1 and 2, would have the second and third highest values, $1,385 and $1,204 per section per year, demonstrates the strength of demand for them.

The average consumers' surplus value for recreational resources used for general outdoor recreation (excluding hunting and fishing) for the entire state was $66 per trip. The total consumers' surplus values for general outdoor recreation resources were largest in Regions 1 and 2, over $55 million and $30 million, respectively. Again, these are annual values.

As noted above, these values are for the use of the resources by Arizona residents; nonresidents were not sampled. Therefore, the values are underestimates and the degree of underestimation could be very great. In 1970 it was estimated that about 70 percent of the recreational activity taking place in the region could be ascribed to out-of-state residents. For Arizona, the proportion was estimated as 75 percent (Four Corners Commission 1972).

The result of ignoring the recreational opportunity costs of disturbed land is likely to be the disturbance of more land than is socially optimal. These opportunity costs will be ignored by the private market since they are not costs to the private firms making the development decisions. Therefore, public intervention, in the form of land use regulation and public land use

---

*Consumers' surplus value is a measure of the surplus satisfaction consumers receive over and above the price they pay for a commodity. Non-discriminating monopolist value is the largest total revenue a single owner of the resources could collect by charging all visitors the same price.

planning, is necessary if a socially optimal pattern of land development and preservation is to be achieved.

Two kinds of opportunity costs arising from two kinds of impacts, direct and indirect, were identified. The direct impacts are often described simply as acres of land or proportion of land occupied or disturbed. Such statements imply that all land is of homogeneous value except for the resource being exploited and, therefore, they have no place in serious discussions of these impacts. Reclamation cannot alleviate the direct impacts of land disturbance on amenity resources and we should not pretend it can. Reclamation may reduce the indirect impacts of an aesthetic nature, but we need to know more about the effectiveness of such efforts.

The information reviewed here indicates that the potential opportunity costs of development, in terms of recreational benefits foregone, are significant. The analytical framework for taking these opportunity costs into account has been developed (Fisher, et al. 1972) and now we must start using it to guide the public intervention that is coming and is needed.

REFERENCES

Bureau of Land Management. 1972. *Public Land Statistics*. Washington, D.C.: U.S. Dept. of the Interior.

Dials, George E. and Elizabeth C. Moore. 1974. The cost of coal. *Environment* 16(7):18–37.

Encyclopedia Americana. 1974. National Forests. International ed. 19:731–734. New York: Americana Corp.

Everhart, W. C. 1972. *The National Park Service*. New York: Praeger.

Fisher, A. C., John V. Krutilla, and Charles J. Cicchetti. 1972. The economics of environmental preservation: a theoretical and empirical analysis. *The American Economic Review* 4:605–619.

Four Corners Commission. 1972. *Development Plan*. Washington: U.S. Dept. of Commerce.

Gibson, L. J. and Richard W. Reeves. 1972. The spatial behavior of camping America: observations from the Arizona Strip. *The Rocky Mountain Society Science Journal* 9(2):19–30.

Martin, W. E., Russell L. Gum, and Arthur H. Smith. 1974. *The demand and value of hunting, fishing, and general rural outdoor recreation*. Technical Bulletin 211, Agricultural Experiment Station, Tucson: The University of Arizona.

National Park Service. 1973. *Public Use of the National Parks*. U.S. Travel Data Center. Washington: U.S. Dept. of the Interior.

U.S. Dept. of Agriculture. 1973. *Agricultural Statistics*. Washington.

19.   Anticipating Amenity Conflicts

Lawrence Royer

THE ARID LANDS of the southwestern United States provide the location for some of the classic confrontations over amenities and recreational use. Historically, these conflicts focused upon the surface disturbance of the land and the consequent loss of natural landscapes, recreational, and scenic opportunities. This pattern of confrontation will continue. Surface disturbance will elicit a response from groups and individuals who perceive that their utility in what is being disturbed is threatened.

One of the problems for the mining professional and other individuals involved in the planning, operation, use and management of reclamation efforts on disturbed land is the problem of anticipation. What, indeed, is the amenity that is being disturbed? How *intense* will be the reaction to the proposed disturbance? In other words, what is the *impact* of the disturbance? From whom will the reaction originate? Are methods available that will monitor or assess the nature of the anticipated conflict?

In these situations, "reclamation," "rehabilitation," and "mitigation" are not synonymous with "restoration." Proposals to reclaim lands to be disturbed by mining are analogous to proposals to "add-on" wildlife and recreational benefits to water projects. Ingram (1971) has shown that in water policy, mutual accommodations of this nature do not always overcome substantial opposition and the conflict remains. In no manner should this fact detract from the importance of examining the "reclamation" of disturbed lands in the southwest. Instead, it posits an additional perspective to the analysis.

To bluntly state the problem, there is no reason to obscure the fact that opposition to surface disturbance by mining will arise. It will not disappear and any self-respecting mining company or federal agency should gird itself for this conflict. There are well-known strategies for engaging the confrontation. A good case study from a university's Department of Political Science can provide details for the uninitiated. But the source of the opposition should not be confused. Its genesis is in our inability to fully restore disturbed lands rather than in the nature of the reclamation or mitigation of the disturbances. The purpose here is not to explain how the objections can be neutralized, but to discuss the meaning and nature of these objections.

The framework for analysis is the concept of amenity resources. The concept embraces two words — amenity and resource. What is an amenity? Universities and researchers in the western United States that study outdoor recreation have ignored the term for many years, but are now beginning to pay the word more than a cursory lip service.

There is a gradual awakening to the fact that the phenomenon labeled as outdoor recreation is a difficult phenomenon to identify. It is even more difficult to describe. In 1969, the National Academy of Sciences in its state of the art study suggested that the researcher should address this enigma. The Academy's advice was generally ignored. But as recreation research efforts seem to become more and more devoted to those real situations where conflicts — conflicts between recreationists and others — are the norm, it is also becoming apparent that the concern is with multiple phenomena. Perhaps it is really "stretching it" if it is attempted to lump these concerns under an all-embracing category of outdoor recreation.

Those of us in recreation research have also been at fault by making recommendations the major objective of our recreation research. We have endeavored to identify the "shoulds." To understand and anticipate conflicts is not of the same order as positing remedies for conflicts. The researcher possesses no inherent superiority as a "recommender." The practitioner can do as well or better. The researcher can examine and describe, and he can help the practitioner anticipate his problems.

The word amenity is understood and appreciated in England. There the word has a sufficient meaning and connotation to stand on its own. In the United States the term is ambiguous and vague. Many subtle meanings attach to the word. The term is confusing enough to render it inoperable in many instances. The British would say that we have abused the word beyond meaning. However, if we attach the word to the idea of a natural resource, we have salvaged it and the concept becomes useful.

We may begin with the definition of amenity offered by a British planner.

> Amenity is not a single quality, it is a whole catalogue of values. It includes the beauty that an artist sees and an architect designs for; it is the pleasant and familiar scene that history has evolved; in certain circumstances it is even utility — the right thing in the right place. . . . (Atkisson and Robinson 1969)

Another definition of the British meaning states that:

> Few terms are so loosely employed in England, yet every usage carries some overtone of its Latin root, connoting pleasure. "Amenity" serves to denote almost any kind of interest in a place and any inherent value that transcends purely economic considerations. It is attached to whatever seems to need protection. In one town the "amenities" are historical buildings and the flavor of the past, in another open spaces, in a third views and vistas, elsewhere facilities for recreation or access to nearby points of scenic interest. The . . . quality of the term gives the public a sense of common cause against those to whom land is merely a marketable commodity. (Lowenthal and Prince 1964)

Webster included "the attractiveness and esthetic or non-monetary value of real estate" and "an area or location that provides comforts, conveniences, or attractive surroundings to residents or visitors" in a close parallel to the British meanings.

Note that the meanings continually refer to place or location upon the land, to values above and beyond economic considerations, and to the preservation or maintenance of these values. Note also that the value mixes vary with location and that they are not necessarily esthetic or landscape values nor are they outdoor recreation values. They may, however, include these kinds of values. The Grand Canyon is an amenity. But so also are the Kennecott open pit mine at Salt Lake City and the ranching life style in the Decker-Birney coal field in Montana. The image of the cowboy riding off into the sunset on a natural prairie in Montana is an amenity image. Eliminate either the prairie or the possibility of the cowboy and the amenity is destroyed. The Piceance Basin deer herd in Colorado is a big game resource. But it is more than that. The hunting of the deer herd can be eliminated and yet the amenity remains.

The term amenity assumes much greater interpretive power for our purposes if we attach a meaning of utility to the word. This is exactly what occurs when an amenity becomes a natural resource. Grand Canyon National Park is an amenity resource. The Decker-Birney ranch life remains just an amenity although it nearly became an amenity resource during the closing moments of the 93rd Congress.

Why? Some theories of natural resources propose that a natural resource is created by our culture (Spoehr 1956). It is a cultural artifact which is defined and given value or utility by our technological culture, economic culture, or political culture. Thus Grand Canyon National Park is defined as a resource because our political culture has defined it in statute. The act of reservation from the public domain was an exercise in the creation of an institutionally defined natural resource. The definition of the coal natural resource can change with new economic or technological situations. Strippable reserves change with the development of stripping technology and reserves are now economically defined in terms of sulfur-Btu content. Amenity resources can be envisioned in much the same way. The strip mining bill legally recognizes the ranching life style on the northern plains as an amenity resource through its owner consent provisions. A resource must possess utility to be a resource but technological or legislative assignment of utility is just as valid as an economic assignment of value.

In Montana, Forest Service recreation researchers have begun to apply this concept to their investigations of wilderness recreation. Using the concept of wilderness as an institutionally defined amenity resource, these researchers have derived the most operational definition of recreation carrying capacity to date (Frissell and Stankey 1972). Research results have been much less rewarding for so-called backcountry areas lacking the stature and definition of wilderness as a resource (Lucas and Stankey 1974). The concept may be applied to subjects and problems other than carrying capacity.

The significance of placing the somewhat esoteric idea of an amenity within the context of a natural resource is that the amenity can then be examined from some of the more traditional modes of analysis that surround natural resources. These modes include the idea of renewable versus non-renewable, the idea of utility, and the idea of fixed location. Whether the stature of a natural resource creates an "amenity right" is another question (Goldie 1971).

When resources are envisioned as amenities, it is also seen that as natural resources they are in the same league as the mineral natural resources. They do not fall in the category of biological natural resources such as forests or range. For example, minerals and amenities do not renew themselves. They cannot be grown or sustained as can forests, range, or wildlife. They are both non-renewable natural resources. Once gone, they cannot be replaced. Secondly, both minerals and amenities are locationally specific. They are indeed "where you find them." Thirdly, each "deposit" of minerals or amenities is unique. You cannot generalize an amenity resource any more than you can generalize a mineral lease or patent. They are both highly individualistic situations. It may also be ventured that mineral resource situations are so individual because of the process of discovery. Amenity resources are unique because each situation possesses a unique mix of amenity values. In either case, there are no standard "species" such as is the case for forest or wildlife resources.

It might be well to clarify the relationship of amenity resources to recreation resources and to landscape resources. Amenity resources may have recreational values, landscape values, or aspects of both. Research indicates that certain landscape types are esthetically superior and that a consensus can be obtained on the psychological preferences for these landscape types. In these instances, the landscape type is analogous to a species and we have already indicated that amenity resources do not possess this characteristic. Recreational resources are generally substitutable. We can create new facilities in new locations if we remove an existing facility. The economic utility of a recreational resource can be measured by the number of visits to a location or the amount spent to visit. However, Grand Canyon National Park remains an amenity resource regardless of whether a recreationist sets foot in it. And, it is impossible to build a new Grand Canyon as a recreation facility.

As indicated earlier, outdoor recreation is a difficult concept to define. Perhaps we have been lumping too many phenomena under the category. In the preceeding chapter, a picture is painted of a recreation situation that would not be duplicated elsewhere. Recreation in the southwest is described as an amenity resource condition. The condition described is more susceptible to analysis if it is envisioned as an amenity resource situation rather than an outdoor recreation situation. The Forest Service takes pains in its *Manual* to distinguish the wilderness resource as a non-recreational resource. The same should be done for the extensive national park resource in the southwest that was described. But until it is, the impacts upon amenities will continue to be described under esthetics or recreation categories in environmental impact statements. And speculation will continue on why all of the furor over a place occurs when it has a low recreation visitation figure or it is stated that the site can be revegetated and recreational facilities can be built upon it after surface mining.

We can anticipate the amenity conflicts on disturbed land if we consider amenity resources to be analogous to mineral resources. Mineral resources are non-renewable and have a fixed location. So also are amenity resources. It is impossible to lessen the conflict between amenity preservation and surface disturbance by suggesting that reclamation of the surface will mitigate the impact upon the amenity. The amenity is exhausted by the disturbance and it cannot be renewed or replaced. You can expect the conflict to continue unabated until the decision to surface mine or to not mine is made. The decision has a certain air of immutability about it. Given this perspective it can be seen that the expansion of an existing mine is of an entirely different order than the opening of a new mine. The conflict will be much greater in the latter case.

If either the mineral or amenity resource is considered as a "location" rather than an "activity," then other implications for anticipating conflict or assessing impact become clear. One of the objectives of the environmental impact assessment process is to generate information about alternatives to the proposed action. The implication is that a good close look at alternatives may show that an alternative will suffice while lessening the impact. A similar idea is applied in land use planning legislation when "areas of environmental or local concern" are to be identified. We then "plan around them" in a flexible manner. In the case of either minerals or amenities this concept does not work. It is not appropriate because of the geographic specificity of these two resources. It is not nit-picking about some vague theory of resources when we speak of fixed location. Donald Schwinn (1974) of Kennecott Copper Corporation has succinctly raised the point in discussing the federal land use bills.

> The . . . general concern of the mining industry with the proposed legislation is that it wholly fails to recognize the peculiar nature of mineral resources and the special considerations that are required as a result. The most important of the several peculiar characteristics of mineral resources in this context is that they must be recovered *where* they exist or they cannot be recovered at all. Land use planning, by its very nature, implies some measure of choice — of alternative. This element is lacking insofar as mineral development is concerned, except in the absolute sense of to mine or not to mine, which is, I submit, a choice of an entirely different quality. We do, in other words, have a considerable latitude about where we put urban development, transportation, airports, auto junkyards and the like. Society does not generally suffer greatly so long as we do permit necessary uses somewhere in the vicinity of the need. And if society is to benefit generally from mineral development at all, the maximum amount of land must be kept constantly available for mineral exploration and development. For if you can't explore, you can't find; and if you can't develop what you find, you won't explore. The bills under discussion do not even recognize these factors, much less take them into account.

This is a perceptive assessment of the peculiar nature of minerals as a natural resource. As Schwinn points out the location nature of minerals resources is often not considered.

Translate "mining industry" to "preservationist lobby" and "mineral resources" to "amenity resources." The Sierra Club once entitled a book about the wilderness amenity as the *Geography of Hope* (Krutch 1967). *A Sense of Place* (Gussow 1971) is the title of a book published by Friends of the Earth. The principle is the same for amenity resources and the defenders are just as intense. Minerals and amenities are "where you find them." When they are in the same place, anticipate conflicts that cannot be resolved, mitigated, or compromised. Geographic specificity limits our choices. It closes options. It makes for either-or decisions. Yet as Schwinn indicates, we fail to recognize this factor when we draft legislation or conduct an environmental assessment. Unless this point is recognized in environmental impact statements about surface mining, sooner or later it will become a point of litigation. Certainly it has inspired litigation that addresses other issues of law.

At the beginning of this chapter the question was asked, "What, indeed, is the amenity that is being disturbed?" It is a difficult question. It must be answered from the perspective of a region such as the southwest, Appalachia, or the northern plains. Some of the answers are provided in the previous chapter. Scenery of the type that yields excellent results for the amateur using color film in an Instamatic camera is part of the southwestern amenity resource mix. It is possible to use the National Park System as the type locality and as an indicator for other areas. Blue sky and red rock record well and surface mining accompanied by mine mouth generation absolutely destroys this amenity resource.

Vastness is another quality of many of the southwestern amenity resources. Vastness can be equated with the situation of large contiguous land areas possessing relatively few highly visible artifacts of Anglo culture. In the southwest, a major change in the appearance of a road map is indicative of the loss of an amenity resource. The location of new surface disturbances or of attendant transport networks such as transmission lines, railroad lines, or exploration roads violates this amenity.

Amenities also possess cultural and historical elements. Monument Valley is an amenity resource not because it is scenic but because it is both scenic and inhabited by the Navajo. Examine the works in an art gallery in Santa Fe or Tucson or the latest "cowboy art" book in a Phoenix bookstore and the significance of this human element in the southwest landscape can be appreciated. Or browse through a year's sample of *Arizona Highways*. In the northern plains, examine what is depicted in the best selling Charlie Russel prints. It is a grassland-rimrock landscape with a mounted man and either buffalo or cattle. Surface disturbance destroys these amenities. We apparently identify with the amenity in a national sense.

At the beginning of this chapter it was asked, "From whom will the reaction originate?" Amenity resources are created by political elites — groups that have access to the political system. Amenities policy is formed in the same manner as most other policies in this country. It has been said that the smaller the group, the better represented its values are in the democratic process. Surface mining impacts upon the value systems of certain organized interest groups. These groups are generally the traditional preservationist organizations such as the Sierra Club or Audubon Society. They now include regional organizations and legal defense groups. It is from these groups that conflict and confrontation emanate. They have the ability to react and take political action.

Few, if any, environmental impact statements assess or monitor the impact of a major significant action upon these organized groups. Impacts do occur upon other segments of our society. In an assessment study of a proposed shale oil pipeline, the travel routes of non-resident summer tourists to Utah's national parks were analyzed (Utah Environmental & Agricultural Consultants 1973). These visitors could be identified, and it was known that their objective in traveling was to consume amenities such as national parks. The pipeline could potentially affect these amenities and impact visitors who had utility in them. But these visitors are not politically active, and we would have missed the point if we had assumed that conflict would have originated from them. If amenity conflicts are to be anticipated for disturbed land, those groups must be identified who place value upon the amenity resource and who also possess the power to do something about it. It is suggested that the best way to understand these groups' definitions of amenity resources is to read their literature and publications and not to monitor visitation to national parks. A dialogue with them is a better way of anticipating the intensity of reaction to a specific surface mining proposal than is a public hearing or a public relations effort. A specific surface disturbance proposal may not violate an amenity resource. The resource may not exist at that location. Dialogue is the manner in which this information may be obtained.

If the two resources exist at the same location, then conflict will emerge. The exhaustive non-renewable nature of an amenity natural resource is similar to a mineral resource. In either case, impacts are difficult to mitigate and reclamation does not address the issue. These are truly either-or situations — they cannot be placated. They are the source of conflict. Yet we often fail to recognize this basic source of conflict and we seldom include this impact in our assessments. A failure to understand amenity resources can become the basis for much of the misunderstanding about the planning, operation, and management of reclamation efforts on disturbed land in the southwest.

REFERENCES

Atkisson, A. A. and R. M. Robinson. 1969. Amenity Resources for Urban Living, pp. 179–201 in *The Quality of the Urban Environment*. Harvey S. Perloff (ed.) Baltimore: The Johns Hopkins Press.

Frissell, S. S., Jr. and G. H. Stankey. 1972. Wilderness Environmental Quality: Search for Social and Ecological Harmony, pp. 170–183 in *Sound American Forestry*. Washington, D.C.: Society of American Forestry.

Goldie, L. F. E. 1971. Amenities rights — parallels to pollution taxes. *Natural Resources Journal* 11:274–280.

Gussow, A. 1971. *A Sense of Place: The Artist and the American Land*. San Francisco: Friends of the Earth/Saturday Review Press.

Ingram, H. 1971. Patterns of politics in water resources development. *Natural Resources Journal* 11:111–118.

Krutch, J. W. 1967. *Baja California and the Geography of Hope.* San Francisco: Sierra Club.

Lowenthal, D. and H. C. Prince. 1964. The English landscape. *Geographical Review* 54.

Lucas, R. C. and G. H. Stankey. 1974. Social carrying capacity for backcountry recreation, pp. 14–23 in *Outdoor Recreation Research: Applying the Results.* St. Paul: North Central Forest Experiment Station.

National Academy of Sciences. 1969. *A Program for Outdoor Recreation Research.* Publication 1727. Washington, D.C.

Schwinn, D. E. 1974. Remarks. *Natural Resources Lawyer.* 7:271–277.

Spoehr, A. 1956. Cultural Differences in the Interpretation of Natural Resources, pp. 93–102 in *Man's Role in Changing the Face of the Earth.* William L. Thomas, Jr. (ed.) Chicago: University of Chicago Press.

Utah Environmental and Agricultural Consultants. 1973. *Environmental Setting, Impact, Mitigation and Recommendations for a Proposed Oil Products Pipeline.* Denver: Colony Development Operation.

20. Archaeological Studies on Land Modification

R. Gwinn Vivian

CONCERN HAS BEEN EXPRESSED RECENTLY for re-establishing certain of our resources and the aesthetic values attached to some of them after land has been disturbed by the extraction of ores and other practices that alter and modify the land's surface. Funds expended for studies of vegetation, wildlife, rangeland, aesthetics and recreation reflect the extent to which industry in the United States is willing to consider re-establishing or creating favorable wildlife habitats, productive acreage, and interesting landscapes from disturbed land. Public support for reclamation of disturbed land is increasing and it is becoming evident that effective reclamation planning and management will become an even more important and necessary business in the future.

Reclamation efforts are necessarily directed toward the re-establishment of renewable resources. Given the proper conditions and enough time much disturbed land can be made suitable for vegetation, habitats can be created for wildlife, and recreational areas can be developed in reasonably aesthetic surroundings. Some resources, such as ore bodies, however, cannot be re-established. Once exploited, these non-renewable resources are lost for any future use, thereby depriving persons of their benefits. For example, recent increased consumption and consequent depletion of oil and natural gas reserves in the United States have stimulated much of the present "energy crisis."

There is another crisis in the United States today, and for that matter in much of the world, but it is a crisis that commands far less attention than that concerned with our dwindling energy reserves. This crisis has been termed the "archaeological crisis," and like the problems that face persons involved with energy development and use, the archaeological crisis has been created as a result of a depleted resource. Archaeological or cultural resources — the material objects and places that have been made or modified by man and the information associated with them — are like mineral resources for they are non-renewable. It is as impossible to gather scientific archaeological data from an area that has been strip mined as it would be

[183]

to recover the original minerals from that area after it was mined. Once removed, whether by natural causes or land modification, archaeological resources are lost forever. The public is confronted then with a situation in which the exploitation of some "critical" non-renewable resource leads to the loss of another non-renewable resource — archaeological remains. Unfortunately, this is a situation that cannot be remedied by reclamation efforts.

Two questions may be raised. First — is the problem of archaeological site loss worthy of consideration? Obviously, the economy and well-being of the citizens of the United States are not dependent on the management of a decreasing archaeological resource base. It would be fair to say that nine out of ten citizens are familiar with the depletion of our oil reserves, the exploration for coal deposits, and the increasing need for copper, iron and other metals. Yet scarcely one out of ten is aware of the fact that within the next twenty-five years a major portion of the prehistoric and historic remains in the United States will be lost as a result of both major and minor land modification projects. There is an impression that the average American has placed the maintenance of his automobile, the heating of his home, and the purchasing of more efficient goods above the protection and preservation of important aspects of his cultural heritage. And yet, there are indications that Americans, even in the face of an energy crisis, are re-evaluating the process of our technological society and the effects of technology on the natural and cultural environments. The environmental or conservation move- ment in the United States is not abating, and the predictions that environ- mental legislation could be drastically curtailed have not been substantiated through any major action on state or federal levels. Even with an increasing inflation rate people are concerned about the environment. A recent article in the *New York Times* reported that, "The consensus of recent public opinion surveys, according to the Council on Environmental Quality, is that citizens still place a high valuation on environmental improvement activi- ties, even at considerable cost. Accordingly, while the national economy may occasion some government belt-tightening in areas related to environ- ment, in the absence of a more explicit link between the two, any major slow- down in environmental programs as an inflation remedy, seems unlikely."

Federal concern for cultural aspects of the environment has also been notable within the past few years. The National Environmental Policy Act of 1969, for instance, specifically directs that the preservation of historic and cultural values should be a national goal. An Executive Order issued in 1971 requires all federal landholding agencies to inventory their properties and locate all cultural resources on lands administered by them. The Archae- ological and Historic Conservation Act of 1974, which passed virtually unanimously in the House and Senate, provides that up to one percent of construction monies on federal, federally assisted, or federally licensed projects can be used for the preservation of cultural resources disturbed by those projects. Within the past years, the number of states that have established antiquities legislation for the protection of cultural resources has doubled and there are currently only four states without some type of antiquities statute. Other examples of conservation and protection measures expressed for archaeological and other cultural remains could be cited, but the point should be clear that Americans do care about the history and pre-history of their country. Furthermore, the loss of the non-renewable resources that make up the greater share of our cultural heritage is a problem that most citizens believe must be dealt with in land modification planning.

The second question is: Should this problem be considered by persons involved with the reclamation of disturbed land if the solution does not lie in reclamation? If efforts to preserve a portion of archaeological data, like measures taken to reclaim disturbed land, are viewed as a practice that can provide benefits to the public even if the preservation of cultural resources precedes reclamation efforts, then this is a problem to be dealt with at this time. In addition, if this is an aspect of resource development that must concern industry as a result of protective legislation and the benefits of positive public relations, then it is a subject that should be reviewed and discussed by archaeologists and representatives of industry.

A substantial body of federal legislation governs the protection of cul-tural resources. The most important is reviewed below and is followed by a summary of its applicability to industry. The basis for virtually all cultural resources mandates in the United States is the Antiquities Act of 1906. This act established concern for archaeological and historic sites on a national level and directed that the exploration of sites on federal land would be through permit only from the secretaries of the Interior, Agriculture or War. The Antiquities Act generated two additional pieces of legislation that expanded the principles set forth in the 1906 mandate. The Historic Sites Act of 1935 declared a national preservation policy for important cultural resources for the inspiration and benefit of the people of the United States. The subsequent National Historic Preservation Act of 1966 furthered this concept and noted that "the historical and cultural foundations of the Nation should be preserved as a living part of our community life and development in order to give a sense of orientation to the American people." The National Environmental Policy Act of 1969 established concern for all resources and required the preparation of environmental impact statements on many projects where resources would be impacted. Cultural resources are defined in this act as one aspect of the environment that must be considered in the development of impact statements. The Archaeological and Historic Con-servation Act of 1974 directs federal agencies to notify the Secretary of the Interior whenever they find that a federal, federally assisted, or federally licensed project may cause irreparable loss or destruction of significant scientific, prehistoric, historical, or archaeological data. This act also allows funding for the recovery of threatened archaeological resources. Finally, Executive Order 11593, which states that "the Federal government shall provide leadership in preserving, restoring, and maintaining the historic and cultural environment of the Nation" directs federal agencies to exercise caution in transferring, selling, demolishing or substantially altering any federally owned property that might contain significant cultural features.

The importance of this legislation for industry centers primarily on its relation to the development of federal lands, and in some instances on fed-eral assistance or licensing for development. A significant portion of western lands is administered by the Bureau of Land Management, the Forest Service, the National Park Service and the Bureau of Indian Affairs. Under the NEPA, impacts on these lands must be mitigated to the fullest extent possible, including mitigation of threatened cultural resources. Executive Order 11593 prescribes an inventory of federal lands for cultural resources prior to transfer or sale, thus insuring that this procedure will occur before or in conjunction with environmental impact statement studies. The Executive Order further stipulates that all cultural properties must be evaluated in terms of significance and those properties that qualify for inclusion in the National Register of Historic Places must be nominated. Section 106 of the National Historic Preservation Act requires federal agencies with jurisdiction over cultural properties to determine the effect of any actions that might alter or otherwise change a cultural property that is on, has been nominated to, or qualifies for the National Register. The determination of effect can include a formal review process. Cultural resources on federal property are provided strong measures of protection, therefore, through the operation of complementary federal legislation and Executive Orders.

State level legislation for the protection of cultural resources is not as comprehensive as the federal mandates. Nonetheless, increasingly stringent requirements are being developed by western states for the preservation and maintenance of cultural heritage values. Enactment of the California Environmental Quality Act essentially carries the provisions of the NEPA to the state level, including the requirement to assess the effects of projects on archaeological and historic sites. Passage of similar acts in other western states is considered by most to be only a matter of time. All of the western states now have some form of antiquities legislation. Arizona pioneered in this field and the Arizona Antiquities Act of 1960 stipulates that no archaeological, paleontological, or historical feature situated on lands owned or controlled by the State of Arizona, or any agency thereof, should be defaced or otherwise altered without authorization of the Director of the Arizona State Museum. This authorization normally is provided through a system of archaeological clearance reports. The size of state owned or controlled acreage in the western states varies, but is large enough to require consideration in land modification projects.

If compliance with legislation is required for many projects, and compliance includes the mitigation of adverse impacts on cultural resources, can industry fulfill the requirements? It has been the experience of most archaeologists that the developers of land modification projects are concerned about complying with statutes prescribing the protection of cultural resources. Relationships with mining and utility companies, in particular, have been good, and a number of important archaeological projects have been carried out in the past few years in Arizona in cooperation with several firms. For example, an agreement between the Peabody Coal Company and Prescott College developed during the early phases of geologic exploration of Black Mesa on the Navajo Reservation led to the systematic survey of the region for archaeological remains and the excavation of those remains prior to land stripping. Since 1967, when work was started, more than 300 prehistoric and historic sites have been located and over 25 of them have been excavated. Analysis of materials recovered and publication of the results of both survey and excavation have been funded by Peabody. Similar agreements between the Arizona State Museum and CONOCO for work on state and federal lands in the Florence area, and with Cities Service Company for survey, excavation and site preservation in the Globe-Miami area have resulted in important cultural studies in these regions. In addition, several institutions in Arizona have cooperated with the Hecla Mining Company in producing archaeological studies on mining lands on the Papago Indian Reservation.

It has also been the archaeologists' experience, however, that developers often are not totally aware of federal legislation and guidelines governing cultural resources. Just as importantly, developers have not been advised of the archaeological requirements and procedures that are necessary for compliance with national, state and local cultural resource mandates.

To assist in solving these problems, the archaeological profession is presently taking steps to safeguard both the resource base of our cultural heritage and the interests of sponsors who contract for the professional services of archaeologists. At a recent series of seminars sponsored by the Society for American Archaeology and funded by the National Park Service, guidelines were developed that include the minimal qualifications for persons working as contract archaeologists. A more immediate solution to these problems has been taken in a number of states by institutions that have created special programs for preparing archaeological studies for environmental impact statements, developing mitigation proposals and other legislated requirements on land modification projects. Currently four institutional programs in Arizona are designed specifically for contract work.

These institutional programs, both in Arizona and elsewhere in the southwest, are designed to accomplish more than excavate archaeological sites and recover historic and prehistoric specimens. Staff members can provide summaries of pertinent legislation and legislative guidelines relating to cultural resources on projects, suggest procedures for complying with statutes and serve in a liaison role between industry and federal agencies, in addition to developing the necessary data and recommendations for environmental impact statements and other required studies. Furthermore, they can suggest management programs that will comply with legislation and yet meet the needs of business. For example, the Cultural Resource Management Section at the Arizona State Museum has developed a program for a mining company that preserved intact two archaeological sites of National Register quality without curtailing the company's plans and without increasing the costs of cultural resource mitigation. In fact, cultural mitigation costs were reduced in this case. It has been the experience of most companies involved in land modification projects that early investigation of impacts on cultural resources can result in savings in both time and money.

The archaeological profession, through the Society for American Archaeology, has been in the process of developing guidelines for the preparation and evaluation of archaeological reports developed for land modification needs. These guidelines attempt to equate the planning stages of projects with the various levels of archaeological research required by legislation and the profession. Archaeological work and the resultant reports are geared toward satisfying the sponsor's needs for archaeological clearance as well as providing the public with the benefits of archaeological research. Five general stages of project development have been defined and include: (1) Regional Plan, (2) Preliminary Project Planning, (3) Alternative Design, (4) Final Design, and (5) Project Execution. The archaeological activities necessary at these stages are: (1) *Overview* of known cultural resources and identification of inadequacies of knowledge in the region; (2) *Assessment* of cultural resources in a more defined project area, recommendations for further study, and possible archaeological field checks; (3) *Preliminary Field Study* of alternative project locations, including laboratory and limited field research; (4) *Intensive Field Study* of the selected project area including collection of reliable samples of data, description and evaluation of cultural resources, determination of the impacts, and development of mitigation proposals; and (5) *Project Specific Mitigation* including, but not limited to, excavation, avoidance, and preservation. It is recognized that no single project necessarily progresses through all five stages or in the order specified, but the stages do reflect an orderly development of study if cultural resources are to be affected. The development of standardized procedures for archaeological work and standardized requirements for archaeological reports should also result in sponsor savings in time and money.

The benefits that can be derived from conducting archaeological studies on land modification projects are not all related to savings in time and money. Like the rewards derived from the reclamation of impacted land, many of the benefits of cultural studies can be measured in terms of the public's attitude toward industry. If it can be demonstrated that industry is concerned about the preservation of segments of our cultural heritage, as they are concerned about reclaiming land, public acceptance of land changing projects can be made easier. It is a fallacy to believe that the results of archaeological work interest only the trained archaeologist. The support of legislation protecting antiquities, the increasing numbers of avocational archaeologists, the popularity of the past in the media, as has been demonstrated by the sales of such books as *Chariots of the Gods,* clearly demonstrates that the average citizen is interested in past and recent cultures. The lack of public concern over the loss of cultural resources through land modification projects is primarily the result of a lack of knowledge about this loss.

There is clearly an opportunity for industry to capitalize on its efforts to preserve important aspects of our cultural heritage through the support of archaeological studies while at the same time complying with protective legislation. Publication of the results of this type of work through the popular media, recovery of important collections for use in public exhibits, and development of technical analyses of data for the advancement of the field of anthropology, among other results, can provide examples of industry's efforts to work toward the public good. If developed effectively, these efforts can become as visible to the public as reclaimed land.

21. Measuring Public Preferences for Scenic Quality

Terry C. Daniel

MAN HAS BECOME increasingly aware of and concerned about the impact of his actions upon his environment. The mutual interdependence of man and his environment is now widely recognized, to the extent that "environmental quality" has become a dominant social theme. Public concern has been expressed in the form of national policies and legislative actions directed at conserving and more effectively managing natural resources and at insuring some degree of protection for the essential aspects of our physical-biological environment. The National Environmental Policy Act of 1969, for example, attempts to ". . . assure all Americans safe, healthful, productive, and aesthetically and culturally pleasing surroundings." Thus, it is recognized that man's well-being depends upon both tangible and intangible aspects of his environment and that responsible management of the environment requires a concern for all elements and their interrelationships.

Aesthetic experiences are among the most important, but perhaps least understood of the intangible benefits provided by our natural environment. These experiences tend to be highly subjective and may depend as much upon characteristics of the experiencing individual as upon features of the environment. All resources are defined and their values determined by human needs and desires; aesthetic resources represent a particularly intimate and direct interaction between man and his environment. While aesthetic resources are generally recognized as essential to a quality environment, the subjective and intangible character of these resources has posed difficult problems for environmental planning and management. "Aesthetics" remains poorly defined and, as a consequence, aesthetic resources have not been represented in any consistent fashion in efforts to manage our environment. If aesthetic resources are to be systematically included in environmental management and planning they must be more precisely defined and identified and their current status must be determined. Further, methods must be developed for predicting the impact of various management decisions and actions on aesthetic resources.

An important aesthetic resource that has been identified, both by public concern and by law, is the scenic beauty of our landscape. While there is as yet no generally agreed upon definition of scenic beauty and no consistently used method for evaluating it, several potentially useful methods have been developed for dealing with this aesthetic resource (Arthur, Daniel and Boster 1976; Krutilla 1972; Redding 1973). One approach, the Scenic Beauty Estimation (SBE) method (Daniel and Boster 1976) has been tested extensively in the context of forest landscape management problems. The method provides objective, quantitative indices of landscape scenic beauty based upon the perceptual preferences of public observers.

Although applications have for the most part been limited to forested (ponderosa pine) landscapes, the Scenic Beauty Estimation method may be extended to the evaluation of other natural and man-influenced landscapes. In particular, the SBE method offers a means for evaluating the scenic impacts of mining and other land-use activities on southwestern landscapes and for assessing the effectiveness of subsequent reclamation efforts.

A MODEL OF LANDSCAPE BEAUTY

Scenic beauty, like all aesthetic values, arises from an interaction between man and his physical environment. The scenic beauty of a landscape is neither entirely "in the eye of the beholder" nor is it entirely attributable to inherent characteristics of the landscape. Rather, the beauty of a landscape must be inferred from evaluative (aesthetic) judgments by human observers in response to their perceptions of the characteristics of the landscape.

The SBE model explicitly recognizes this dual-component nature of scenic beauty. As Fig. 21.1 indicates, an evaluative judgment (for example, "That is beautiful, I like it.") is assumed to reflect the relationship between the aesthetic criterion or standards being applied by the observer and the perceived beauty of the judged landscape. The *perceived beauty* of the landscape represents perceptual effects that are assumed to result primarily from physical properties of the landscape. The *aesthetic criterion* is established by the observer/judge and its value is assumed to be determined by factors such as the observer's previous experiences with similar stimuli (landscapes) and by the more immediate circumstances surrounding the making of the judgment.

An important implication of this model is that scenic beauty judgments are, by themselves, necessarily ambiguous. That is, a reaction such as "That landscape is ugly!" could be the result of unaesthetic properties of the landscape, or it could indicate that the observer/judge is applying particularly stringent criteria. The effects of such interactions between criteria and perceptions are very much apparent in every day judgmental situations. Consider for example, the factors that might affect your "aesthetic appraisal" of an exotic dish under two different judgment situations: (1) you are in an expensive restaurant, entertaining an out-of-town guest and he asks with apparent disgust "How is your————? Mine is rather poor," as opposed to: (2) being served the identical dish by your new mother-in-law (or new boss, or principal stockholder) who asks "How do you like the————? I prepared it especially for you." Presumably the actual taste of the dish has something to do with your reported judgment in either case, but. . . .

Responsible managers and planners whose actions have effects on the environment must be reactive to public judgments whether these judgments are communicated directly, in the market place or in the political arena. By explicitly including aesthetic concerns the National Environmental Policy Act legislatively mandates that (presumably) public aesthetic judgments will be an important consideration in any plan or action having significant environmental impacts. For the reasons described above, the separate roles of observers' criteria and their perceptions must be recognized if managers' reactions to expressed public judgments are to be maximally appropriate and effective. Attempts to alter the physical characteristics of the landscape (as by reclamation efforts, for example) may do little to alleviate negative public appraisals if those negative judgments are primarily the result of unrealistically high judgmental criteria; especially if those criteria have been established on the basis of factors entirely separate from the actual features of the landscape. On the other hand, management options must be selected and implemented with due regard for their effects on the perceived beauty of the landscape if even minimal aesthetic criteria are to be met.

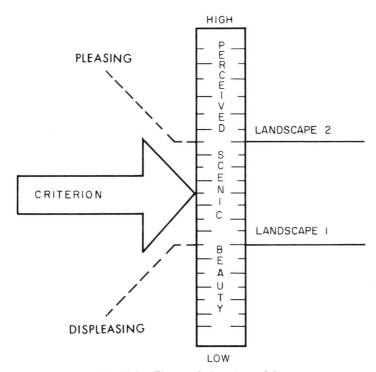

Fig. 21.1.   The scenic beauty model.

The importance of distinguishing between the criterion and the perceptual components of public aesthetic judgments is clear, but how are they to be distinguished? In most situations neither the perceived beauty of the landscape nor the aesthetic criteria being applied by the observer can be determined directly. An aesthetic judgment can, however, be interpreted indirectly by a systematic analysis of judgments made in response to several different landscapes. The SBE method, described in detail by Daniel and Boster (1976) provides a procedure for quantitatively estimating the perceived scenic beauty of different landscapes (or landscape treatments) independent of the particular aesthetic criteria being applied by the observer/ judge. The resulting "Scenic Beauty Estimate" is a standardized, relative index of the perceived beauty of landscapes. The SBE value for any landscape may be derived independently for single observers and/or combined to yield an overall index for any specified group of observers.

Extensive testing of the reliability, validity, and utility of the SBE index has been carried out in the context of scenic beauty evaluations of forested landscapes (Boster and Daniel 1972; Daniel et al. 1973). Several recent extensions and applications of the method suggest that it may be applied to a much wider range of problems (e.g., Angus and Daniel 1974, Stull et al. 1974). Quantitative assessment and prediction of public perceptions of landscape disturbances (e.g., from mining or other activity) and reclamation efforts should be possible in both forested and desert regions. Before considering these potential applications, it will be useful to review briefly the basic procedures that have been employed. Only a general outline of these methods will be presented as a complete description is available in Daniel and Boster (1976).

THE SBE METHOD

The major impetus for developing the SBE method was the desire on the part of forest managers to have some more effective means of integrating aesthetic resources with other, more tangible resources in comprehensive land use plans. Many of the resources that must be considered are (relatively) objectively measured and their values are expressed in quantitative terms (for example, board feet of lumber, acre-feet of water, pounds of forage). The relative costs and gains for these resources can be determined for alternative management plans and can be taken into account when deciding among alternative actions. Aesthetic resources, on the other hand, are notoriously difficult to define and have not been measured in any consistent fashion. Determining aesthetic costs and gains, then, has not generally been possible, at least not with degrees of precision comparable to the more tangible resources. As a result, aesthetic resources generally have not been systematically included in land use decisions. Quantitative indices of public perception of landscape scenic beauty can provide an important step toward resolving this problem.

Before public perception of the landscape can be determined, the public must be taken to the landscape, or the landscape must be taken to the public. For a variety of reasons, most of them economic, it is usually more feasible to pursue the latter course. In the majority of applications of the SBE method the landscape has been brought to observers in the form of color slides (photographs). This is an effective and efficient procedure.

Several important questions must be answered before appropriate photographic representation of a landscape can be accomplished. First, from what vantage point do observers usually view the landscape to be represented? Some landscapes may be viewed primarily from scenic vistas or viewpoints elevated above and at some distance from the landscape. Other landscapes will be experienced primarily from an "on-the-ground" level. Most, of course, can be viewed in several ways. Photo samples should be taken to represent appropriate views. For the majority of SBE applications, the on-the-ground vantage point has been chosen as the landscapes of interest are usually viewed in that fashion from foot trails, roads, and the like.

After the vantage point is determined, the next question is "Which of the virtually infinite number of possible pictures do you take?" An often chosen approach to this problem has been for the investigator to select "representative" scenes. The obvious danger in this procedure is that the investigator's choices may be influenced by his own biases and thus the validity of his representation is endangered. Some objective, unbiased procedure is certainly preferable. A technique that has proved useful in our applications is what we have dubbed the "random-walk" procedure. Briefly, each random walk begins at an arbitrarily selected point within the landscape that is to be sampled. The orientation of the first photograph is determined by referring to a random number table that ranges from 1 to 360 (degrees). A camera, held at eye level, is focused along the selected compass heading and the sample photograph is taken. Another random heading is then selected from the table and, by following this heading for a predetermined distance such as 100 paces, a second sample point is located. The second photograph is then taken along yet another random heading. This procedure is repeated until the desired number of sample photographs has been obtained. Other methods have also been used. The essential feature of all methods, however, is that some unbiased (random) process determines where and in what direction photo samples will be taken.

Slide representations of each landscape (usually ranging from 15 to 30 slides per treatment or area) are shown to groups of observers and their individual perceptual judgments are recorded for each slide. The slides are shown one at a time in a completely random order and every observer rates each slide on a 10-point scale which ranges from 1 ("extremely ugly") to 10 ("extremely beautiful").

Because scenic beauty ratings may depend to a large extent upon the aesthetic criteria adopted by the observer and on some rather arbitrary decisions he may make about using the 10-point scale, ratings that are obtained are ambiguous. As indicated earlier, the computation of SBEs involves procedures designed to separate these effects and yield a standardized, relative value indicative of the perceived scenic beauty of the represented landscape. These procedures have been described in more detail by

Daniel and Boster (1976). Variations are also discussed in Angus and Daniel (1974) and in Wheeler, et al. (1971). Separate SBE values may be computed for each observer for each landscape (based on the 15 to 30 slides rated for each area). If a single group value is desired, the individual values may be averaged for each landscape of concern.

RELIABILITY AND VALIDITY OF SBE

The results of eight separate applications of the SBE method to evaluate six different landscape types are presented in Fig. 21.2. The left panel presents obtained average SBEs based on random walk photo samples taken in the fall of 1971 and the right panel represents average SBEs for the same landscapes resampled on two separate occasions in the summer of 1972. All six landscape types (the same in all eight experiments) were ponderosa pine forests located near Flagstaff, Arizona. Each represents a different vegetative treatment (see Daniel et al. 1973 for a description of the areas).

While all the experiments reveal very consistent differences in perceived scenic beauty among the six landscapes, the primary concern here is with the consistency of the results from one experiment to the next. In spite of the fact that the four experiments in each set (left panel or right panel of the figure) differed in the number of slides sampled, the specific slides sampled, and in the observers making the judgments, the pattern of the results is remarkably uniform. Some apparent differences between the 1971 and 1972 samples probably reflect the fact that the 1971 sample was taken in October, after the summer rainy season, while the 1972 samples were both taken in June, a much drier period prior to the rains.

Generally, the results of these eight experiments do indicate that the SBE method provides stable, reliable measures of scenic beauty. Values for the SBE index stayed relatively constant in spite of substantial changes in the slide samples and observers. At the same time, the SBE index consistently revealed quantitative differences in the perceived scenic beauty of the sampled landscapes.

While the reliability of the scenic beauty index is an important concern, it is also very important to determine whether the index is a measure of what it purports to measure, i.e., is it a valid measure of landscape beauty? A major validity concern is whether the values obtained accurately indicate the scenic beauty of the landscape, or only the beauty of the *slides* sampled from the landscapes. To answer this concern, two separate groups of observers (a student group from Northern Arizona University and a more diverse sample of the public from Phoenix, Arizona) were bussed to the actual landscape sites and asked to rate them on the 10-point beauty scale. These observers walked around in and judged the same landscapes that had been represented by color slides in previous experiments.

Comparisons of SBE values based on color-slide samples with those based on in-the-field judgments revealed essentially a perfect correspondence. These data, graphed in Fig. 21.3, indicate that color-slide based values may serve as valid stand-ins for judgments that would be offered by observers who actually visit the sites.

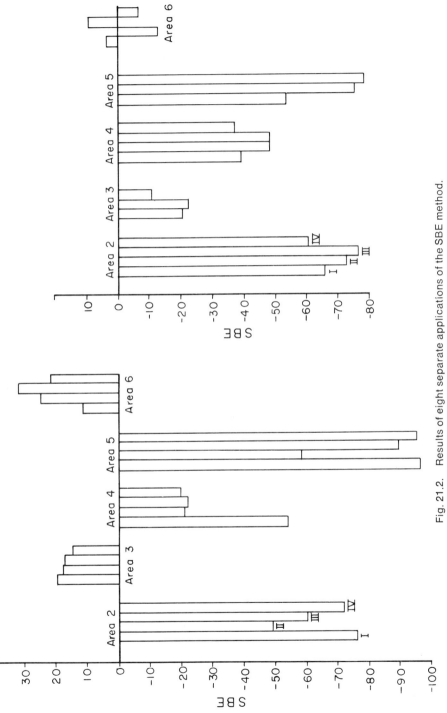

Fig. 21.2. Results of eight separate applications of the SBE method.

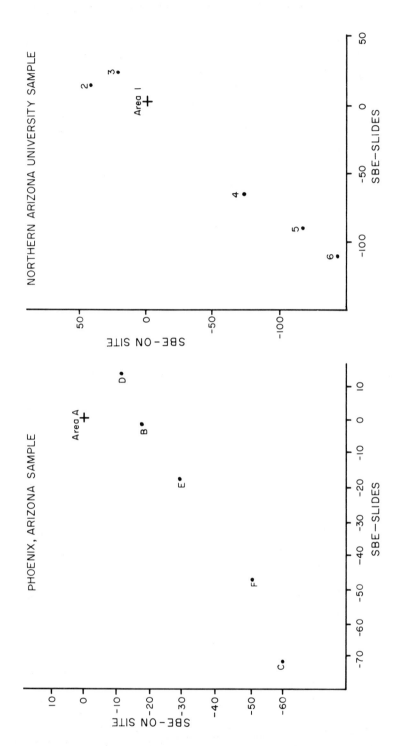

Fig. 21.3. Comparison of SBEs obtained for photo samples and actual visits to the same forest sites.

Other experiments have also been directed at establishing the validity of the SBE procedures as a scenic beauty measurement. Two studies compared SBE values against other methods purporting to measure scenic beauty and, in both instances, the SBE values compared quite favorably with the other measures. Further studies have shown the method to yield stable indices of scenic beauty in spite of changes in the context in which the measurements are taken. While the validity of any measurement system can never be completely assured, the studies indicated above, along with the wide range of other problems to which the SBE method has been successfully applied, do provide substantial support for the argument that the SBE is a valid and useful index of landscape beauty.

APPLICATIONS OF THE SBE METHOD

The basic procedures and analytic techniques described above have been applied to a number of forest-landscape management problems. In each application modifications and extensions of some aspects of the methodology were required, but the SBE index is a central component common to them all. A brief outline of the essential features of several of these applications will illustrate the potential of the SBE method as a landscape management tool.

Evaluating Forest Harvest Procedures. It is generally agreed that different harvest prescriptions have different scenic impacts on the forest landscape. Further, while scenic values are often very low immediately after a harvest, they may recover in time. The SBE method has been applied in this context to assess quantitatively and to compare several harvest prescriptions in terms of their immediate and long-term scenic impacts. Thus, the manager's ability to determine an optimum harvest solution, in regard to all affected forest resources, is greatly enhanced.

Comparing Public Groups. Throughout this chapter, the terms public perception and public preferences have been used somewhat loosely. It is important to realize that the "public" is composed of a number of different sub-groups, each having (potentially, at least) different perceptions and preferences about forest landscape aesthetics. Landscape management decisions that are applauded by some of these sub-groups may produce very negative reactions on the part of others. In recognition of this diversity in "the public," the SBE method was applied to determine differences and similarities in the scenic beauty perception of 26 selected public interest groups. Six management options for ponderosa pine forests were presented to each group. The results of this experiment (Daniel and Boster 1976) revealed more than expected agreement among the diverse groups assessed. Exceptions to this pattern of agreement were understandable in terms of each groups' special interests. For example, range interests (cattle growers and range management groups) generally preferred more open timber stands than did the other groups. Because the impacts of specific forest management decisions are often greater for some groups than others (because of differences in locations and patterns of use), it is important that differences in the perception of identifiable public groups be recognized and systematically assessed.

Scenic Beauty Mapping. Often landscape treatments (and/or their effects) will be different from one site to another within a management area. These differences (depending, for example, on soil conditions, vegetation types, slopes, etc.) may be represented by resource maps or response units that can be overlaid to identify areas where specific treatments are required. For example, steep slopes with unstable soils and low concentrations of harvestable trees may be excluded from a timber sale. Scenic values may also vary from site to site across the landscape, and these differences should be considered in the formulation of any management plan. Ideally scenic beauty differences should be represented in such a way that they could be "overlaid" with other resource/response indicators.

An application of the SBE method to this problem produced the "Scenic Beauty Map" pictured in Fig. 21.4. The procedure was to locate a large number of sample points (in a grid pattern) across the landscape. Randomly oriented photographs (slides) were taken from each sample point, the slides were judged by groups of observers, and SBE values were obtained for each sample point. The origin of the SBE values (the zero point) was arbitrarily established by the average perceived beauty for the entire landscape. Thus, positive SBE values indicate higher than average beauty, and negative values indicate below average beauty. Conventional contouring techniques were then used with SBE values serving as "elevations." The resulting Scenic Beauty Map reveals the topography of scenic values across the landscape and allows for the identification of specific high and low regions. When overlaid with other resource data, this map should provide valuable assistance in the specification of landscape treatments and uses and in the evaluation of the impacts of various management actions on the scenic resource.

Predicting Scenic Beauty. While the ability to measure the scenic beauty of existing landscapes is an important and essential development, there remains a great need for predicting the scenic consequences of management actions in advance. Because the SBE is a reliable quantitative measure of scenic beauty, it may serve as the basis for systematic prediction models. For example, we have attempted to predict scenic beauty (SBEs) in terms of measurements of manageable landscape features. While this work is not yet complete, indications are that public perception of scenic beauty may be predicted by linear combinations (regression equations) of physical landscape features. In one application, almost 90% of the variance of scenic beauty values (for ponderosa pine forests) was explained by a combination of four manageable features: amount of downed wood, density of trees, variation in diameter of trees, and condition of ground cover. Another study showed that approximately 60% of SBE variance could be explained by four timber inventory factors: amount of slash/acre, number of trees > 20″

SCALE

8 chains

N

Fig. 21.4.   Scenic beauty contour map.

DBH/acre, percent crown canopy, and cubic feet/acre of trees 6–10″ DBH. These statistical predictions need further refinement and must be empirically verified and validated, but the indication is clear; scenic beauty can be predicted for specified observer populations on the basis of measured characteristics of the physical landscape.

The results of extensive research devoted to developing, testing, and applying the SBE method are very encouraging. Reliable, valid, and useful indices of landscape beauty can be provided for general or specific observer populations. Costs of the application of the method are minimal, whether gauged in terms of time or dollars. Certainly, further refinements and extensions of the method are desirable and are being explored, but the present effectiveness of the method should not be overlooked. The usefulness of the SBE method has been demonstrated in a variety of practical and important landscape management problems.

IMPLICATIONS FOR MINING AND RECLAMATION

All mining operations have environmental impacts. Many of these impacts have serious consequences for the quality of water, air, and other essential elements of the environment. The mining industry has generally been sensitive to and concerned about these dangers and methods have been developed and implemented that may avoid or alleviate many of these adverse effects. Mining also may have significant impacts on aesthetic resources. Roads, construction sites, mining and milling plants, products, and wastes can all affect the scenic attractiveness of the landscape. As in other land use contexts, these aesthetic impacts have been very difficult to assess and attempts to avoid or reduce them have only been marginally successful. The SBE method provides a conceptual approach and a technology that can greatly enhance the effectiveness of efforts to deal with these aesthetic impacts.

An important implication of the SBE model of landscape scenic-beauty judgments is that aesthetic impacts can only be understood in terms of *both* their physical effects on the environment *and* the standards or criteria being applied by the observers of those effects. If public reactions are the product of criterial factors established by concerns that are unrelated to physical-perceptual impacts on the landscape, attempts to change those reactions by altering the physical features of the landscape (e.g., reclamation) may have unintended effects. At best under these circumstances, reclamation efforts may have minimal effects on public judgment. At worst, if viewed as an attempt to cosmetically disguise negative impacts, they may intensify negative reactions. Perceptual elements (related to physical landscape features) must be distinguished from judgmental criterion components (established by the observer) if aesthetic impacts are to be accurately assessed and appropriate reclamation efforts are to be designed and implemented.

The SBE method can provide separate estimates of perceived scenic beauty and observer judgment criteria. Thus, where aesthetic impacts can be attributed to the perceptual effects of specific landscape features, appropriate alterations in procedures or reclamation efforts could be implemented to alleviate these impacts. If, on the other hand, public reaction can be traced to criterion or judgmental standard components, some other course of action may be indicated. In some instances, efforts may best be directed at informing or educating the public (or other relevant observer-groups) about the factors that underlie or necessitate the apparent impacts on the environment. That is, some effort to justify the objectionable aspects of the mining operation may be the only appropriate (or possible) recourse for the mine operator.

Objective measures of mining impacts on aesthetic resources, especially measures that distinguish between perceptual and criterial components, should be of great value in planning operations and reclamation efforts. Precise quantitative evaluations of sites prior to the beginning of any mining activities could provide appropriate bases for subsequently determining the aesthetic impacts of a mine and its associated effects. The absence of such preliminary data would make it difficult or impossible to assess the extent or nature of changes (positive or negative) that should be attributed to the mining activity. The SBE type of approach would make it possible to determine the aesthetic effects of specific features of mining operations (roads, building, tailings, etc.) so that plans for new operations and/or reclamation efforts can be directed accordingly. Certainly, where several reclamation programs may exist as options (each with different costs) quantitative measures of their effectiveness, namely the scenic-aesthetic dimension, should be very helpful in choosing among the alternatives.

There are, no doubt, a number of other areas in which a quantitative assessment of public scenic (and perhaps other aesthetic) preferences would be very useful. All of these remain, for the most part, untried and untested in the context of mining and reclamation. However, the success of the SBE method in dealing with very similar land use problems in the forest management context provides substantial encouragement for attempting applications in the mining context. Aesthetic impacts, especially scenic impacts, of mining in the "wide open spaces" of the southwest are likely to be great (sometimes affecting hundreds of thousands of viewers for many, many years). If these aesthetic costs are to be properly considered in the context of other, more tangible costs and gains, some reliable and valid (quantitative) assessment of aesthetic impacts would seem to be essential.

REFERENCES

Angus, R. C. and T. C. Daniel. 1974. Applying theory of signal detection in marketing: Product development and evaluation. *Journal of the Agricultural Economics Association* 56(3):573–577.

Arthur, L. M., T. C. Daniel and R. S. Boster. 1976. Scenic Beauty Assessment: A Literature Review. USDA Forest Service Research Paper.

Boster, R. S. and T. C. Daniel. 1972. Measuring public response to vegetative management. Proceedings of 16th Annual Arizona Watershed Symposium.

Daniel, T. C. and R. S. Boster. In press. Measuring scenic beauty: The scenic beauty estimation method. USDA Forest Service Research Paper.

Daniel, T. C., L. Wheeler, R. S. Boster and P. R. Best, Jr. 1973. An application of signal detection analysis to forest management alternatives. *Man-Environment Systems* 3(5):330–344.

Krutilla, John V., ed. 1972. *Natural Environments: Studies in Theoretical and Applied Analysis*. Baltimore: Johns Hopkins University Press.

Redding, Martin J. 1973. *Aesthetics in Environmental Planning*. Washington, D.C.: U.S. Environmental Protection Agency.

Stull, J. W., R. C. Angus, R. R. Taylor, A. N. Swartz and T. C. Daniel. 1974. Rich flavor discrimination in ice cream by theory of signal detection. *Journal of Dairy Science* 57:1423–1427.

Wheeler, L., T. C. Daniel, G. Seeley and W. Swindell. 1971. Detectability of degraded visual signals: A basis for evaluating image retrieval programs. Optical Sciences Center Technical Report. No. 73. Tucson: The University of Arizona.

22. Balancing Resource Extraction and Creative Land Development

Fred S. Matter

THE THEORETICAL POTENTIAL in a balancing mechanism is as obvious as are the stumbling blocks that fall between the theory and its actualization. On the one hand men are constantly reshaping the surface of the earth looking for better ways to accommodate their packaged dwelling units into what remains of the existing natural landscape. On the other hand, another group of men are creating giant landscapes purely as by-products of their search for mineral deposits below the earth's surface (Fig. 22.1). One group is trying to mold a setting, frequently on a far less successful scale, that is similar in many aspects to what another group is literally throwing away. The builders are scraping off the vegetation of the foothills, paying high premiums for the right to build subdivisions of little plateaus, each one with a picture window view of the valley below. The mining companies are creating foothills with unobscured views of the surrounding areas and then trying their best to disguise their efforts with cosmetic attempts at revegetation. The irony is completed by another set of subdividers developing housing in a floodplain, probably best suited for agricultural purposes, directly below the mining areas thereby setting up the primary source of tension between the inhabitants of the area and the constantly expanding waste dumps above.

Is it an impossible vision of harmony through creative land use planning to speculate on the combined efforts of disparate groups of land developers, including both miners and community builders, working together to create a landscape that serves the best purposes of both interests? Are there, in fact, a variety of interests that could benefit from the application of a creative overall plan directed toward the redevelopment of the mining areas? Could some of these interests derive direct economic benefits from redevelopment in addition to satisfying the more pressing environmental issues that have aroused the general public? Could some of these potential economic benefits be realized in a short-term time frame applying to the current operations of the mining companies or are they all of a long-term more speculative nature? Also, if such an overall, three-dimensional land use plan is developed for a particular mining area, are there transferable ideas that could benefit the study of other situations, particularly mining areas that are just in the beginning stages of development? Finally, given a positive answer to any or all of these questions, what is the best method to use in the study of a three-dimensional redevelopment plan for a large area combining natural topographical features, constantly expanding mining operations, and all other man-made conditions in the immediate surroundings?

Fig. 22.1. Composite view of mining operations,
Sierrita Mountains, southern Arizona.

A. Duval blasting
B. Twin Buttes pit
C. Pima tailing pond

D. Pima tailing revegetation
E. Anamax Twin Buttes conveyor
F. Twin Buttes edge of tailing pond

These are some of the questions that were asked at the beginning of a multidisciplinary team study conducted at the University of Arizona on the mining operations in the Sierrita Mountains–Santa Cruz Valley region immediately south of Tucson. The conclusions and conceptual designs drawn from this study were published by the university in Fred S. Matter's extensively illustrated report entitled *A Balanced Approach to Resource Extraction and Creative Land Development Associated With Open-pit Copper Mining in Southern Arizona* (1974). Most of this chapter is drawn from the report as are the figures and photographs used to describe both the methodology and the design conceptions.

The crux of the entire work, as has been suggested, was the creation of a positive attitude toward the land redevelopment possibilities that are inherent in the operational nature of a large open-pit copper mine. These possibilities were analyzed for both long-term and short-term benefits and were also calculated to aid both the interests of the surrounding communities and those of the mining companies themselves. The first step in the investigation was the construction of a large-scale relief model of the entire Sierrita mining area including the Santa Cruz River valley bordering it on the east. This model, constructed at $1'' = 1000'$ with $100'$ contour intervals covered approximately 25 miles by 20 miles in scaled area. Included in the model were six open-pit mines: the Mission and San Xavier mines operated by the American Smelting and Refining Company, the Twin Buttes Mine operated by the Anamax Company, the Pima Mine operated by the Cyprus Pima Mining Company, and the Sierrita and Esperanza mines operated by Duval and the Duval-Sierrita Corporation. The pits of these mines were cut into the relief model at the scale of the then current operations, according to detailed contour maps supplied by the individual companies. All other landforms created by the mining operations, including waste rock areas and tailing ponds, were constructed to scale in movable segments which could be easily removed to facilitate the use of the same base model for the investigation of other landform design schemes arranged in alternative configurations. The procedure that followed this basic technique involved the construction of a number of alternative land use, landform configurations manipulating the shapes, sizes, and layouts of waste dumps, tailing ponds, and holding basins. Each scheme was designed to investigate a particular set of ideas about the overall efficiency of current mining operations, the environmental impact of these operations and their adaptability for future redevelopment patterns. A number of general guidelines were followed in the conceptualization of these preliminary models including the maximum consolidation of trucking patterns for economical waste haulage, the utilization of gravity flow for tailing disposal, and the ability to reshape existing dumping configurations through gradual pattern modification without resorting to moving any of the predisposed material.

The results of this first phase of the study produced an overall land use, landform model that combined the best ideas from each of the alternative schemes into a single composite design encompassing the entire study area (Fig. 22.2). In general, the achievements of this overall plan dealt with short-term benefits to both mining operations and environmental concerns stemming from maximum operational consolidations and minimal land dis-

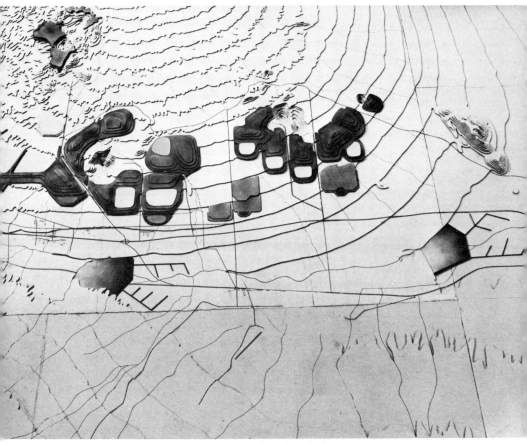

Fig. 22.2.   A long-term land use design.

turbance. The success of this initial phase, in comparison with the previously existing expansion plans of the various individual mining companies, was revealed by an increase in the efficiency of both the land coverage and haulage distances involved in waste disposal. This increase in efficiency would seem surprising in view of the competitive nature of the ore extraction process except for the inclusion of a very fundamental and overriding factor in the local mining picture. That factor is the highly complicated landownership and land leasing situation that surrounds all of the operating mines. The intricacy of this situation can only be suggested by listing a few of the major landholders in the area. A majority of the land involved is owned by the State of Arizona, other large portions are owned or controlled by the Papago Indian Reservation, the Farmer's Investment Company, various agencies of the federal government, numerous privately-owned development companies, ranching outfits, the City of Tucson, and each of the mining companies. Further complicating the landownership picture is the extremely critical issue of groundwater withdrawal rights, tied by Arizona State law to the surface ownership of land. In this area, the competing interests of the City of Tucson, the Farmer's Investment Company, the Papago Reservation, and the mining companies are involved in unresolved litigation that threatens to become more intense with each additional expansion of water needs. All of this is situated in a state-designated "Critical Groundwater Basin" within the semiarid region of southern Arizona.

Given this diversity of highly competitive and conflicting ownership rights in close proximity to one another, the existence of inefficient land usage in the absence of a coordinated overall land use plan becomes understandable. The investigative study identified each one of the current conflicts and proposed a system of trade-offs between parties to consolidate the existing patchwork of disconnected ownership into a systematic workable overall pattern. Without question, the implementation of such an idealistic scheme, involving private interests with long histories of highly individualistic operating methods and built-in antagonisms could be a long drawn-out procedure. In the normal course of large-scale land-planning histories, such a complicated picture of negotiations between diverse parties could well spell the end of the projected plans. The major deterrent to such a failure would traditionally lie in the satisfaction of the economic self-interests of each of the competing groups. An additional and perhaps more pervasive deterrent in the current mining scene has to do with impending legislation for mined land rehabilitation on both the federal and various state levels. The worst effect of this legislation would be to force the mining companies to try to return their mined land to a pre-existing "natural" condition according to a single governmental guideline without deviation for local characteristics or for positive redevelopment possibilities.

The second phase of this study was inaugurated in order to anticipate and perhaps to aid in beneficially modifying and improving such legislation. This phase was to define more vividly the long-term benefits that would accrue to each of the interested parties if a systematic land use pattern were to be implemented as the first step in an extended consideration of the possibilities for the eventual redevelopment of the entire mining area. This plan was to be pursued in harmony with the projected patterns of development in the adjacent areas of the Santa Cruz Valley, in particular, the retirement community of Green Valley, and the City of Tucson. An additional benefit projected from the extension of the study had to do with the prevailing psychological attitude toward mining operations in general. In the 1970s attitudes reflecting current and past mining practices regarded the closing down of a mining area as a disaster involving not only a magnitude of economic problems to the servicing communities but also a major blight on the landscape in the form of abandoned waste areas and tailing ponds. These land scars would take hundreds of years to be reabsorbed into the natural setting if left to the slow recovery processes of the desert environment. The prevalence of such an attitude, combined with the sure knowledge of a continually depleting ore body, leads to a mixed set of values toward the entire cycle of exploitation of any mineral deposit, especially one in close proximity to other existing urban or rural developments. If, on the other hand, the long-range disposition of the wasted areas surrounding a mining operation could be shown to have positive additional development potential, potential that would be a benefit to existing communities and other surrounding interests, the attitude toward such operations could become far more positive. In fact, the economic and the ecological improvements held in reserve for such future activities could be of value not only to the adjacent parties, but also to the mining companies themselves. Given these long-term possibilities, such a short-term consideration as the development of a consolidated land use plan should be easier to implement. Certainly the long-term factor becomes an added incentive in the complicated process of finding a way for a combination of interests to work together.

With this goal in mind, the study team began the process of collecting a wide variety of ideas that could be used in the redevelopment of the mining areas. In this search for ideas all parts of the mined landscape were considered. The main focus centered around the use of a modified waste rock dumping pattern, with a covering of overburden stockpiled from the initial phases of the excavation process, as the foundation for a satellite community development plan. The locations of the dwelling units in such a community would be fitted into the slopes of the waste areas, each unit having an unobstructed view of the Santa Cruz Valley to the east and the north. At intermediate stages between the slopes, flat benches would provide horizontal circulation spaces for individual automobile access to each house. On top of the overall configuration a larger flat plateau would serve as the location for community services, larger building sites and recreation spaces. On the other side of the housing slopes, additional benches could serve as agricultural plots with industrial sites on the lower edge of the complex facing the pit areas to the south and west (Fig. 22.3). It has been suggested that the full functioning of these redevelopment proposals would occur only after the operations of the mining companies had ceased. It is possible, however, utilizing a carefully staged development plan to allow limited construction to begin on sites that are widely separated from the active mining areas. One of the characteristics of the overall development plan was that the waste dumps were arranged in a series of semi-independent foothills, each of which was capable of being completed to its final configuration before dumping began on the next in the series. This procedure facilitated a staged redevelopment process as well as insuring a finished community with naturally sized subdivisions of neighborhood hillside groupings. The total complex was then linked with major access roads tucked in the valleys between residential areas to minimize the conflict between high speed traffic and slow moving quiet neighborhood systems.

Following this general concept, two larger relief models with 50' contours and 1" = 400' horizontal scales were built by the study team. Each model represented a variation of the community development idea constructed around a major open-pit copper mine. The technique involved in the construction of the finished landforms was similar to that of the original model. Twenty-year dumping projections supplied by the individual companies were used to calculate the volumes of the long-range earth forms which were again built as removable sections for study purposes. The pits were cut into the relief models, again using twenty-year projections. As a result of this design conceptualization, the advantages inherent in the utili-

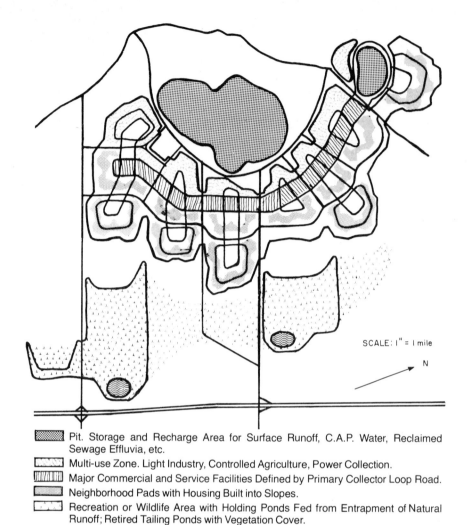

SCALE: 1" = 1 mile

N

▓ Pit. Storage and Recharge Area for Surface Runoff, C.A.P. Water, Reclaimed
Sewage Effluvia, etc.

▨ Multi-use Zone. Light Industry, Controlled Agriculture, Power Collection.

▨ Major Commercial and Service Facilities Defined by Primary Collector Loop Road.

▨ Neighborhood Pads with Housing Built into Slopes.

▨ Recreation or Wildlife Area with Holding Ponds Fed from Entrapment of Natural
Runoff; Retired Tailing Ponds with Vegetation Cover.

Fig. 22.3.   Long-term re-use scheme: satellite city

zation of a completely pre-controlled landscape for three-dimensional devel-
opmental purposes, as opposed to the rough surface treatment of existing
natural landscapes for the same ends, was vividly illustrated (Fig. 22.4).

The particular example just described is only one type of development
that would benefit from such overall control. In addition, the projected
configuration of waste dumping was investigated for the large-scale entrapment of surface water runoff from the adjacent mountain ranges. The south-
facing slopes of both the dumps and the pits were also found to be ideal for
the arrangement of solar collecting devices for energy production. The controlled and dramatic elevation change created by the open pits, in combination with an upper level water storage area, was found to make an excellent
location for a pumped storage system balancing peak and off-peak power
usage in the utility system for the entire region (Fig. 22.5). These and a

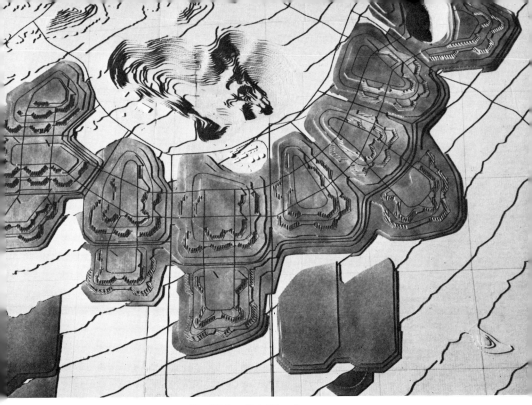

Fig. 22.4.   Satellite city.

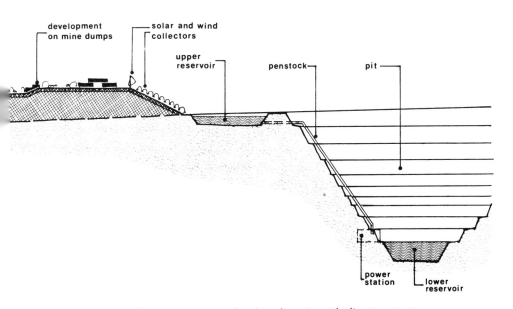

Fig. 22.5.   Section through redeveloped waste and pit area: energy
collection and storage system.

number of other redevelopment ideas working in concert or functioning independently would serve the interests of both the existing communities in the mining area and any new developments planned on the waste areas themselves. Other general conceptualizations could easily be developed for different mining areas depending on the particular climate and the context in which the operations were located. Recreation, wildlife refuges, and income-producing agricultural production could well be the central foci out of many possible reuses to which the mined landscape could be adapted.

The particular project described here is only a beginning for further investigations, both in the collection of new ideas for redevelopment possibilities and in the more exact detailing of the conceptual designs projected thus far. Engineering studies of dump stabilities, hydrological studies on surface-water catchment, the sealing of tailing ponds to decrease the possibility of groundwater pollution, the resulting increase in the amount of water available to holding ponds for multiple use in recreation, and efficient recycling of water to the mill are all areas that need further investigation. It is clear, however, that without the achievement of the first-phase goals of consolidated land use and landform patterns, the hope of ever arriving at a stage in which such comprehensive redevelopment could take place is impossible. The foundations for any future reuse scheme are being built on a daily basis by the operations of the producing mines. These foundations are being distributed in amounts that stagger the imagination when compared to even the most extensive land development schemes known in the world today. The existing relocation of approximately 2½ billion tons of earth compounded by the addition of one million tons each working day from the six operating mines in the Sierrita area creates a man-made landscape that rivals the foothills of the mountain ranges surrounding the Tucson basin. If these man-made foothills were created in such a way to support viable satellite communities of the future, what is now a negative factor in the landscape could become a long-term asset to the entire region (Fig. 22.6).

It is evident that current attempts to produce such a desirable change of status, although made in good faith, are not comprehensive enough and, in some cases, are misdirected in the definition of their ultimate goals. The monumental scale of such earthmoving and land disruption can never be disguised as an apparently natural condition by the revegetation of an exposed slope. Neither does the prospect of trying to recreate the original condition associated with the land prior to the beginning of mining operations hold much realistic hope for these disturbed areas. In addition to the prohibitive costs of once again moving the massive amounts of earth back to the pits, the leftover mess that would be the result of such a tremendous operation would still be an aesthetic and ecological blight for years to come. Current legislative attempts to provide for such conditions in association with strip coal mining are completely inappropriate for the type of open-pit mine frequently encountered in the semiarid regions of the southwest.

Finally, there is the always intriguing argument that most of the landscapes existing outside of the designated wilderness areas of the National Park System are all landscapes that have been significantly modified by

Fig. 22.6.   Pima Mine waste dumps, southern Arizona.

the presence of mankind in ways more pervasive than is visible to the casual observer. A return to the natural condition of a disturbed area, therefore, frequently means a return to the man-made condition before the most recent revision. If one accepts this theory, it makes even less sense to pursue legislation that is based on a purely convened image of "naturalness" which in all probability did not exist in the original state. Whatever the philosophical argument, however, the most telling case for comprehensive redevelopment will probably remain an economic one. The theoretical basis for this approach, as was stated at the beginning of this chapter, is clear and compelling. An industry that is highly equipped and skilled in the business of moving massive amounts of earth, that accomplishes this task in a manner that pays for itself through the value of the ore thus extracted, is in the process of creating large-scale man-controlled landscapes. These landforms could easily serve as the replacement for the destruction of more natural areas poorly designed to meet the needs of an ever expanding housing and community development industry (Fig. 22.7). If the waste by-products of the former process can be shaped to serve as the foundations for the needs of the latter, if the ultimate developer of the man-made landscape finds that his primary site development costs are paid for by the calculated dumping patterns of the operating mines, at a cost which is less than that of modifying existing foothills, are not the difficulties involved in

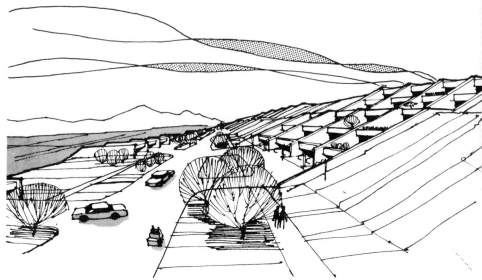

Fig. 22.7.   Housing on redeveloped waste areas.

the coordination and design of such an idea worth pursuing? Given the additional stimulus of improved short-term environmental impacts, improved operating efficiencies through landownership trade-offs, and a new positive image for the mining companies as long-term contributors to a regional development scheme, perhaps the theory can start to become a reality before the chances are lost forever beneath the weight of billions of tons of wasted opportunities.

PART V

REVEGETATION TECHNIQUES

RECLAMATION ON MINED LANDS most often implies revegetation to restore sites to at least their former level of productivity. Problems vary across a wide spectrum of conditions including toxicity, infertility, salinity, high or low pH, sodic or alkaline materials or some combination thereof. In the arid southwest, water is the key factor that determines the success or failure of reclamation efforts. It is alternately a blessing and a threat, and the problem is to maximize the one and minimize the other. Much of the reclamation research in the southwest is directed toward this goal. Surface treatment for reducing the erosive forces of water and for retaining it on site for use by vegetation is one of the most effective methods. Richard L. Hodder describes several techniques that have been successful in Montana.

Quick results are often a major thrust of reclamation projects. Irrigation is frequently suggested as the means of obtaining this in an arid environment. However, there is some controversy over its ultimate effectiveness, perhaps due to the lack of extensive research and experience in its use in rehabilitating southwestern mine lands. Dale DeRemer and Dan Bach describe trickle irrigation, which appears suitable to many of the rehabilitation problems in the region. Earl F. Aldon reports the results of field and laboratory experiments with the system in New Mexico as well as some additional work on plant establishment.

Slopes of mining waste are often steep and not easily accessible to planting machinery. Handseeding is expensive. Thus, considerable attention has been given to hydroseeding. Burgess L. Kay describes extensive experience with hydroseeding in a variety of situations applicable to seeding and planting mining wastes.

K. D. Dean and M. B. Shirts of Utah discuss the problems of toxicity, infertility, pH and alkalinity, and how stabilization may be obtained by physical, chemical, vegetative, and combination methods. Kenneth L. Ludeke describes similar problems as well as the approach the Cyprus Pima Company has taken in southern Arizona to revegetate copper mine tailings.

23. Dry Land Techniques in the Semiarid West

Richard L. Hodder

CONCEPTS OF RECLAMATION have changed drastically, primarily because of intense interest and concern from opposing factions. Until the late 1960s, reclamation in the semiarid west invariably meant irrigation of dry lands to make them productive. Reclamation emphasis has since focused primarily on treatment of surface mined spoils, particularly from mining coal.

Stringent regulations have been devised by state and federal agencies to eliminate the mistakes characteristic of the past. Attitudes and the economic situation have changed drastically, for by the early 70s the challenges of land reclamation were limited far more by lack of ingenuity than by lack of funds. Approaches previously considered impractical and unrealistic may be acceptable under new reclamation circumstances.

Since 1967, a major concern of the Montana Agricultural Experiment Station has been in the stabilization and revegetation of bare areas caused by strip mining for coal and highway construction. Associated problems of land stabilization related to these two types of disturbance are similar in many respects and dissimilar in others. Roadside cuts and fills are often composed of materials similar to mine spoils. However, mine reclamation problems are usually considered different from roadside stabilization problems because mine spoils may be graded, mixed, manipulated, farmed, and so forth, whereas roadside problem materials must be accepted as designed by highway engineers. Essentially the same basic criteria apply to all revegetation of severely disturbed sites.

Restoration of bare areas, especially in the dry western states, must start with site stabilization which first demands physically stable gradients. Once physical stability is achieved, the surface soil may be temporarily stabilized with mulches, annual species, mechanical surface manipulation such as pitting, listering, and finally with permanent vegetation.

The successful establishment of vegetation is dependent not only on supplying the essential prerequisites of plant nutrients, water and adaptable species, but in providing these materials in adequate amounts at the critical time. This crucial combination of factors naturally occurs infrequently in the semiarid west.

Dry land planting techniques have to do directly and indirectly with broadening the limits of the critical period. This may be accomplished with innovations and new systems used in the establishment of permanent vegetative cover. New techniques usually hinge on the most limiting factor, the availability of soil moisture. Several effective methods to circumvent limiting moisture are familiar; yet some innovations with great potential may be entirely new to many. ·

Impeding runoff, increasing infiltration, and reducing evaporation are means of increasing available soil moisture. These accomplishments may be achieved by gradient reduction, cultivation, mulching, topsoiling, buffering, surface manipulation, compaction relief, and special seeding techniques.

Reducing gradients of slopes to a degree that will provide mechanical stability and be conducive to the operation of farm machinery is essential. With slope reduction, runoff may be reduced, infiltration increased, and opportunity for seedbed preparation greatly improved. Five-to-one slopes, as required by Montana regulation since 1973, are sufficiently level to be farmed. Crop trials on such slopes have included winter and spring wheat, barley, speltz, and oats. Hay and forage crops tested include alfalfa, sweet-clover, sainfoin, cicer milk-vetch, and grass-legume mixtures. Annual species, such as cereal grains, may provide the surest and quickest means of obtaining additional organic matter in the surface of spoils to promote stability and increase available moisture.

Cultivation will be considered briefly, since its effects are generally understood. It is sufficient to say that a cultivated layer on the surface improves infiltration, breaks up capillary action, reduces evaporation, and thus conserves moisture available to plant roots.

Mulching with various materials is a technique used to increase available soil moisture. It encourages infiltration by impeding runoff, lowering soil temperatures, reducing evaporation, and minimizing raindrop splash, surface puddling, and sealing.

Topsoiling usually increases infiltration because of the organic material content of topsoil and its friable, relatively coarse aggregated texture. Topsoil is typically removed by scrapers which may transport and deposit it directly on reshaped spoils or stockpile it for later use. Many topsoil values are lost when it is stockpiled. If, however, topsoil is applied directly and with proper timing, some sod is effectively transplanted, and many native plants that cannot practically be obtained by seeding are introduced to the new spoils site in this way.

Topsoil is scarce or nonexistent in many areas, and occasionally it is excessively salty. Topsoil for each disturbed site must be tested for quality before it is used. Organic matter in topsoil is an effective erosion deterrent necessary to protect this valuable natural resource from the erosional forces of wind and water.

Topsoiling introduces some disadvantages and hazards if improperly timed or handled. It is by definition loose and friable, and is a natural storehouse of seed of undesirable, as well as desirable species. The challenge, then, is to control immediate erosion, subdue fast germinating weeds, and encourage establishment of desirable perennial species. As of 1975, more research is necessary to accomplish this feat effectively so that the potential benefits of topsoiling may be more fully realized.

A thick buffer layer of light-textured soil material below topsoil has been found effective in increasing absorption of surface flow, thus increasing soil moisture reserves, and when combined with the use of annual stabilizing plant species may minimize erosion problems and encourage perennial cover establishment. However, this is possible only when plant nutrient needs are satisfied. Because spoils are typically void of certain essential nutrients, it is not a question of whether or not to fertilize, but rather when and at what rate. Optimum application rates must be determined by testing at each site. Fertilizing in spring following seedling emergence is effective.

Surface manipulation utilizing several configurations has been tested to study water retention capabilities. The configurations included deep chiseling, offset-listering, gouging, and dozer basins; all have been compared for efficiency in encouraging seed germination and rapid establishment of seedlings, erosion control, and compaction relief.

Deep chiseling is a surface treatment that loosens compacted soils for a depth of 6 to 8 inches. The process creates a series of parallel surface furrows on the contour of the spoils which effectively impede water flow and markedly increase the infiltration rate. Chiseling forms a cloddy seedbed which is ideal to receive broadcast seeded native species mixtures. Some benefits of chiseling are relatively temporary because erosion from low intensity storms may be sufficient to fill the cultivation slots, and thus the high initial water absorption rate diminishes. This treatment is effective on relatively flat slopes during the first spring to help establish a vegetative cover, which, when adequate, will create a lasting erosion control system.

Off-set listering (Fig. 23.1) is a surface configuration consisting of

Fig. 23.1.   Off-set listering is an effective surface manipulation treatment in reducing saltation and impeding runoff.

Fig. 23.2. Gouging (right) is more effective than deep chiseling (upper left) in reducing saltation, controlling water erosion, and increasing infiltration.

alternately arranged elongated pits approximately 6 inches deep and 4 feet long. The design is especially effective in reducing saltation and in impeding runoff.

Gouging (Fig. 23.2) is a surface configuration composed of series of depressions approximately 10 inches deep, 18 inches across and 25 inches long. This pattern is amenable to gradual slopes and flat areas. It is especially effective in conserving runoff from moderate intensity storms and in causing differential melting of snow in the winter, thus retaining snow moisture that would otherwise be lost. It, too, creates a cloddy seedbed ideal for broadcast seeding.

Dozer basins are large depressions designed to accomplish goals similar to those of terracing but without the characteristic precision, hazards, and expense of the latter technique. Dozer basins are usually about 2 feet deep and 15 feet long, and are placed on the rough contour interval of about 30 feet. This spacing is adequate to allow for combinations of treatments which will improve water retention capacity and overall efficiency (Fig. 23.3). Precipitation intercepted within each mini-drainage accumulates in the basin bottom in quantities sufficient to thoroughly saturate the basin limits. This increased soil moisture availability assures the establishment of an initial stand of vegetation during the first growing season, which can ultimately spread between basins to provide a complete cover.

Compaction caused by the reshaping of new spoils with heavy equipment may be severe and limiting to root penetration, plant development, and water infiltration. Precipitation that is not readily absorbed by spoils contributes to increased runoff and resultant erosion. If soil moisture is augmented via special treatments, the heaving action caused by freezing and thawing of increased soil moisture may be the most economical manner of relieving compaction. Mechanical ripping is immediately effective, but

Fig. 23.3.   A combination surface manipulation treatment composed of gouging (right) and dozer basins (left).

it is an expensive approach to eliminating compaction. Research has been initiated to indicate optimum depth and spacing of ripping for maximum effectiveness in vegetation establishment.

Special seeding techniques or seeding designs are interesting methods of increasing availability of soil moisture. We have used a modification of an approach developed by D. H. Heinrichs of the Canadian Swift Current Research Station, Saskatchewan. This technique utilizes a system of cross-seeding intended to increase total forage production and beef production, and to significantly extend the productive life of a grass-legume stand. Grasses may be seeded in drill rows at two foot intervals in a direction crosswise to the prevailing wind and over-seeded by drilling legumes in rows at 90 degrees to the grass seeding. The resulting raised grid pattern that develops is effective in intercepting and retaining both snow and rain, thus increasing available soil moisture.

Seeding occasional multiple rows of tall, stalky, wind resistant species of grasses, such as tall wheatgrass, crosswise to prevailing winter winds creates a snowfence effect that has increased snowmelt and soil moisture reserves at the Eastern Agricultural Research Center at Sidney, Montana. This seeding technique is applicable to dryland and semiarid reclamation systems.

Snow fences have been used successfully to increase moisture accumulation on dry spoils. Dr. Morton May at Laramie, Wyoming, has worked considerably with this approach. Unfortunately, snow fences are relatively expensive, sometimes difficult to place effectively, and troublesome to relocate or remove from the site. Nonetheless, in certain situations snow fences may have a very real place in reclamation.

There are two obvious approaches to making stored soil moisture more readily available to plants: extend plant roots down to where the moisture is, or bring the moisture up to where the roots are. Some schemes used to accomplish these goals are intriguing.

Two planting innovations with great potential in placing plant roots purposefully deep into soil where moisture is more available are the planting of developed tubelings and the use of supplemental root transplants.

Tubelings are usually seedlings of woody species planted and nursery developed in 2-ply paper cores or tubes. The tubes are typically 2½″ in diameter and 2 feet long. The paper core is reinforced with a ½″ square mesh plastic sleeve. When the root system develops and extends from the bottom and sides of the tube, the tubeling is ready for transplanting. A powered soil auger is used to drill tubeling-sized holes in the field. Tubelings are inserted, sealed around the top and abandoned without further care or maintenance. Species found particularly adaptable to this technique include honeysuckle, three-leaf sumac, sand cherry, chokecherry, old man wormwood, and Siberian peashrub.

Supplemental root transplanting is accomplished by first carefully removing a pair of interconnected seedlings of a rhizomatous shrub species. The top of one seedling is pruned off at the crown leaving two root systems connected to the uncropped seedling. The horizontally connected root system is then planted in a vertical attitude, the root system of the cropped seedling being placed at the deeper depth, hence adjacent to deeper soil moisture reserves. The connected intact seedling is placed in a normal manner in the drier surface soil level. Snowberry, wild rose, sea buckthorn, and many sageworts are responsive to this technique.

A condensation trap consists of a deep planting basin containing a regular setting-out stock seedling planted in the basin center. The entire basin is covered with a plastic sheet and heeled in around the edge to contain a large amount of air. The foliage of the plant is guided up through a hole in the plastic sheeting. Rocks are placed on the tarp around the plant to provide protection and to weight the plastic, keeping it taut in a funnel form. Condensate collecting on the underside of the plastic sheet trickles down to the plant location and effectively irrigates it (Table 23.1, Fig. 23.4).

This brief review of techniques being used to alter and extend the critical factors involved in revegetation work under semiarid conditions should suggest that the potential for stabilization and revegetation of disturbed land may be very real. Similar or new methods are expected to develop in the future. They will require intensive field testing, and should therefore be submitted to research organizations with suitable facilities and expertise for proper evaluation. As mentioned previously, adequate field testing is a prerequisite if more effective techniques are to be developed. If effective methods are not developed, it may be necessary to resort to limited irrigation for successful reclamation. In the semiarid west the feasibility of such irrigation has not yet been proved.

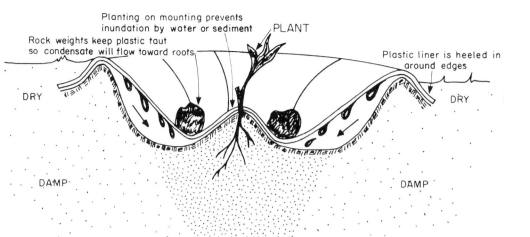

Fig. 23.4. Condensation trap.

TABLE 23.1
Temperature and Moisture Data From Condensation Traps at Logan, Montana

Outside Basin Temperature (F)		Date of Reading	Inside Basin Temperature (F)*		Time Period		Plastic Color	Condensate Collected (ml.)
Min.	Max.		Min.	Max.	hours	days		
57	88	7-25-69	—	—	21½		Clear	73
		7-25-69	—	—	21½		Black	20
51	85	7-26-69	50	108	24		Clear	63
		7-26-69	52	103	24		Black	16
56	95	7-29-69	48	120	72		Clear	403
		7-29-69	50	110	72		Black	144
55	89	7-30-69	47	118	24		Clear	88
		7-30-69	46	102	24		Black	16
56	89	8- 1-69	49	119	44		Clear	181
		8- 1-69	51	113	44		Black	84
42	100	8-15-69	44	134		7	Clear	1,109
		8 15-69	41	122		7	Black	319
44	96	8-22-69	34	131		7	Clear	2,011
		8-22-69	38	114		7	Black	1,016
37	101	8-30-69	42	128		8	Clear	2,142
		8-30-69	32	126		8	Black	1,568
31	92	9- 5-69	24	122		6	Clear	2,062
		9- 5-69	22	110		6	Black	1,242

Note: Mean annual temperature, 46.5 F; mean annual precipitation, 10.3 in.; elevation, 4,035 ft.
*Temperatures recorded on basin bottom in shade.

24. Irrigation of Disturbed Lands

Dale DeRemer and Dan Bach

SOME OF THE REASONS to reclaim disturbed lands are: to provide stabilization from wind, water, and gravity erosion; to meet environmental regulatory standards; to improve public relations; and to enhance aesthetics. An additional, but important, reason may be to develop a greater production on the area than existed before mining.

Some disturbed lands may be made available to lease for agriculture. However, it must be made certain that the agricultural products will not contain toxic materials such as lead or copper which would render the food unfit for consumption. Most agricultural crops can be grown successfully on tailings materials. Watermelons, fruits, and nut trees have also been successful.

After the problem is defined and goals are established, there are often several alternative approaches for vegetating disturbed lands. The plan used must determine if irrigation is to be used, and, if so, what type. If irrigation is used, some investment is needed. Irrigation is necessary if one or more of the following conditions exist: if rainfall is less than 12 inches (this could probably be raised from 12 to 30 inches or perhaps higher if other problems exist), if rainfall is undependable, and/or if slopes are steep. On slopes, if water is applied by either rainfall or irrigation, water can go into the soil for plant use, pass through the plant root zone and become unavailable, or run off and cause erosion. The usable water that remains in the root zone, "effective moisture," is generally less on steep slopes.

Toxic materials can be found in various places around the world. In Arizona, many tailings materials have a pH ranging from about 1.5 to 3.5. These materials are often high in many of the toxic heavy metals, particularly copper. The only effective way to solve this problem is to leach the material out of the plant root zone.

If land is bare and remains that way for a period of time, then one or more of these problems exists. An example is a tailings berm of alluvium which had been barren for 1 to 3 years. It is a case in point, having two problems: steep slope and low rainfall. A certain amount of rill erosion took place due primarily to some rain, although rainfall in the area is in the the neighborhood of only 3 inches annually. A quick examination of a chronological period at this berm will reveal what can happen when a system of revegetation is properly applied.

In 1970, a drip system was installed. Water was applied through the system to wet and leach the root zone for a period of time. Native plants

[224]

of one-gallon size were planted. The emitters were placed at random along the hose line, but, unfortunately, they could be randomized in only one direction. After about 75 days, the plants showed good growth. Some had doubled in size. Wild tobacco reached 6 feet in height in about 3½ months. Chilean mesquite trees reached 6 feet in height with an 8 to 10 foot diameter drip line spread in about 2½ to 2¾ years.

At the beginning of the third year, vegetation covered 60 percent or more of the berm, and blended well into the desert, when viewed from a distance. That the plants were lined in one direction was not particularly evident, and the berm was quite green compared to what is seen on the desert floor. When water is withdrawn expectations are that there will be some sloughing of vegetation and some loss of the vibrant green color. The berm received between 20 and 40 lbs. of actual nitrogen per acre per year for a period of three years, totaling 60 to 120 lbs. No further fertilizer applications will be made after irrigation is discontinued.

Although irrigation has been around for more than 6,000 years, only two types lend themselves to today's needs — sprinkler systems and drip irrigation. Both have been developed in the last 30 years.

SPRINKLER SYSTEMS

Sprinkler systems are very effective, and have a number of advantages: they give full area coverage, are durable, and are easily portable. The portable types can be quickly assembled. Sprinkler systems come in all types, sizes, and shapes to meet a very wide variety of needs. They have a high water requirement. If water is scarce or expensive, this is a major consideration; the application rates of most sprinkler systems are often higher than the infiltration rate of tailings soils. This is a serious problem on sloping ground. However, on some soils it may be avoided if the sprinkler system is properly designed. Proper design is crucial to proper operation. Initial cost of sprinkler systems may be high. Since sprinkler systems cover large areas, they may also create weed problems.

Sprinklers have been successful in a number of applications in reclamation of disturbed lands, but only when the system was properly designed by an experienced and responsible sprinkler system designer. A 40 percent slope can't be irrigated by spraying with a firehose, and toxic copper can't be leached out of the soil with a high-application-rate sprinkler. Therefore the sprinkler system must be very carefully suited to the job that it has to do. After the system is installed, the facility requires someone who is experienced and has success in the use of the particular system used. This is essential to avoid costly trial and error mistakes. He could perhaps be a local farmer who has successfully used sprinkler systems.

DRIP OR TRICKLE IRRIGATION

Drip irrigation provides water slowly over a long period of time in order to maintain soil moisture at optimum levels. Irrigating with a five, six, seven, eight, or ten-day frequency, as may be done with sprinklers or furrow irrigation, provides soil moisture levels varying from saturation to near wilting point between irrigations. At the time of irrigation, the soil moisture level is too high and oxygen is excluded from the root zone. Then there is a period of time when the soil moisture is just right. As the plant and surface evaporation use soil moisture, a period occurs when soil moisture becomes limiting. The soil moisture cycle is approximately a third of the time too wet, a third at optimum, and a third too dry for good growth.

The concept which has made drip irrigation successful is to irrigate every day or every other day to maintain the soil at optimum soil moisture levels. Drip irrigation employs a device that will emit water slowly and a plastic hose to deliver the water.

Drip irrigation began in the late 1950s. Dependable types of equipment have been developed during this time. The water discharge from a given orifice is about one gallon per hour.

ADVANTAGES OF A DRIP IRRIGATION SYSTEM

Less water is used than with most any other type of irrigation system. Generally, experience has shown that water use will be 5 percent to 40 percent of that used by other irrigation systems, a savings of 60 to 95 percent.

Irrigation of a tailings berm 30 feet high (58 feet on the face) and 2 miles long will require a flow of 13.4 gal. per minute, about the flow obtainable from a household faucet. Thus, only very small quantities of water and relatively small quantities of fertilizer are needed. The water is applied at a rate far less than the infiltration rate, eliminating runoff. Drip irrigation is a very efficient leacher. As water moves into the soil, the soluble materials (salts and toxic elements) move with it, out of the root zone, resulting in a relatively clean region for plant roots. Usually the irrigation system is operated for a period of about two weeks before planting in order to leach the root zone.

Fertilizers and other pesticides can be applied through the system very efficiently. A rule of thumb is to use only about 25 percent of the amount of fertilizer needed with other irrigation methods.

Generally, a drip system is less expensive than many other types of available systems. Although prices are rising, particularly in the plastics industry, technology is improving to keep costs low. New materials are being developed that will do certain jobs far less expensively than ever before.

Drip systems are easily automated from a controller-timer or soil moisture sensing device to maintain the proper amount of water for plant needs (Figs. 24.1, 24.2, and 24.3).

SOME DISADVANTAGES OF DRIP IRRIGATION

The system is new, and experience is needed in its use. One of the problems in the past has been improper design and improper use, although most growers have been successful in its use. The lifespan of the components usually ranges from one to twenty-five years, depending upon the type of system and how it is used. The plastic materials are designed to be used in the sunlight and are resistant to sunlight decomposition.

Fig. 24.1. Selected desert types on tailings berm south of Tucson which have been drip irrigated for 18 months.

Fig. 24.2. The author checks *Atriplex semibaccata* planted 13 months before the picture was taken. The grey-white tailings material supported no growth prior to drip irrigation.

Fig. 24.3.   Thirty-month-old mesquite trees and selected brush types have grown to 9 feet in height and 3-inch trunk caliper with drip irrigation and proper fertilization on a 40 degree slope.

Another disadvantage is the inability to achieve full area coverage. For grass or lawns, the cost of going full coverage is high. However, the state of the art is improving rapidly and is close to the point where it is economically feasible to go nearly full coverage; at least in the neighborhood of 2′ x 4′ spacings. After extensive testing by the manufacturers, it appears very promising.

Filtration is perhaps the biggest problem and is absolutely essential. A drip system should never be installed without a good filtration system. Without it, the system will not function well.

Animal damage is a problem in some areas. It may be best solved by keeping the area clean of materials that harbor animal pests.

Irrigation is an effective way to facilitate revegetation of problem areas. On steep slopes, soil materials with low infiltration rates or toxic soils, drip irrigation will likely be most effective. Flat areas requiring full coverage should be sprinkler irrigated. Mine operators anticipating revegetation projects would do well to seek out expertise in the areas of irrigation design and operation.

25. Reclaiming Coal Mine Spoils in the Four Corners

Earl F. Aldon and H. W. Springfield

THE FOUR CORNERS AREA CONTAINS some of the most important coal resources in the west. Much of the coal can be mined by stripping. Large-scale stripping in northeastern Arizona and northwestern New Mexico has been underway since the early 1970s.

Research on mine spoil reclamation began in 1972, the year New Mexico passed a Coal Surfacemining Act. That year, the Rocky Mountain Forest and Range Experiment Station entered into cooperative agreements with Utah International (Navajo Mine) and Pittsburg and Midway Coal Co. (McKinley Mine) to conduct experiments on reclaiming spoils in the Four Corners. Results are based on experiments conducted on spoils from these two mines, both in New Mexico.

The Navajo Mine, about 20 miles southwest of Farmington, operates in the Navajo Fruitland coal field of the San Juan basin. The climate is arid. Average annual precipitation is only 6 to 7 inches. Summer is wetter than winter; spring and fall are the drier seasons. Temperatures reach extremes of −25° and 105°F. Elevation ranges from 5,100 to 5,600 feet. The sandy soils, which comprise half the area, support an open grassland with a scattering of shrubs. Principal grasses are galleta *(Hilaria jamesii)*, alkali sacaton *(Sporobolus airoides)*, and Indian ricegrass *(Oryzopsis hymenoides)*. The shaley, saline soils and thin breaks and badlands support mainly low-growing shrubs, including shadscale *(Atriplex confertifolia)* and Nuttall saltbush *(Atriplex nuttallii)*.

The McKinley Mine, 20 miles northwest of Gallup, operates in the Gallup Mesaverde coal field. The climate is semiarid. Annual precipitation averages 11 to 12 inches, a third of which falls as high-intensity rains in July and August. Another third comes as snow from December through March. Dry, windy weather prevails during the spring season (< 1 inch of precipitation from April to July). Elevation ranges from 6,800 to 7,300 feet. Temperatures reach extremes of −35° to 95°F. Pinyon *(Pinus edulis)*-juniper *(Juniperus monosperma)* woodland is the dominant vegetation, occupying the plateaus, benches, mesas, rocky breaks, and steeper slopes. Associated with the trees are several shrubs, such as mountain mahogany *(Cercocarpus montanus)* and cliffrose *(Cowanica mexicana)*, and a sparse

[229]

herbaceous understory, including squirreltail *(Sitanion hystrix)*. The relatively narrow valleys filled with moderately deep alluvium support big sagebrush *(Artemesia tridentata)* with an understory mainly of western wheatgrass *(Agropyron smithii)* and blue grama *(Bouteloua gracilis)*.

Results will be discussed under the following topics: spoil amendments, irrigation techniques, water harvesting, direct seeding trials, and mycorrhizae.

SPOIL AMENDMENTS

The first studies were designed to answer the questions: Will plants grow in mine spoils? Do organic amendments help? Is fertilizer necessary?

The first experiment with representative spoil material from the McKinley Mine showed emergence and early growth of mountain rye *(Secale montanum)* and fourwing saltbush *(Atriplex canescens)* was not improved by incorporating organic amendments (manure, bark, straw, or sawdust at 10 tons/acre). Applying a complete fertilizer (10-5-5) at rates of 800 or 1,600 lbs/acre did not affect emergence, but resulted in slightly taller seedlings 5 weeks after seeding (Aldon and Springfield 1973).

A follow-up experiment with McKinley Mine spoils was conducted in the greenhouse for 90 days in 1973. Three spoil materials were used: (1) old spoil, from mining in 1968; (2) raw spoil from mining in 1971; and (3) raw spoil from 1972 mining. For comparison, topsoil obtained from under big sagebrush was included. Texture was classified as clay loam and pH was 7.2–7.3 for all four materials. Fertilizer (10-5-5) was added at 800 or 1,600 lbs/acre. Results showed fertilizer did not improve seedling emergence of fourwing saltbush or western wheatgrass, but significantly improved their height and yield. Topsoil was not significantly better as a growth medium than either 5-year-old spoil or the younger spoil material. The results of this experiment suggest no advantages from applying fertilizer at the time of seeding, as emergence and survival were not enhanced. Moreover, in this experiment, the 1,600 lbs/acre rate showed practically no advantage over the 800 lbs/acre rate of application.

A greenhouse experiment was completed with two representative spoils from the Navajo Mine. Alkali sacaton and fourwing saltbush were seeded separately. Nitrogen was applied at 0, 80 (N_1), and 160 (N_2) lbs/acre, alone, and in combination with phosphorus at 0, 80 (P_1), and 160 (P_2) lbs/acre. Fertilizer was applied either at the time of seeding, or one or two months after seeding. Seedling emergence was not affected by fertilizer treatment or time of fertilizer application. Yields varied, however, according to species, spoil material, and fertilizer levels and combinations. Saltbush yielded more when fertilizer was applied at the time of seeding, whereas alkali sacaton yielded more when fertilizer was applied one month after seeding. Best production of saltbush in 90 days came from the N_2P_1 and N_1P_2 treatments in one spoil, and from the N_1P_1 and N_1P_2 treatments in the other spoil. Maximum yield of sacaton came from the N_2P_2 or N_2P_1 in one spoil and from any of four treatment combinations (N_1P_1, N_1P_2, N_2P_1, or N_2P_2) in the other spoil. In this second spoil, the lower level of fertilizer (80 lbs./acre) produced as much alkali sacaton as the higher level (160 lbs./acre).

IRRIGATION TECHNIQUES

Irrigation probably is essential to insure establishment of perennial plants at the Navajo Mine due to the extremely low and erratic precipitation. At present, research and planning for reclamation at Navajo are based on the premise that irrigation will be used the first year only. The question is, then, what happens when no water is supplied the following year? Can the plants survive, or is it necessary to provide supplemental water by some other means? Several field studies and one laboratory study were conducted to help answer these questions.

Two types of artificial floodways (steep slopes and gentle slopes) were installed at the Navajo Mine in July 1973. For the steep-slope floodways, the spoil piles were left intact and a 20-foot-wide floodway was bulldozed between the piles. For the gentle-slope floodways, the surrounding areas were graded to slopes less than 5 percent before the bulldozer bladed the floodway. Test plots were installed in two steep-slope and two gentle-slope floodways. Alkali sacaton was seeded: (1) broadcast into a straw mulch, (2) in shallow furrows, and (3) in collars. Three-month-old transplants of fourwing saltbush also were planted in three ways: (1) in an entire plot mulched with straw, (2) with straw around the base of each plant, and (3) with paraffin surrounding each plant to harvest water. All plots were sprinkler irrigated the first year. The test in 1974, and in subsequent years, was to learn whether the adjoining slopes supply enough runoff to maintain the plants in the floodways. Unfortunately, precipitation totaled only 3.27 inches from September 1973 (when irrigation was terminated) to September 1974. Virtually no water ran off the slopes into the floodways. As a consequence, many plants died. Despite the dry weather, however, about three-fourths of the fourwing saltbush plants survived through the second growing season with growth as shown in Table 25.1.

TABLE 25.1
Survival and Growth of Fourwing Saltbush Without Irrigation

Date of Measurement	Survival (%)	Height (in.)	Diameter (in.)	Size Index (Ht. x Diam.)
August 1973	98	6.2	5.2	32.21
September 1974	76	26.4	19.8	522.7

Alkali sacaton at a seeding rate of 144 seeds/sq. ft. produced 13 seedlings/sq. ft. the first season. Only 38% of these survived through the second season. Nevertheless, the overall average of 5.0 seedlings/sq. ft. could be considered a good stand in view of the very dry weather.

Drip irrigation also has been studied. A small-scale test of this technique was conducted in September 1973 at the Navajo Mine. Low-cost plastic pipe, equipped with emitters that slowly drop water on a small area, was installed in two artificial floodways. A western wheatgrass or fourwing saltbush transplant was planted at each emitter. Plants were watered each week for a month, then watering was terminated. Precipitation from September 1973 to May 1974 totaled 2.51 inches, and from May to October

1974, only 0.76 inch. Plants were severely stressed during the hot summer. Despite the dry weather, 73% of the fourwing saltbush and 68% of the western wheatgrass plants survived. The saltbush plants, which were just 3 to 4 inches tall at planting time, measured 20 inches tall and 12 inches in diameter a year later. Western wheatgrass culms averaged 18 inches high. Survival and growth of unirrigated plants were significantly less.

In 1974, a 40-acre demonstration planting was made at the Navajo Mine, based largely on results of the 1973 tests. Twenty acres were drip-irrigated at four watering rates and another 20 acres were sprinkler-irrigated at four rates. Alkali sacaton, galleta grass, fourwing saltbush, and Indian ricegrass were seeded singly at emitter spots under the drip system, whereas they were seeded in various mixtures under sprinkler irrigation. Gypsum, bottom ash, and topsoil were laid down in strips across the graded spoils to test their effectiveness as spoil amendments. Results of these plantings were to be evaluated after two growing seasons.

In addition to these field studies, a growth-chamber study was conducted to determine the frequency and amount of water necessary for emergence and growth of fourwing saltbush and alkali sacaton under simulated field conditions. Seeds of the two species were seeded in spoil from the Navajo Mine. Water was applied through drip emitters under five different watering regimes that varied from one to five times weekly and totaled 600 to 1,500 mls. per week. Seedling emergence and growth were good under all except the 600 mls./week regime. Plant shoots and roots were harvested and weighed after 19 weeks. Both species showed excellent growth response from applying 400 mls. three times a week. Consumptive water use was significantly different by treatments and by weeks. Use was highly dependent on total weekly amount applied. Use rose steadily for the first four weeks of the test, then leveled off at rates commensurate with the amounts of water applied. Further work on determining the water requirements for other native species offers promise for maximizing efficient use of water when establishing plants by irrigation on mined areas.

WATER HARVESTING

Harvesting water to help grow plants under arid conditions is not a new idea, but the practice is seldom used in reclaiming mine spoils. Theoretically, the amount of water available for plant growth can be increased severalfold by water harvesting. Our principal concern has been the feasibility of water harvesting to improve the survival and growth of shrub transplants.

In a small field test last summer, ground paraffin and black polyethylene were used around fourwing saltbush transplants to catch rainfall. The effective water-collecting area was 4 sq. ft. Paraffin was applied at the rate of ¾ lb./sq. ft. The paraffin melts into the soil at 120°F and forms a surface coating which repels rain. Runoff from summer storms added an average of 0.75 inch more moisture to the plants than in the untreated plots over the 2+-inch summer precipitation (Aldon and Springfield 1975). Growth of the transplants reflected the increase in available moisture. The plants, only 2 in. tall at the start of the experiment, were the following heights 75 days later: untreated, 6 in.; paraffin, 9 in.; polyethylene, 14 in.

Water harvesting was also investigated on spoils at the McKinley Mine. Fifty transplants of Siberian peashrub *(Caragana arborescens)* were planted in shallow basins in August 1973. Paraffin was applied around some plants and black polyethylene around others. No precipitation fell during the first 2 weeks after planting, but all plants survived. Precipitation from November 7, 1973 to October 30, 1974 totaled only 7.46 inches, far below average. Nearly all plants survived through the second growing season regardless of treatment. Plants that received extra water, however, grew larger (Table 25.2):

TABLE 25.2

Size of Caragana aborescens in Water Harvesting Basins After Two Growing Seasons

Treatment	Height (in.)	Diameter (in.)	Size Index (Ht. x Diam.)
Polyethylene	16.0	6.2	99.2
Ground paraffin	9.7	4.5	43.6
No treatment	8.9	3.1	27.6

In addition to paraffin, a silicone spray emulsion was tested on spoil material to evaluate its potential for increasing runoff. Both strength and water repellence of the silicone crusts were improved by increasing either concentration or rate of application of the emulsion. If the crust became fractured, however, water repellence was greatly reduced. Although all the wax treatments produced a soft and easily disturbed crust, they gave better water repellence than the silicone, whether the wax was disturbed or not.

Both treatments aid plant establishment on coal mine spoil. They are now undergoing large scale pilot tests at the Navajo and McKinley mines. Areas shaped at 6:1 and 12:1 ratios of water yield to planted area have been treated with paraffin and silicone. Survival after the first growing season shows a slight benefit from shaping the area and treating with either paraffin or a silicone spray. Again more time will be needed to assess the value and longevity of the treatments under field conditions.

DIRECT SEEDING TRIALS

The requirements for seed germination and seedling establishment of fourwing saltbush are known (Springfield 1970). It was seeded directly on outslope terraces at the Navajo Mine in March of 1973. The area had a wet winter and early spring, so residual soil moisture was good. Seeds were planted when temperatures were optimum — 65°F daytime and cold nights. A problem was to hold the moisture until roots could reach greater depths. Treatments consisted of a straw mulch and white portland cement slurry mulch sprayed alongside the seeded row. The plots were covered with wire cages to protect the plants from animal damage. Germination was good, but heavy rains deposited sediment over much of the upper terrace. Survival and growth of plants on the lower terrace (15 plots), however, was good (Table 25.3):

TABLE 25.3

Fourwing Saltbush Emergence and Survival Under Different Mulch Treatments

Treatment	Seedling Emergence (% of seeds planted)	Seedling Survival (% of seedling emergence)
No mulch	38.2	71.2
Portland cement mulch	22.4	80.4
Straw mulch	39.4	82.7

Unfortunately, sheep and goats from the adjoining rangeland invaded the area and destroyed the plants.

On other spoils at the Navajo Mine, fourwing saltbush seeds were hand broadcast on a small low-lying area during snowy, wet weather in March 1973. Measurements showed good stored moisture to a depth of 4 feet. In May, more than 400 seedlings were seen. By September 1973, 205 remained alive. A year later, 58% of these plants had survived. They averaged 22.3 inches tall and 17.2 inches in diameter.

MYCORRHIZAE

Endomycorrhizal associations are found on members of many plant families. These symbiotic fungal associations with the plant's roots benefit plant growth by increasing nutrient absorption, including normally unavailable phosphorus, reducing internal plant resistance to water flow, and improving water uptake. The advantages of these attributes in semiarid environments are obvious.

A study of mycorrhizae was undertaken when it was found that transplants of fourwing saltbush grew better if some soil from beneath a growing shrub was added to the potting mix. Follow-up studies indicated plants grown on inoculated soil grew better and accumulated more phosphorus than plants grown on sterile soil (Williams, Wollum, and Aldon 1974). These laboratory findings were then tested in the field on spoils at the McKinley Mine. Transplants growing on inoculated and uninoculated (sterile) soil were planted on a steep outslope of 3-year-old spoil. Survival after 1 year was 37% on inoculated and 22% on uninoculated soil. Since many plants were buried with sediment or frost heaved, the test was repeated in 1973. After two growing seasons, plants grown on inoculated soil have survived and grown better than those on the uninoculated soil (Table 25.4):

TABLE 25.4

Performance of Fourwing Saltbush in Relation to Mycorrhizae

Treatment	Survival* %	Height (in.)	Diameter (in.)	Size Index (Ht. x Diam.)
Control	84	10.8	8.4	90.7
Mycorrhizae	95	16.4	14.1	231.2

*Excluding plots covered with sediment.

Further studies will be made on species of fungi to use, and to see whether low-cost methods can be devised to inoculate soil on a pilot basis.

Seventeen important field-grown shrubs have been examined for endomycorrhizae. Endomycorrhizal plants include fourwing saltbush, winterfat, mountain and Utah serviceberry, true mountain mahogany, apache plume, rock spirea, bitterbrush, fendlerbush, mock orange, gambel oak, snowberry, big sagebrush, and skunkbush (Williams and Aldon 1976). These shrubs would be of value in mine spoil revegetation efforts.

COOPERATIVE STUDIES

A substantial share of the research effort of this project is being conducted by outside institutions where additional skills are available to conduct studies on these complex problems. Research currently under way at the University of New Mexico, New Mexico State University, New Mexico Institute of Mining and Technology, University of Arizona, and Arizona State University includes: germination and moisture requirements of saltgrass *(Distichlis stricta)*; endomycorrhizal associations on important shrubs; prediction of surface lithology on spoil banks; composition and rate of natural revegetation of coal mine spoils; techniques for reducing copper tailings pollution; and microbial establishment in mine spoil material.

CONCLUSION

Results from two years of research on reclamation of mine spoils in the Four Corners have been encouraging. There have been a few disappointments, of course, since the environment at the two mine sites is harsh, particularly at the Navajo Mine. The establishment of a good stand of fourwing saltbush merely by broadcasting the seed during a March snowstorm was very successful as was the relatively high survival of seeded alkali sacaton and transplanted fourwing saltbush, despite only ¾ inch of rain throughout the second growing season.

Two years of research on such a complex, difficult problem, however, do not permit far-reaching, firm conclusions. Vegetation establishment will be possible but somewhat difficult and costly, due to climatic uncertainties, rather than to any toxicity of spoil material. Native plant species seem to offer the best possibility for establishment and survival, but reliable seed sources must be developed, and work is needed to determine their germination requirements. Good results have been obtained using fourwing saltbush, alkali sacaton, and western wheatgrass in revegetation efforts. It is unreasonable to expect "front lawn" plant densities on reclaimed lands, however. Natural densities of plants on areas to be mined vary widely by location, season of year, and year to year. Reliable guidelines are needed to know what plant density a particular site can support with natural precipitation.

The problem at the two mines is to establish a stand from direct seeding, or by transplants, but a greater problem is to keep the plants alive and growing. This problem is more acute at the Navajo Mine, where irrigation must be used to obtain initial establishment the first year, and plants must undergo severe stresses the next year, especially if the weather is dry.

It is possible to seed directly and get a "good" catch of fourwing saltbush if very specific guidelines are followed. In 1973, however, below-average precipitation caused heavy losses in these stands. Accordingly water harvesting techniques are being investigated so that dense stands can be maintained regardless of precipitation. These stands could provide seed that would disseminate to adjacent areas and germinate during favorable years.

The replacement of "topsoil" on mined areas is being explored by several investigators, but the value of the practice in the southwest remains to be seen. Blowing sands from surface soil can bury small seedlings or, worse, abrade or even cut off the stems at ground level. Populations of microflora on mined areas need to be re-established. Replacing surface material may accomplish this if the material is handled properly. Mycorrhizal inoculations are being investigated as a way of speeding up the return of microflora.

A second conclusion is that supplemental irrigation will be necessary for stand establishment when annual precipitation is less than 8". Still to be worked out are minimal amount and frequency of watering needed, since irrigation water, always in critical supply in the southwest, is regulated by complex water rights.

The necessity for using supplemental water has been determined through experience in establishing stands under natural precipitation on relatively dry sites. Stands can be established with natural precipitation, but successes are dependent upon conditions that may occur briefly each year, or perhaps only one year in ten or longer. For example, alkali sacaton can be planted with fair success (80% survival) on wild lands if exacting requirements are followed. These requirements are met in a single 2-week period per year (Aldon 1975). Success of plantings outside this period drops off markedly toward zero. Reclamation efforts must be almost continuous if environmental concerns and state and federal laws are to be followed. Supplemental irrigation makes this continuing effort possible by starting in the spring with cool-season germinators and proceeding through the growing season with warm-season species of both grasses and shrubs.

A final conclusion is the absolute necessity of proper management of reclaimed areas. A few head of domestic livestock at the wrong time can eliminate all reclamation efforts in a very short time and have done so. In addition, rodents must be managed and wildlife must be watched to prevent them from destroying stands before plants are self sustaining. There is little knowledge about insect or disease problems that may attack stands, but these, too, may cause problems. An advantage of using native vegetation is that it may be better able to withstand these influences.

Acknowledgment. This work was conducted in cooperation with Utah International, Inc. and Pittsburg and Midway Coal Co. We appreciate the assistance these companies have given in furnishing study areas, mechanized equipment, other facilities and manpower for conducting the field investigations.

The research reported here is a contribution to the SEAM program. SEAM, an acronym for Surface Environment and Mining, is a Forest Service program to research, develop, and apply technology that will help maintain a quality environment and other surface values while helping meet the nation's mineral requirements.

REFERENCES

Aldon, Earl F. 1975. Establishing alkali sacaton on harsh sites in the southwest. *Journal of Range Management* 28(2):129–132.

Aldon, Earl F. and H. W. Springfield. 1973. Revegetating coal mine spoils in New Mexico: a lab study. USDA Forest Service Research Note RM-245. Fort Collins, Colorado: Rocky Mountain Forest and Range Exp. Stn.

Aldon, Earl F. and H. W. Springfield. 1975. Using paraffin and polyethylene to harvest water for growing shrubs, pp. 251–257 in *Proceedings of the Water Harvesting Symposium,* Phoenix, Arizona. March 26–28, 1974. U.S.D.A., Agricultural Research Service ARS w-22.

Springfield, H. W. 1970. Germination and establishment of fourwing saltbush in the southwest. USDA Forest Serv. Res. Paper RM-55. Fort Collins, Colorado: Rocky Mountain Forest and Range Exp. Stn.

Williams, S. E. and Earl F. Aldon. 1976. Endomycorrhizal (vesicular arbuscular) associations on some arid zone shrubs. *Southwestern Naturalist* 20(4):437–444.

Williams, S. E., A. G. Wollum and Earl F. Aldon. 1974. Growth of *Atriplex Canescens* (Pursh) Nutt. improved by formation of vesicular-arbuscular mycorrhizae. Soil Science Society of America Proceedings 38(6):962–965.

26. Hydroseeding and Erosion Control Chemicals

Burgess L. Kay

A NUMBER OF PRODUCTS are advertised to control erosion or assist plant establishment. These are not miracle or all-purpose products that will always produce results on a single site or on all sites. Rather, they are specialized products which must be used correctly to be of value. Common misuses include: applying them where they are not needed; applying them at a rate too low to be of value; using the wrong type of product; or not following instructions. They are often added as "insurance" or to give the impression that the responsible agency is doing a knowledgeable and thorough job, when in fact it may be spending money foolishly.

First, the engineer must design the job for proper management of water. Flow must be diverted from adjacent surfaces via structures, and the length of the slope must be limited. All structures, temporary or otherwise, must be in place before the first rain. No chemical or fiber can cancel those needs.

In seeding, the plant species must be the correct ones for the area, plantings must be at the proper season, and the fertilizer must be the one needed. No product is an improvement over covering the seed with soil. But seed cannot always be covered — particularly on steep and rocky slopes. To avoid applying seed to the surface and letting it take its chances, there is a recognized need for fibers and chemicals, keeping in mind they are no substitute for good agronomic practices.

HYDROSEEDING AND HYDROSEEDERS

Hydroseeding or hydraulic seeding is the application of a slurry of seed and water to soil. The slurry may also contain wood fiber and fertilizer (hydromulching). Hydromulching is an excellent method of applying seed and fertilizers to steep areas, where the wood fiber holds the seed and fertilizer in position. On short or gentle slopes the fiber also provides protection against erosion until the vegetation is established.

In hydroseeding, damage to grass seed has been excessive with certain practices and machines. Particularly harmful are centrifugal pumps and bypass agitation. In contrast, gear-type pumps combined with paddle agitation appear very satisfactory. The following study illustrates the problem.

Procedure. Seed of Topar pubescent wheatgrass (*Agropyron tricophorum* (Link) Richt.) was circulated continuously in two hydroseeders for up to 120 minutes. Samples of the slurry were collected from the nozzle at 1, 10, 20, 30, 60, and 120 minutes. Four seed lots of 50 seeds each were germinated at room temperature on inclined blotters for 21 days. Tests were conducted both with and without wood fiber (Conwed fiber at 50 lbs./150 gals. water).

Machine no. 1 used a 3 x 3-inch Gorman-Rupp self-priming centrifugal pump (Model 12 D1). The slurry was agitated by bypassing it through this pump back to the main holding tank. This involved constant recycling of the slurry.

Machine no. 2 used a gear pump with rubber-covered gears (Model 2500/IB Bowie) and agitation by paddle only. Thus, the slurry went through the pump only a single time, and much more slowly than in machine no. 1.

Results. Soaking seed in water and immediately pumping it increased the number of naked caryopses from 5% to 15%. This initial loss can apparently be expected from any hydroseeder and is perhaps a small price to pay for the versatility of the seeding technique. Damage beyond that point, however, is of concern.

Seed damage with machine no. 1 was severe, although wood fiber reduced the damage (Fig. 26.1). Thus, without fibers, germination was

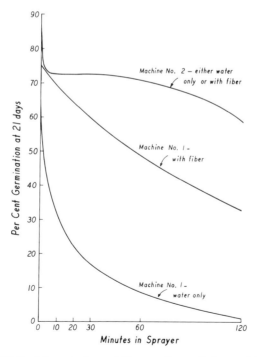

Fig. 26.1.   Effect of time and wood fiber on Topar Pubescent Wheatgrass seeds in hydroseeding machines.

reduced to 10% by 60 minutes in the pump, and to 1% by 120 minutes. With fiber, germination was 49% for 60 minutes and 33% for 120 minutes.

Damage in machine no. 2 was not significant through 60 minutes, and germination declined from 73% to 59% between 60 and 120 minutes. Added wood fiber had no apparent effect.

In hydroseeding, the following precautions should be taken:

1. Use a gear pump with paddle agitation if possible. If a centrifugal pump is used, limit the delivery time (moment from placing seed in water until tank is empty) to 20 minutes or less, and use fiber.

2. Do not put seed into water until just before the start of seeding.

3. In seeding loose soils on benches or slopes where the seed would cover naturally, or on straw-mulched areas, the seed should be broadcast by hand, or use a hydroseeder with a gear pump *without fiber*. Do not use centrifugal pumps to apply seed without fiber. If a centrifugal pump is used, apply 500 lbs. wood fiber/1,500 gals. water/acre. Fiber can be applied over seed with either pump if that is felt to be important.

The grass seed used in these tests was relatively large, but the same principles apply to other seeds. Seed damage could be particularly important with shrub or tree seeds, which are commonly larger, sometimes of irregular shape, and always expensive. Slow-release fertilizers, which depend on their particle size for their slow-release properties, may suffer similar destruction.

FIBERS

A covering of mulch is desirable to hold the seed in place. An excellent product used in hydroseeding is wood fiber applied at 1,000 to 3,000 lbs./acre with the seed and fertilizer in a water slurry; it will stick in place on near-vertical surfaces for extended periods (even over one year). It contains a green dye, which both helps monitor its application and gives it a pleasing appearance. This is sold commercially as Conwed or Silvafiber.*

Fiber's most important function is to hold the seed in place, although it also has a true mulch effect in modifying the environment, particularly holding moisture. This extra moisture often induces earlier and better germination and has resulted in better seedling survival. Increasing the rate of fiber within the range of 1,000 to 3,000 lbs./acre not only increases the number of plants but also increases erosion protection.

Three species of trees have been used recently in the manufacture of "hydromulch" — aspen, alder, and hemlock. Hemlock appears to last longer on the slope but is more difficult to apply and is apparently disappearing from the market. All seem to give comparable protection while they remain on the slope, and all remain long enough to establish plants.

Waste paper, recently substituted for virgin wood fiber, does not hold seed in place as well and gives less erosion protection.

A fiber made from corrugated paper boxes (PFM), marketed on the east coast, seems better than waste paper but inferior to virgin wood fiber.

---

*No endorsements are implied in this chapter.

Other fibers have been tried. Dairy waste fiber (DWF), washed from milking-parlor floors, shows promise but does not stick to the slope as well as virgin wood fiber and will require a chemical additive. Ground straw, ground newsprint, recycled office waste, rice hulls (whole or ground), and cubed alfalfa were not satisfactory.

ORGANIC SEEDING ADDITIVES

Organic seeding additives are generally advertised to hold fiber in place, promote germination, hold moisture, and retard erosion. Most sales literature acknowledges that fiber should be used with the product. Bio Binder, Ecology Control M Binder, Kelgum, Petroset SB, Terra Tack (Guar), Terra Tack III, and Verdyol Super have been tested.

It soon became evident that virgin wood fiber generally holds very well by itself. Under extreme wind conditions, higher rates of fiber application proved to be as good as adding a chemical. Many of the products did not provide additional erosion protection, and one of them gave poorer results than using the fiber alone. One product (Ecology Control M Binder) at 100 lbs./acre has held fiber in place, improved grass stands, and reduced erosion better than fiber alone in some of the more severe tests (sandy soil, steepest slopes, most wind and rain). A new product, Terra Tack III, shows promise at 40 lbs./acre. Both products seem compatible with commercial fertilizers. Some others are not.

Some of the other products might prove more effective if the optimum rate could be determined although manufacturers do not always know what an effective rate is. They generally recommend a rate far below the most effective level. Often they are reluctant to recommend what might be an effective rate because of the high price, knowing the customer would rather purchase a low rate, hoping that it will work, rather than pay a higher cost for an amount which may. Lacking a control area without the chemical to show that the product was not needed and may actually be of no value, the customer is often satisfied because the plants appear to grow well. Thus, everyone is happy, and the product continues to be sold. No law says it must be effective to be offered for sale.

PLASTIC EMULSIONS

Plastic emulsions are used commercially both with and without plant material. They form a crust which is useful for both erosion and dust control. Among these are polyvinyl acetate (PVA) (Aerospray 70, Crust 500, Curasol AE, Enviro, MGS, Terra Krete, and Soil Bond), or co-polymers (PVA plus acrylics), and a co-polymer of methacrylates and acrylates (Soil Seal). All retard soil erosion if used in sufficient quantity. They are most effective in intermediate stages of construction or where plant materials are not desired, as when trees or shrubs are to be transplanted. Again, they are commonly used at rates too low to be effective, used at too high dilution rates, or are used where they are not needed.

The use of plastic emulsions as seeding additives is questionable. They form a crust, an obvious deterrent to establishment of the small-seeded plants commonly used in erosion control. Tests show that they may delay germination, may reduce total germination, and may reduce overall establishment. A salt burn is common on grass seedlings growing with PVA products, particularly if fertilizer is used. Such chemicals are used to shed water from the slope, which may be fine for erosion control but which can make the environment too dry for plant establishment on steep sites with low water-holding capacity and low total precipitation.

If the plastics are used with seed, the slurry should also contain wood fiber. The plastics will not usually hold the seed or fertilizer in place without fiber. The fiber also adds to the effectiveness of most plastics in controlling erosion. The seed rate should be doubled in an effort to compensate for the undesirable effect of the plastic. In special situations the plastics may retain moisture and improve germination, particularly if used without fertilizer. However, the conditions which consistently produce this effect have not been defined.

The optimum average rate of the plastics as suggested by Ron Mearns and Tom Hoover of Cal Trans is 1,000 lbs./acre of dry matter for the polyvinyl acetates (750–1,100 lbs./acre on various soils). This value is expressed as lb./acre because the products vary in solids content, making gal./acre misleading, and the contents may not be clearly defined on the label.

The dilution rate is important with plastic emulsions. A dilution rate of five or six parts water to one part PVA is superior to more dilute mixtures. If seed, fertilizer, and fiber are to be included, they should be applied first in a separate operation and allowed to dry because the low volume of a 5:1 mix will not permit the inclusion of sufficient fiber in the hydroseeder. Fertilizer should be applied with the emulsion. Commercial machines can pump from 3% to 6% (by weight) of wood fiber (250 to 500 lbs. fiber/ 1,000 gals. water).

STRAW AND TACKIFIERS

Straw is an excellent mulch for stabilizing soil, encouraging seed germination, and speeding plant growth. However, straw must be incorporated into the soil (disked or rolled), or be held down with a net or a sprayed chemical tackifier. A commonly used chemical tackifier is asphalt emulsion. The following tests investigate the merits of products which may be substituted for asphalt. The results agree with three earlier tests (unreported), and are presented as a summary of all testing with tackifiers.

Procedure. Barley straw was broadcast at 2,000 lbs./acre on the surface of greenhouse flats filled with decomposed granite. The chemicals listed in Table 26.1 were sprayed over the straw. After curing, the flats were inclined on a 2:1 slope and subjected to winds of up to 84 m.p.h. created by a Finn Bantam straw blower. Velocities to 35 m.p.h. were measured with an anemometer, and higher velocities were measured with a pitot tube. Table 26.1 also shows the velocity at which 50% of the straw blew away.

Results. Asphalt emulsion is commonly used in the eastern U.S. at rates of 200 to 500 gals. per acre. The Asphalt Institute recommends 484 gals./acre.

Asphalt is seldom used in California because of the hazard of accidentally getting this black sticky substance on nearby surfaces. In this and other tests, it was observed to soften in hot weather, allowing the straw to slip from the flats. It may have to be heated, as in this test, to permit spraying. The black color probably helps increase soil temperatures and may encourage growth in cool weather. Asphalt emulsion was applied under the product specifications of SS-1 and was an excellent tackifier at 400 and 600 gals./ acre. Six hundred gallons were superior to 400 gallons. Two hundred gallons were not satisfactory.

Terra Tack I is a free-flowing powder produced from the ground endosperm of a natural vegetable gum, guar *(Cyamopsis tetragonoloba)*, and contains gelling and hardening agents. It is applied at 40 lbs./acre in 600 gals./acre or more of water (1,600 gals./acre in this study), and 250 lbs./ acre of wood fiber. When mixed with water and properly cured, it forms an insoluble network. It is colorless and unaffected by heat, and can be removed easily from spills or oversprays. The higher rate tested gave results similar to those with asphalt at 400 gals./acre.

Terra Tack II is a free-flowing powder produced from semirefined seaweed extracts. It is sold as two parts, the alginase and a gelling agent which are mixed with water (750 gals./acre) and fiber (150 lbs./acre). When properly mixed, it polymerizes, and upon application forms an insoluble network of binding membranes which is also nonstaining and easy to clean up. Proper mixing, however, is essential. It must be applied in stringers or lines rather than uniformly since it is too viscous to give complete coverage at the low rate tested. The resulting network will give satisfactory

TABLE 26.1

Effect of Tackifier Products on Wind Stability of Barley Straw Broadcast at 2,000 lbs./acre

Product	Chemical Rate/acre	Fiber (lb./acre)	Wind speed (m.p.h.) at which 50% of straw was blown away				
			Trial 1	Trial 2	Trial 3	Trial 4	Mean
None	—	—	8–10	8–10	8–10	8–10	9
SS-1 Asphalt	200 gal.	—	45	35	40	40	40
	400 gal.	—	84+	75	84+	79	80
	600 gal.	—	84+	84+	84+	84+	84+
Terra Tack I	40 lbs.	250	67	75	63	72	69
	89 lbs.	250	84	81	84+	80	82
Terra Tack II	45 lbs.	150	63	81	58	69	68
	90 lbs.	300	84+	84+	84+	84+	84+
Aerospray 70	50 gal.	—	20	15	20	13	17
	100 gal.	—	30	20	15	25	22
	50 gal.	250	48	45	35	50	44
	100 gal.	250	84+	45	35	50	54
Curasol AH	45 gal.	250	30	40	40	40	38
	90 gal.	250	66	63	64	63	64
	180 gal.	250	84+	84+	75	63	76
Soil Seal	100 gal.	250	84+	72	63	84	76

results under less severe conditions. A higher rate should be used for steeper slopes, heavy traffic, or areas of high wind. A higher rate (1,500 gals./acre water) gave results comparable to those with 600 gals./acre asphalt. At this rate Terra Tack II was superior to all of the nonasphalt treatments.

Aerospray 70 and Curasol AH are white liquid glues (polyvinylacetates). They may be corrosive to equipment if not carefully flushed off. They have considerable binding effect on soil and are excellent for erosion control if used at high enough amounts and at the proper dilution rates. As successful straw tackifiers they must be used with wood fiber, as can be noted from data under Aerospray in Table 26.1. The low rate of 50 gals./acre with fiber was superior to 100 gals./acre without fiber. None of the rates tested gave results equal to those with 400 gallons of asphalt. Under excessively high rainfall, however, these products might prove superior if used at higher rates than tested here. Aerospray was applied as 10 parts water to 1 part product, and Curasol as 11 parts water to 1 of product.

Soil Seal, a liquid plastic (co-polymer of methacrylates and acrylates) is not normally advertised as a straw tackifier. However, it was tested at 10 parts water and 1 part concentrate and proved to be equal to a much higher volume of Curasol.

LEGUME INOCULATION IN HYDROSEEDING

Legumes are important in seeding infertile sites because they can supply their own nitrogen, surviving where other plants might not. Legumes that are common on California road cuts after several years of "natural revegetation" are bur clover, vetch, annual clovers, and lupines.

Legumes receive their nitrogen from root-nodule bacteria which remove it from the soil air, converting it to a form usable by plants. Thus, the bacteria are as important as the plant itself.

Many soils either lack root-nodule bacteria or contain ones that do not fix nitrogen. The use of inoculated seed introduces efficient nitrogen-fixing bacteria of the proper strains. Good legume stands require that the inoculum used be especially prepared for the species or variety of legume planted. Since bacteria native to the soil will be highly competitive with bacteria introduced on inoculated seed, the latter will need protection until the seed germinates. By pellet inoculation, high numbers of live bacteria are concentrated on the seed, and the pellet helps protect them until germination.

Each seed pellet contains a legume seed, the inoculant, and an adhesive and a coating material that influences survival of the bacteria. The operator can prepare the pellets in a cement mixer or on a concrete floor, or he can order inoculated seed from a dealer. (Details of pellet inoculation are given in University of California Agriculture Experiment Station Bulletin 842, "Range Legume Inoculation and Nitrogen Fixation by Root-Nodule Bacteria," and Agricultural Extension Service Publication AXT 280, "Pellet Inoculation of Legume Seed.")

Since there was some question whether the bacteria were washed from the seed in the hydromulching process, the following experiment was designed to measure the effectiveness of inoculation under field conditions.

Procedure. The experiment was conducted on the north coast of California at the Hopland Field Station of the University of California in cooperation with Dr. M. B. Jones. The site had previously been determined to be a problem inoculation area. Seeds of both the "Mt. Barker" and "Woogenellup" cultivars of subclover (*Trifolium subterraneum* L.) were pellet-inoculated with the appropriate *Rhizobium* bacteria at 4 times the supplier's recommended rate, with the UC formula used in pelleting. The pellets were allowed to cure overnight. They were then circulated through a hydroseeder for up to 120 minutes. This machine has a self-priming centrifugal pump (Gorman-Rupp Model 13 DI, Size 3 x 3 in.) and is agitated by passing the slurry through this pump back to the main holding tank. Thus, the slurry was constantly being recycled through the pump. The slurry contained 150 gallons water, 50 lbs. "Conwed" wood fiber, 35 lbs. of treble superphosphate fortified with elemental sulfur (0-35-0-20), and 2 lbs. of clover seed. Samples of the slurry were sprayed on a prepared seedbed after 1, 10, 20, 30, 60, and 120 minutes in the hydroseeder. Application rates per acre were 875 lbs. of wood fiber, 612 lbs. of 0-35-0-20, and 35 lbs. of seed. These were compared with the same batch of pellet-inoculated seeds broadcast dry at 20 lbs./acre and with uninoculated seed also broadcast dry at 20 lbs./acre. The latter two treatments were fertilized with 0-35-0-20 at 500 lbs./acre. Each treatment was replicated four times. Planting date was October 1, 1971. The first effective rain was in mid-November.

Results. On March 2, 1972, the treatments were evaluated. Healthy plants were determined by their vigorous growth, size, and dark green color. Plants inoculated ineffectively or not at all were very small and sickly yellow by comparison. Stands were rated on a 1-to-10 basis, from no plants inoculated to all plants effectively inoculated. Mean data appear in Table 26.2.

Mt. Barker subclover remained effectively inoculated through 30 minutes, and Woogenellup subclover through 60 minutes. It is not known whether the decline beyond those times was due to a washing of the bacteria from the seed or to seed damage by the centrifugal pump. Other studies with this machine indicate excessive damage to seed of pubescent wheatgrass

TABLE 26.2
Effect of Hydroseeder on Subclover Inoculation

Treatment	Variety of Subclover	
	Mt. Barker	Woogenellup
1 minute	10a *	9a
10 minutes	9a	9a
20 minutes	8a	8a
30 minutes	8a	8a
60 minutes	4b	8a
120 minutes	1c	2b
Dry pellet	8a	8a
No inoculation	1c	1b

*10 is excellent inoculation, 1 is none. Values followed by the same letter are not significantly different at the 0.01 level.

(*Agropyron tricophorum* (Link) Richt.). Very few plants resulted from the 120-minute treatment (either inoculated or uninoculated), indicating that seed damage may have been a factor. A gear-type pump would probably result in less seed damage, and would be less likely to wash the inoculum from the seed.

Conclusions. Pellet-inoculated subclover seed will remain effectively inoculated in a hydromulching slurry if delivery time is limited to 30 minutes or less with a centrifugal pump, and probably longer with a rubber-covered gear pump.

PRESOAKING SEEDS AND USE OF GROWTH REGULATORS

Enhanced germination from soaking seeds before planting has been investigated by many scientists. The technique is limited in application, however, because of the difficulty of drilling wet seeds and providing a seedbed environment conducive to continued germination or growth of pre-soaked seeds. Germination is enchanced in only a few species when the seeds are soaked and dried again before planting.

Presoaking would seem a natural part of hydroseeding, which is done with wet seeds. The large volume of water and fiber applied in hydroseeding ensures a seedbed environment conducive to germination, provided that temperatures are not limiting. Hydroseeding contractors feel there is a need to increase the percentage of germination and speed of germination. The contractor would like to see the seeded species germinate before turning responsibility for care of the seedlings over to maintenance departments. Many supposedly bad seeding jobs were probably the result of poor water management during the germination period after the contractor had gone.

Experiments were conducted in cooperation with Dr. James A. Young and Dr. R. A. Evans of the Agricultural Research Service in Reno, Nevada. Presoaking time depends on the phenology of germination of the species being soaked and the temperature of the soaking liquid. If the radicle emerges before hydroseeding, it may be injured when the seeding mixture passes through the pumping system. The hydroseeding mixture is applied with considerable force so that large cut or fill banks can be reached from the roadway.

Field-tests with presoaked seeds for hydroseeding resulted in difficulties with fermentation of the soaking liquid and seeds. High-temperature soaking of short duration with aeration prevents fermentation.

The addition of plant growth regulators to the soaking liquid is a natural extension of the methodology. Plant growth regulators offer the possibility of breaking dormancy, speeding germination, and increasing the size and vigor of seedlings.

Growth regulators that have been evaluated for use in presoaking for hydroseeding are: gibberellin (as the potassium salt of GA_3); kinetin (6-furfurylaminopurine); ethephon ([2-chloroethyl] phosphonic acid); hydrogen peroxide (H_2O_2); and potassium nitrate (KNO_3). The inclusion of KNO_3 and H_2O_2, which are not generally considered growth regulators, may be questioned, but they were used in the same manner as the other materials.

Many different plant species (25 or more) are commonly hydroseeded, and mixtures of five or more species are often seeded together. These mixtures vary from hard-seeded legumes to light-requiring Kentucky bluegrass *(Poa pratensis)*. Obviously, with such diverse species, a variety of soaking times, concentrations, and types of growth regulators may be needed. Possible combinations for several species are still being evaluated.

The most detailed investigations have been with common Bermuda grass *(Cynodon dactylon)*. Seeds of this species germinate at relatively high temperatures (35° to 40°C), and the germination rate is greatly increased by presoaking for short periods. GA_3 solutions at very low rates increase seedling size. The addition of KNO_3 plus GA_3 and kinetin possibly increases total germination and seedling size.

RELATIVE RESULTS AND COSTS

The data in Table 26.3 are taken from a recent study on decomposed granite and are representative of results from a number of other studies. The data are for 1:1 slopes following 11 inches of rain.

TABLE 26.3
Effect of Erosion Control Treatments on Plant Establishment, Soil Loss, and Cost of Treatment

Treatment	lbs./acre	No. seedlings per sq. ft.	Soil lost 100 lbs./acre	Cost $100/acre
Check	—	0	249	—
Fiber	1,000	12	26	4
	2,000	18	12	6
	3,000	14	8	6
Ecology Control Fiber	100 1,000	41	10	8
Straw Terra Tack II	2,000 90	155	5	7–9
PVA Fiber	1,000 500	?	2	9–11

The cost of treatment increases with the effectiveness of both plant establishment and erosion control. The costs in Table 26.3 were estimated by a commercial firm in June 1974 on a base of 10 acres and minimum travel, and include seed and fertilizer.

27. Vegetation for Acidic and Alkaline Tailing

K. C. Dean and M. B. Shirts

THE MINERAL INDUSTRY OF THE UNITED STATES discards 1.7 billion tons of solid wastes annually, and the total accumulation is near 25 billion tons covering over 2 million acres of land. Mineral wastes are second only to agricultural wastes in quantity, and represent nearly 40 percent of the total solid wastes produced in the United States. Almost 40 percent of the mineral discard is fine-size material that requires some sort of stabilization to prevent air and water pollution.

On active mill tailing ponds, air pollution is controlled by keeping the surfaces of the ponds wet either by tailings discharge or by sprinkling. On inactive ponds, more lasting stabilization is required; physical, chemical, vegetative, and combination methods for stabilizing tailings have been developed and are in use.

The Federal Bureau of Mines has been conducting research on the utilization and stabilization of mineral wastes for many years (Dean et al. 1969, 1974; Havens and Dean 1969). In addition, many mining and milling companies and academic groups have worked independently, and during the past few years, cooperatively with the Bureau in evaluating various stabilization methods in the field. A free exchange of information has taken place. Through these individual and cooperative studies methods have been developed that can achieve effective stabilization on most existent milling wastes at costs ranging from $40 to $1,750 per acre of waste accumulation (Dean et al. 1974). Perhaps the greatest problem that still requires resolution is to achieve preferred vegetative stabilization of tailings that are or may become excessively basic, saline, or acidic. This report focuses on research directed to overcoming this remaining major problem.

REVIEW OF STABILIZATION PROCEDURES

The principal methods of stabilizing milling wastes include:

1. Physical — covering tailings with soil or other restraining materials. This includes the use of temporary expedients such as water sprinkling; organic materials with a moderate effective life such as bark or straw; and materials that have a long-term effectiveness such as crushed or granulated smelter slag or country rock and soil.

    Physical methods have been used extensively on many different types of wastes and in differing environments. Water sprinkling is, perhaps, the most used dust control method. Bark was used as a covering and straw was harrowed into the tailings surface at Anaconda, Montana. Granulated slag proved effective as a tailings cover in Arizona, Montana, and Nevada, and soil has been used in many states.

2. Chemical — the use of a material to interact with fine-size minerals to form a crust. Over 70 different materials have been tested in the laboratory by the Bureau of Mines and the more promising have been field tested. These include: lime, pyrite, sodium silicate with or without $FeSO_2$ and $CaCl_2$ additives, various lignosulfonates, redwood bark extracts, amines, acetate salts of amines, dicalcium silicate, bituminous base products, resinous adhesives, and elastomeric polymers. Reagents of the resinous adhesive, lignosulfonate and elastomeric polymer groups appeared to be most effective. Chemical stabilization has not proved as durable as soil covering or vegetation on typical tailings. Nevertheless, chemicals can be used on sites unsuited to the growth of vegetation because of harsh climatic conditions or the presence of toxic substances in the tailings, or in areas that lack access to a soil-covering material. Chemical stabilization is also applicable for erosion control on active tailings ponds. Chemicals can be effectively used on portions of these ponds to restrict air pollution while other portions remain active.

   Complications arise, however, in achieving satisfactory chemical stabilization in that the surfaces of tailings piles seldom are homogeneous. Sections of slimes frequently alternate with sections of sands. The permeability, reactivity, pH, acidity, basicity, and salt content of the surfaces vary considerably. Selected chemicals, however, have been used successfully in diverse environments in Arizona, Colorado, Michigan, and Washington.

3. Vegetative — the growth of plants in the tailings. Consideration of aesthetics and renewability generally favor use of vegetation for stabilization, but the successful initiation and perpetuation of vegetation on fine wastes involve ameliorating a number of adverse factors. Mill wastes usually (1) are deficient in plant nutrients; (2) contain excessive salts and heavy metal phytotoxicants; (3) consist of unconsolidated sands that, when windblown, destroy young plants by sandblasting and/or burial; and (4) lack normal microbial.populations.

Research indicates that, other than the excessive acidity, basicity, or salinity, perhaps the greatest problem to be overcome in establishing vegetation is that of windblown sands. Several procedures are in use for preventing windblown sands from covering or cutting off the growing plants. These include combination methods such as water sprinkling while the plants are growing, covering the tailings with soil or country rock, hydroseeding, using excelsior-filled matting as a cover directly over the tailings, and a combination chemical-vegetative procedure developed by the Bureau of Mines. Sprinkling, soil covering, hydroseeding, and matting have proved useful on various types of wastes, and the chemical-vegetative procedure has proved effective during the past four years on tailings ponds in Arizona, Colorado, Michigan, Missouri, Nevada, and Washington.

COMPARATIVE COSTS

Mining companies have tested many of the outlined procedures for stabilizing and reclaiming mineral wastes. Stabilization costs using various procedures are shown in Table 27.1. In general, costs for reclaiming sloping dike areas have been about 25 percent greater than costs for flat pond areas. The costs given in Table 27.1, given in 1973 dollars, are estimated for a tailings accumulation consisting of 80 and 20 percent, respectively, of pond and dike areas. These costs, although broadly generalized, provide some comparison of different methods.

These data indicate that several methods are available for stabilizing normal mill tailings at costs of less than $400 per acre. However, only continued water sprinkling, soil coverings, soil covering plus vegetation, or other methods designed to neutralize or offset the alkalinity, salinity, or acidity can be effective on tailings containing excessive alkalies, salts, or acids. G. W. Morgan studied the growth potential of different grass and root crops when planted on growth adverse bauxite mined land using soil-cover depths ranging from 6 to 24 inches (Morgan 1973). His conclusion

TABLE 27.1
Cost Comparison of Stabilization Methods*

Type of stabilization	Effectiveness	Maintenance	Approximate cost per acre, dollars
Physical:			
Water sprinkling	Fair	Continual	—
Slag (9-inch depth):			
By pumping	Good	Moderate	350–450
By trucking	Good	Moderate	950–1050
Straw harrowing	Fair	Moderate	40–75
Bark covering	Good	Moderate	900–1000
Country gravel and soil:			
4-inch depth	Excellent	Minimal	250–600
12-inch depth	Excellent	Minimal	700–1700
Chemical:			
Elastomeric polymer	Good	Moderate	300–750[†]
Lignosulfonate	Good	Moderate	250–600[†]
Vegetative:			
4-inch soil cover and vegetation[‡]	Excellent	Minimal	300–650
12-inch cover and vegetation[§]	Excellent	Minimal	750–1750
Hydroseeding	Excellent	Minimal	200–450
Matting[‖]	Excellent	Minimal	600–750[†]
Chemical-vegetative	Excellent	Minimal	120–270[†]

*Based on average tailings. Costs could be revised upwards for acidic tailings requiring limestone or other neutralizing additives.

†Bureau-industry derived costs based upon cooperative stabilization efforts. The remaining data were obtained from industry.

‡Generally used on pond area rather than on dikes. Also, not as effective as 12-inch soil cover when tailings are excessively acidic or saline.

§Substantiated as the optimum economic depth of soil cover when reclaiming bauxite mined lands with soil covers ranging from 6 to 24 inches (Morgan 1973), although a lesser soil cover may be satisfactory on other types of waste materials.

‖Based on placing 3-foot-wide matting at 3-foot intervals over the seeded areas.

was that a 12-inch cover was an economically preferable depth when growing grasses. Based upon his research and the assumption that a 12-inch depth of soil cover would also be sufficient for maintaining adequate vegetative growth on excessively adverse acidic, basic, or saline tailings, the cost of a 12-inch soil depth was selected as a criterion for evaluating alternative procedures. The cost of a 12-inch cover and vegetation as given in Table 27.1 is estimated to range between $750 and $1,750 per acre. Thus, research on alternative stabilizing procedures for excessively adverse tailings would have as a goal a cost per acre of less than $750.

It was assumed for purposes of this report that direct vegetative stabilization on the tailings, rather than use of a soil cover would be preferable. Vegetation improves the esthetics of an area and does not hinder the potential retreatment of tailings as greatly as covering the tailings with foreign materials.

VEGETATION FOR ADVERSE TAILINGS

Practically all mill tailings contain deleterious inorganic salts, lack organic components and essential nutrients, and do not have the physical nature required for sustaining vegetative growth. For many tailings these adverse conditions can be overcome in time by fertilization, gradual buildup of organic and microbial populations by encouraging plant growth, and use of chemicals for binding the surface to prevent blowing of loose sands that cut off or bury established vegetation.

Many tailings also are discharged at pH's above or below optimum for general plant growth (> 8 and < 4) or may contain excessive quantities of salts which complicate initiation or perpetuation of plant growth. Additionally, combination problems can exist in which tailings may be excessively saline, because of the recycling of processing waters, and simultaneously contain sulfide materials such as pyrite that upon oxidation would drop the pH of the tailings to a low level in a relatively short time. Such a tailing is Kennecott's Utah Copper Division milling waste. This material, with a pH of 7.8 when fresh, contains salinity equivalent to 2.4 atmospheres osmotic concentration plus approximately 1.3 percent pyrite. The salts in the tailings cause an osmotic gradient that transfers fluid from the plants and thus dehydrates vegetation shortly after sprouting. Another problem is that if vegetation is planted on these tailings and irrigation of the plants leaches away the salts, the pyrite oxidizes and the pH may drop from 7.8 to less than 3.0 within one month's time. Representative tailings of each of these types have been used in growth tests in the laboratory or in the field.

Background Methods. Several vegetative methods are available for overcoming moderate to severe pH or salinity problems of tailings, including variations in neutralization, seedbed preparation, planting, and watering practices.

A minimum of work was done by the Bureau of Mines on percolation leaching because an apparently satisfactory means of leaching highly acidic tailings has been reported (James 1966). The work by James showed flooding is not an effective means of leaching acidity from tailings because (1) prolonged flooding compacts the materials and produces conditions unfavorable for plant growth, and (2) when flooding is halted, evaporation brings the acid to the surface again. James' investigations indicated that the downward movement of acidity can be encouraged by an extremely fine spray of water which forms a mist over the surface and retards evaporation. If the spray of water applied does not exceed the rate at which it can penetrate the tailings, high acidity can be moved to a sufficient depth to permit vegetative growth within a period of 3 to 4 weeks. If the acid can be moved downward to contact a slime layer with which it reacts, subsequent evaporative movement will not return the acid to the surface. These projections have recently been substantiated by research on the reclamation of saline soils using drip irrigation. Practitioners of drip irrigation indicate that spot irrigation with small amounts of water sufficient to drive the majority of the salts beyond the pickup zone of the plant roots speeds vegetative growth. Agronomic researchers have conducted seedbed preparation and planting research designed to offset salinity by the use of widely spaced deep furrows. Good vegetative growth of grasses and legumes in harsh saline environments has been achieved by planting the seeds in 4- to 6-inch deep furrows spaced at intervals from 12 to 42 inches apart. The wide spacing between furrows reduces plant competition, and the deepness of the furrow permits concentration of the water and directs the salts to the ridges between furrows. Corroborative work by the Bureau of Mines on long mounds of saline tailings, piled 12 to 18 inches high, oriented in an east-west direction, and separated by furrows showed that the greater solar radiation on the south slopes affected salinity (Dean et al. 1974). The salt content on the south slope of the mounds was 30 percent higher than on the north slopes while the plant growth on the north slopes was 17 times greater than on the south slopes.

Considerable neutralization work also has been accomplished by adding: (1) sulfuric acid to irrigation waters for improving saline and alkaline soils; (2) lime and limestone to acidic materials; and (3) various forms of sulfur to saline soils and tailing materials.

Continuing Stabilization Research. The basic objective of current research is to ultilize the background methods already available, supplemented by increasing the organic content and microbial populations of the tailings to overcome the more difficult problems that remain.

As previously noted, fine mineral wastes lack nutrients, humus, and microbial populations, all of which militate against the growth of vegetation. They may also lack the proper surface structure for air and water access and resistance to being blown or washed away. Fertile soil in which vegetation grows most abundantly, unlike mill tailings, usually contains a small but important percentage of organic matter and micro-organisms. Plants can be grown well in pure sand cultures to which have been added a proper balance of all the necessary mineral compounds needed by the plant, but if a little appropriate organic matter, such as leaf mold, is added, the plants grow much better. To prepare synthetic soils from barren tailings, research was conducted using additions of chemicals, sewage sludge, and compost.

Buried Organic Layers. Two types of municipal wastes were tested as additives to tailings. One was a commercial compost made from municipal refuse and the other was sewage sludge. Both are bulky organic materials which, when mixed with tails, yield soil-like textures and improved mineral, air, and water relationships. Plots prepared with Kennecott Copper tailings and 5 to 15 tons per acre of each of the two waste-derived products indicated the additives to be beneficial to both vegetative germination and growth. The sewage sludge produced better growth than the commercial compost, and plots treated with either material demonstrated much better growth when additional fertilizer was added. Subsequently, tests were made on the use of tailings pelletized with sewage sludge to form a surface layer and on the effect of buried layers of sewage sludge on the pH and salinity of the tailings being tested.

A series of tests was made in which 2-inch layers of sewage sludge were placed at different depths in barrels containing an 18-inch depth of Kennecott tailings. Sludge was placed at depths of 3, 7, 11, and 15 inches below the surface of the tailings. In another test series, the layering pattern was unchanged but an additional equivalent of 15 tons per acre of sewage sludge was mixed into the top 3 inches of tailings. Crested wheatgrass, Ranger alfalfa, and rye grain seeds were planted for both series of tests with barley used as the fourth seed in one test series, and a small transplanted tomato was used as the fourth plant variety in the other series of tests. Table 27.2 shows the number of plants germinating per 100 seeds 2 weeks after planting and the number surviving at the end of 10 wccks.

TABLE 27.2
Germination and Survival of Four Species of Plants With 2-Inch Sludge Layers at Various Depths

	Number of Plants*															
	Depth of Sludge Layer Below Surface (in.)															
	Series 1†								Series 1A‡							
	3		7		11		15		3		7		11		15	
Plants	G§	S‖	G	S	G	S	G	S	G	S	G	S	G	S	G	S
Crested wheat-grass	71	4	88	37	73	42	53	28	90	26	80	24	78	15	77	56
Ranger alfalfa	70	8	78	4	81	9	59	6	36	5	41	4	51	9	67	19
Rye grain	83	23	92	39	89	43	83	28	74	64	71	56	78	53	80	60
Barley grain	89	0	97	21	94	14	95	2	Tomato plants 2-inches high planted in this quadrant							

*Germination rate at 2 weeks after planting and survival of plants after 10-week growth.
†Series 1, no sludge admixed into top 3 inches of tailings.
‡Series 1A, 15 tons per acre sewage sludge in top 3 inches of tailings.
§Germination.
‖Survival.

The great variability in germination for the various seeds is not easily explained. Differences between series 1 and 1A, however, can be readily compared by assessing the combined results of germination and survival for the three comparable plants in each series. This comparison shows that in Series 1, of a total 1,200 wheatgrass, alfalfa, and rye seeds planted, 920 seeds germinated and 271 survived after 10 weeks to furnish a 77-percent overall germination and 23 percent survival of the total seeds planted. The results in series 1A show that with sludge present in the top 3 inches of tailings the germination rate was 69 percent, but that the survival rate was 32 percent of the total seeds planted. Addition of sludge to the top 3 inches of tailings thus appeared to be beneficial to overall plant survival.

For evaluation of plant-species response and the effect of sludge-layer depths on germination and survival, the overall combined results from both test series were compared. These comparisons show that the combined germination and survival percentages for wheatgrass, alfalfa, and rye grain were 76-29, 61-7, and 81-46, respectively. The grain and wheatgrass showed major advantages over the legumes. The overall germination and survival of plants with sewage sludge layers at depths of 3, 7, 11, and 15 inches were 71-22, 75-27, 76-27, and 70-33, respectively. These data indicate that the test with the sludge at a 3-inch depth was the poorest of the group. However, the vegetation appeared healthier in plots with the shallow sludge layers. Once the roots penetrated the sludge layer, the plants took on a much healthier appearance and were much hardier than plants with roots only in the tailings.

Comparison of series 1 and 1A plots was made at 10 weeks, after which the series 1 plots were disassembled for other tests. The 1A plots were permitted to grow for 10 months to determine the effect of longevity on the plants and especially the deeprooted tomato plant. The plots were systematically disassembled at 10 months to evaluate the plant and root systems of all the tomatoes. Results of this examination are presented in Table 27.3.

Inspection of the root systems showed that practically all roots for the sludge at depths of 7, 11, and 15 inches had grown and remained in the upper 3 inches of tailings containing admixed sludge. The main root for the 11-inch buried sludge layer had grown only 3.25 inches downward and

TABLE 27.3
Plant and Root Growth of Tomatoes

Sludge position (in.)	Weight of plant and recoverable roots (grams)	Length of plant and main roots (in.)		
		Plant	Root	Total
3	411*	89†	19.5	108.5†
7	193	64.5	54.5	119
11	97	67.5	3.25‡	70.25
15	69	50	32	82

*Includes 70 grams pruned from plant while growing so as to keep plant in bounds of plot.
†Does not include length of plant pruned during growing period.
‡The main root system grew only in the surface layer of sewage sludge and tailings.

TABLE 27.4

Percent Germination and Percent Survival of Plants With Buried Sludge Layers and Pelletized Cover

Binder and depth of sludge layer (in.)	Crested wheatgrass		Ranger alfalfa		Rye grain		Alsike clover		Tomato	
	Germ.*	Surv.†	Germ.	Surv.	Germ.	Surv.	Germ.	Surv.	Germ.	Surv.
Coherex										
4½	79	75	57	49	54	43	29	10	—	1
16½	82	70	67	42	72	48	50	7	—	1
Paracol										
4½	72	66	70	55	63	43	54	17	—	1
16½	79	69	58	38	64	49	49	11	—	1

*Germination rate at 2 weeks after planting.
†Survival of plants after a 10-week growth.

had then sent out lateral feeder roots. The main roots in the plots with sludge layers at depths of 7 and 15 inches were much longer but grew in a circular manner, again only in the presence of the organic admixture. Conversely, the main root system for the plot with sludge at 3 inches grew down into and through the sludge layer to within 2 inches of the bottom of the tailings in the barrel. Data in Table 27.3 clearly show the much greater plant growth attained with the sludge layer at shallow levels.

Pelletization of Tailings. Earlier tests indicated that better root environment and soil aeration were obtained by pelletizing the tailings. Pelletized material also appears well suited for use on tailing pond slopes, because the textured surface provides natural sites for plant growth. Therefore, a test series was made using a 1.5-inch layer of pellets on top of tailings containing sludge layers buried 3 and 15 inches under the normal tailings surface.

The pellets were minus-⅜-plus-⅛ inch size and contained 92.4% tailings, 6.6% sewage sludge, and 1% chemical binder, either Coherex or Paracol TC 1842. Pellets were prepared in a rotary drum pelletizer and demonstrated good structural strength. The pellets contained the equivalent of 10.8 tons per acre of sludge in the top 1½-inch layer, somewhat less than the maximum 15 tons per acre used in the top 3 inches of tailings in previous tests. Approximately 1¼ inches of pellets were distributed over the surface of the plots, the seeds were planted, and the final ¼ inch of pellets were used as a cover for the seeds. Seeds used on each plot included crested wheatgrass, Ranger alfalfa, rye grain, and alsike clover. A 2-inch tomato plant was also transplanted into each plot. Two plots had sludge layers 4½ inches below the top of the pellet layer. The other two plots were made up similarly but with the sludge layers at 16½ inches beneath the surface. Table 27.4 shows the number of plants germinating per 100 seeds 2 weeks after planting and the number surviving at the end of 10 weeks for each plot.

As in previous tests the germination of seeds varied broadly and comparison of results was best made by comparing survival of plants at the conclusion of 10 weeks. This comparison showed that the overall germination for all three seeds in the pelletized tailing test was 68 percent with

TABLE 27.5

Comparison by Percent of Overall Germination and Survival of Plants on Buried Sludge Layer Plots

Plant	No sludge added		Sludge mixed in top 3 inches of tailing		Sludge mixed into pellets	
	Germi-nation*	Survi-val*	Germi-nation	Survi-val	Germi-nation	Survi-val
Crested wheatgrass	71	28	81	30	78	70
Ranger alfalfa	72	7	51	7	63	46
Rye grain	87	33	76	58	63	46
Overall	77	23	69	32	68	54

*Germination rate at 2 weeks after planting and the survival of plants after a 10-week growth.

a survival rate of 54 percent. Germination and survival rates for the specific seeds, in percent, were respectively, wheatgrass 78 and 70, alfalfa 63 and 46, and rye 63 and 46. At the 10-week time interval, few differences existed in the survival of plants in the 4½- and 16½-inch deep buried sludge layer plots or between the two chemicals used to make the pellets. However, visual inspection of the surviving plants showed markedly better plant growth for the shallower layer and for Coherex binder.

All plots reported on in Tables 27.2 through 27.4 were maintained for at least a 10-week interval, the results were compared and are summarized in Table 27.5.

The addition of sludge to the tailing, particularly in pelletized form, increased the survival rate of plants. Despite daily waterings for the first month and twice weekly watering thereafter, the bulk of the pellets were still coherent at the end of the 10-month period. The surface pellets had disintegrated to some extent, but those immediately under the surface retained their cohesive pelleted forms for the entire period.

Combination of Pelletization and Buried Organic Layers. A series of tests employing a combination of buried organic layers together with pelleted surface materials were established to determine the effects upon vegetative growth and pH values of the component materials. Four plots were established using Kennecott tailing that had a pH of 7.8 when first obtained but which had decreased to a pH of 6.6 after several months' storage. Two replicate plots were established with a 2-inch sewage sludge layer placed under 15 inches of tailing capped by a 1½-inch layer composed of 92.4% tailing, 6.6% sludge, and 1.0% chemical binder. The second two replicate plots were the same, except that the 2-inch sewage sludge layer was placed under 3 inches of tailing and capped by pellets.

Each plot was planted with a tomato and crested wheatgrass, alfalfa, and rye grain seeds. The plants were allowed to grow for 13 months, by which time most of the wheatgrass, alfalfa, and rye had matured and died and then the plots containing living tomato plants were dismantled and examined. Examination showed the following: (1) The root systems of all plants were principally located in the pelletized portion of the tailing, and (2) placing the sewage sludge layer at 3 inches prevented to a great extent

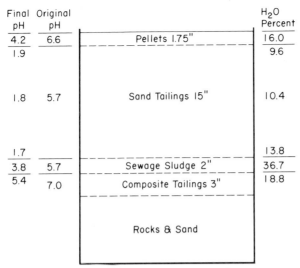

Fig. 27.1. Effect of a Buried Sewage Sludge Layer 16¾-inch Depth on the Oxidation of Tailing Sulfides.

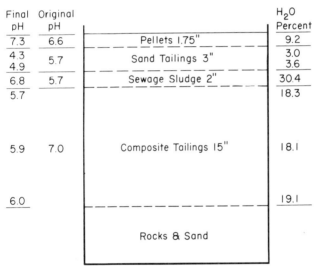

Fig. 27.2. Effect of a Buried Sewage Sludge Layer 4¾-inch Depth on the Oxidation of Tailing Sulfides.

the oxidation of pyrite in the tailing as witnessed by an average tailing pH range of 3.9 to 6.1 and a sludge pH of 6.8 for the 3-inch-deep layer as compared with a tailing pH range of 1.7 to 2.0 and a sludge pH range of 3.6 to 4.0 for the sludge layer at 15-inch depth. A comparison of Figs. 27.1 and 27.2 shows the difference in pH change and moisture content within the tailing materials. Proper placement of the sewage sludge layer to control water and gas movement appears exceptionally promising for controlling acidification of pyrite-bearing tailings.

Costs of Using Sludge as Pellets. As previously noted, several tested physical, chemical, vegetative, and combined vegetative methods have been evaluated as to stabilization effectiveness, required maintenance, and cost. The costs for the preferred vegetative methods have ranged from $100 to $1,750 per acre. Based upon the success of laboratory testing, costs were also estimated for the buried sludge layer, pelletization, and a combination method by the Bureau of Mines. Cost evaluations were made utilizing laboratory data and after consultation with firms having land-moving, transportation, and pelletizing capabilities. The following figures were basic to the Bureau evaluations: Cost of sludge, $1 per ton; hauling of sludge, $0.42 per cubic yard, based on a 7-mile haul; amount of sludge per acre, 130 tons of buried layers and 11 tons for pellets; and a cost of seeds, fertilizer, and planting of $50 per acre. Calculations on laying of the buried sludge were made using the following criteria: (1) An agriculturally developed method using a broad flat plow for lifting the soil and a device for spreading a layer of buried material under the long plow, and then allowing the lifted soil to drop back and bury the layered material; and (2) use of a conventional wheel scraper and spreader method. The mixture of pelletizing consisted of 92.4% tailing, 6.6% sludge, and 1.0% chemical binder. Pelletizing was calculated as being done in place on the tailing using a recently designed high-speed light-compaction roll pelletizer which can produce pellets at an estimated cost of $0.80 per ton. All sewage layer costs were based on application to flat pond areas only. These costs, although broadly generalized, provide comparison for the different methods and are presented in Table 27.6.

TABLE 27.6
Cost Comparison of Auxiliary Stabilization Methods

Type of stabilization	Effectiveness	Maintenance	Approximate cost per acre, dollars
Buried sludge layers	Excellent	Minimal	$405–810
Pellet cover (1½ inches)	Excellent	Minimal	350–660
Combined buried sludge plus pellet cover	Excellent	Minimal	735–1,470

The buried sludge layer and pelletized cover methods still are unproved in the field and the actual costs incurred from application of these procedures to differing sites may range to double the lower calculated figure. However, if the calculated cost range of $735 to $1,470 per acre is approximately correct for the combined buried sludge and pelleted surface method, this procedure is comparable on a cost basis to that using a 12-inch soil cover and vegetation, and tests may be warranted to evaluate the relative effectiveness of both procedures for obtaining vegetation on adverse sites. If the use of pellets is not mandatory to achieve satisfactory vegetative stabilization, then the lowered cost for application of only buried sludge layers and vegetation makes the method more competitive.

FIELD TESTS ON ACID NEUTRALIZATION

Holden Village tailings in the State of Washington are representative of aged, abandoned copper tailings in which pyrite has oxidized to produce acidic tailings with a pH of 3.6. The surface of these tailings has become indurated. These tailings were used to test the growth effectiveness of six different plants on raw tailing and those treated with lime, Milorganite sewage sludge (reference to specific company or brand names is made for identification only and does not imply endorsement by the Bureau of Mines), and fertilizers. An initial test plot was established in May 1973 and a second plot in June 1974.

Due to the relatively isolated location of the tailings requiring either air or water transport (via Lake Chelan) the option as to materials used and the acreage to be treated was limited. Therefore, the initial plot established was limited to one-eighth of an acre and the second plot to 1.54 acres. Although it was known that the plots were subject to severe winds, transportation costs precluded the use of a combined chemical-vegetative approach for the first planting. A bulldozer was used in both tests to break the surface of the tailings to permit plantings.

The first plot was divided into eight sections, aligned in a crosswind north-south direction, and dolomitic lime, in varying amounts, was applied. Also, one-half of the total Milorganite used was applied only to the south half of each of the receiving sections, as shown in Table 27.7. The sections were then roto-tilled and planted in furrows 2 to 3 inches deep (Table 27.8). Fertilizer was broadcast over the plot at the rates of 60 lbs. ammonium

TABLE 27.7
Treatment on Holden Village Tailing, Washington

Section No.	Added, tons per acre	
	Lime (dolomitic)	Milorganite
1	15	5
2	5	5
3	5	10
4	Control	No treatment
5	15	10
6	None	15
7	5	None
8	15	None

TABLE 27.8
Seeds Planted on Holden Village Tailing, Washington

Row No.	Seeds	Seeds per foot
1	Regar brome	110
2	Crested wheatgrass	161
3	Resistador alfalfa	137
4	Orchard grass	133
5	Tall wheatgrass	109
6	Alta tall fescue	105

nitrate and 100 lbs. treble super phosphate per acre. The remaining one-half of the Milorganite was then applied to the applicable sections of the north end of the plot to compare such a top dressing with the intermixed application on the south end of the plot.

The plot was examined after 3 months, and although it showed that the almost constant west to east wind had transported appreciable amounts of the treated and untreated tailing from section 1 toward section 8, considerable growth had taken place and the hardiest plants had 8 to 10 inches of root and aboveground growth. A true correlation of plant growth with lime and sewage sludge treatment could not be achieved in that the wind had transported some of the amendments with the tailing. However, in general, the addition of lime and sewage sludge appeared beneficial in promoting growth. The greatest evidence of their beneficial use was provided by taking the pH of simulated soil samples in the various sections. These samples were taken on the south of the sections at a 1-inch depth and at the hardpan level at depths varying between 8 and 12 inches. Similar samples were taken on the north half of the sections with an additional third sample taken from a 3-inch depth. The results of these pH measurements are given in Table 27.9.

TABLE 27.9
Holden Village Tailing Soil pH, August 1973

Section No.	Lime, tons/acre	Milorganite, tons/acre	South pH		North pH		
			1-in.	Bottom	1-in.	3-in.	Bottom
1	15	5	4.1	2.8	6.4	5.2	3.6
2	5	5	4.3	3.5	7.2	5.7	4.0
3	5	10	4.3	2.8	7.3	6.0	4.0
4	0	0	2.9	2.6	4.1	3.6	3.5
5	15	10	7.1	2.7	6.3	5.1	3.7
6	0	15	5.1	3.6	7.2	3.8	3.8
7	5	0	4.3	3.5	5.2	3.5	3.2
8	15	0	4.6	4.5	5.2	4.3	4.0

These results show that the top 1-inch of tailings to which lime and/or Milorganite was added had a pH of at least 4.1 and ranged to a high of 7.3. The north half of the plot to which the Milorganite was applied as a top dressing consistently produced higher pH values than the south half in which the Milorganite was intermixed with the tailings. The consistency of this effect was especially noticeable nearer the surface on which the Milorganite was placed, thus indicating a possible buffering action by the organic sludge.

Repeat observations in June 1974 showed that plant growth had increased markedly in the 10 months since the first inspection. Section 5, to which were added the largest quantities of lime and sewage sludge, demonstrated superior growth. Measurement of pH on the plot in 1974 showed that the pHs were, with few exceptions, higher than in 1973, indicating that considerable additional neutralization of acid had taken place. Only three of the 16 sections tested in 1974 had a pH less than 5 (4.1, 4.7, and 4.8) and these were in the south half of the plot with intermixed sludge. The pH in the north half sections with surface sludge ranged from a low of 5.6 to a high of 7.6. These were again considerably higher pH values for this type of treatment.

An evaluation of plant growth in 1973 and 1974 indicated that in order of performance the plants ranged from an excellent stand of tall wheat-grass to progressively less growth from crested wheatgrass, brome grass, tall fescue, orchard grass, and alfalfa. The alfalfa and orchard grass growths, respectively, were less than one-tenth and one-sixth that of the tall wheat-grass.

Insufficient time had elapsed by 1975 for proper evaluation of the plot established in 1974 except to note that the plant count made 3 months after planting was proportionately better than for the earlier planting. This was attributed principally to the use of chemicals for preventing sand movement by the winds. The Holden Village testing proved, however, that a relatively rapid vegetative growth can be achieved in excessively acidic tailings by judicious use of neutralizing agents together with previously developed stabilization techniques.

REFERENCES

Baines, S. S. and K. N. Singh. 1966. Utilization of Solar Radiation in Desalinization of Ridged Plantbeds on Saline Soils. *Nature* (5069)212:1391–1392.

Dean, Karl C., Richard Havens and M. W. Glantz. 1974. Methods and Costs for Stabilizing Fine-Sized Mineral Wastes. Bureau of Mines RI 7896.

Dean, Karl C., Richard Havens and Kimball T. Harper. 1969. Chemical and Vegetative Stabilization of a Nevada Copper Porphyry Mill Tailing. Bureau of Mines RI 7261.

Gates, David M. 1965. Radiant Energy, Its Reception and Disposal. *Meteorological Monograph* (28)6:1–26.

Havens, Richard and Karl C. Dean. 1969. Chemical Stabilization of the Uranium Tailings at Tuba City, Arizona. Bureau of Mines RI 7288.

James, A. L. 1966. Stabilizing Mine Dumps With Vegetation. *Endeavor* (London) (96)25:154–157.

Morgan, G. W. 1973. Crop Productivity as Affected by Depths of Topsoil Spread for Reclaiming Bauxite-Mined Lands in Jamaica — I. Agricultural Division, Alcan Jamaica Ltd., Mandeville, Jamaica, W. I. Presented at Conference on Soils of the Caribbean and Tropical America, Trinidad, January 1973. Available for consultation at Salt Lake City Metallurgy Research Center, Salt Lake City, Utah.

28. Tailing Reclamation

Kenneth L. Ludeke

AN ENVIRONMENTAL RECLAMATION PROGRAM was initiated by Cyprus Pima Mining Co. in 1968 to evaluate the existing ecological and physical conditions associated with copper tailing pond stabilization. The elevation of the tailing ponds to be vegetated ranges from 3,000 to 3,250 feet. The climate is cool in the winter and hot in the summer with considerable variation in seasonal precipitation. Average annual precipitation ranges from 8 inches for a poor year to 13 inches in a good year. Maximum precipitation occurs in July and August for the summer rainfall and in February and March for the winter rains. The year can be divided into two growing seasons: spring and summer, beginning in April and continuing through November, with an average frost-free period of 240 to 248 days; and winter, beginning the latter part of November and continuing through March with temperatures ranging from 24°F. to 72°F. Summers are hot, dry, and windy with high evaporation rates.

Vegetation surrounding the mine is characterized by mixed desert grasses with scattered saltbush and other drought-tolerant shrubs. Extensive cactus species occur throughout the desert area surrounding the mine.

Irrigation may be a critical factor in vegetative stabilization of tailing slopes in Arizona, primarily because of dry summers. Native vegetation is limited to adapted, well-rooted, and drought-tolerant species. Serious problems are encountered where tailing pond reclamation requires establishment of vegetation on the slopes. Erosion due to rapid rainfall and precipitation losses from the tailing slopes not only result in surface disturbances that require backfilling, but also restrict vegetative establishment and growth by limiting soil moisture.

Until recently, few effective procedures had been developed to reduce the rate or amount of surface runoff from 1.5:1 tailing slopes. Consequently, tailing erosion and surface water losses were severe. Several new approaches to alleviate these conditions were evaluated. They included:

1. Preparation of a smooth, loose seedbed for planting, accomplished by dragging a "Klodbuster" over the tailing slope, filling in the smaller erosion crevices, and smoothing the larger rills.

2. Application of barley straw mulch to insulate against heat and cold, to provide organic matter through decomposition, and to protect the slope against rainfall impact.
3. Application of sewage effluent and composted manure as a basic soil conditioner and micronutrient supply.
4. Use of hydroseeding.

These treatments adequately reduced the rate of runoff from tailing slopes at the Pima mine.

SEEDBED PREPARATIONS ON TAILING SLOPES

To evaluate the relative merits of "capping" the surface slopes of a tailing pond a study was conducted in 1970. Initial results indicated that the capped slope produced a greater amount of plant material than untreated tailing slopes. This was expected since the untreated tailing slopes have no fertility, organic matter, or micronutrients essential for plant growth. However, a similar study revealed that mulch, fertilizer, and composted manure added to the raw tailing slope, produced considerably more plant material than the best capped tailing treatment.

The study treatments were as follows:

1. An undisturbed tailing slope hydroseeded with seed and fertilizer.
2. A six-to-eight inch layer of overburden placed on top of a tailing slope and the area hydroseeded.
3. Barley straw incorporated into the tailing slope with sewage effluent and composted manure. The area hydroseeded with seed and fertilizer.
4. Barley straw incorporated into the tailing slope with sewage effluent and composted manure. The area hydroseeded with seed, fertilizer, silva-fiber mulch, and two additional applications of nitrogen fertilizer (50/lb. acre N at each application).
5. Control (hydroseeded with seed mixture but no fertilizer or irrigation were added as part of its treatment).

The treatment test areas were hydroseeded at a rate of 100 lbs./acre of Arivat barley on November 4, 1971. As part of the treatments, five tons of barley straw were incorporated into each plot along with 1,000 lbs./acre of sewage effluent and ten tons of composted manure. Treatment No. 4 had the same amounts of manure, sewage effluent, fertilizer, and barley straw per acre; however, it also contained 1,200 lbs./acre of silva-fiber in the hydroseeding mixture. Immediately after seeding, test areas 1–4 were fertilized at a rate of 200 lbs./acre, 100 lbs./acre, and 50 lbs./acre of available nitrogen, phosphorus, and potassium, respectively. On June 1, 1972 the test areas were sampled by harvesting all vegetation present. The data were analyzed to determine the sources of variance among and within the treatments.

Results demonstrated that treatment No. 2 produced a relatively poor stand of barley (Tables 28.1 and 28.2) due partly to the lack of spring precipitation and poor seedbed preparation. Many test areas with low densi-

TABLE 28.1

The Effect on Barley Grain and Barley Forage of Various Hydroseeding Treatments and Fertilizer Evaluation

Test	Barley Straw (tons/acre)	Sewage Effluent (lbs./acre)	Compact Manure (tons/acre)	Silva-fiber (lbs./acre)*	Fertilizer†	Plant Height (cm.)	Forage Yield (gms./0.0001 acre) Green Weight	Forage Yield (gms./0.0001 acre) Oven-dry Weight
1	5	1000	10	0	Yes	22	62	31
2	5	1000	10	0	Yes	53	297	115
3	5	1000	10	0	Yes	65	582	265
4	5	1000	10	1200‡	Yes	87	788	387
5	5	1000	10	0	No	0	0	0

*Added by hydro seeding.
†Fertilized, dry, 200 lbs./acre of N, 100 lbs./acre of P and 50 lbs./acre of K.
‡Two additional applications of N fertilizer (50 lbs./acre) during the growing season.

TABLE 28.2

Arivat Barley Yield From Various Hydroseeding Treatments

Treatment	Average Number of Heads Per Unit Area	Average Number of Seeds Per Head	Average Weight of 2500 Seeds (Mg.)	Average Grain Weight Per Plot (g.)
1	9	11	1851	43.31
2	33	25	2308	65.52
3	98	62	7823	152.34
4	150	83	9784	183.22
5	0	0	0	0

ties were infested with Russian thistle. Treatment No. 1 gave the lowest grain and forage yields. Treatment No. 3 gave fair to good results. Treatment 4 produced the highest yields (Fig. 28.1). Apparently the seed was protected and had germinated within the silva-fiber. Excellent infiltration occurred due to the presence of mulched barley straw and adequate organic matter. Nutrient in fertilizer present from the sewage effluent, composted manure and fertilizer produced a dense cover. The control treatment was unproductive (see also Chapter 31).

Results indicate that any treatment which improves the surface of the tailings slope to increase infiltration, improve fertility, maintain high moisture levels and adequate organic matter, will produce high yields of grain and forage.

The barley plant used in this experiment was selected specifically for the tailing treatment study, and it was intended to be used for long-term vegetative stabilization. A mixture of annual grasses, perennial grasses, cacti, and perennial shrubs would normally be selected with native and introduced plants for a maintenance-free vegetative cover.

Fig. 28.1.   High plant density obtained by mulch, hydroseeding, and fertilization.

PLANT SELECTION CHARACTERISTICS

Saline-alkali tailing slopes could result in very serious revegetation exchangeable sodium often occur in southern Arizona when drainage is impeded and surface evaporation is high. Various natural soluble salts, especially calcium (Ca), sodium (Na), and magnesium (Mg) contribute to soil salinity and result in high pH values. Alkalinity is caused by excessive sodium ions, which tend to lower pH. Both alkalinity and salinity can damage plants or limit their growth.

Native plant species growing under saline-alkali conditions in southern Arizona are especially adapted and drought resistant. Most of the salt stresses on nature are due to Na salts, particularly NaCl. The mechanism of salt tolerance in plants is intricate and allows some species to exist under the most saline-alkali conditions.

Saline-alkali tailing slopes could result in very serious revegetation reclamation problems. At Cyprus Pima Mining Company, the slopes are high in sodium and other salts. Therefore, a necessary step in reclamation includes research in the selection of plant species with high salt and sodium tolerances.

A species testing program was designed with the objective of determining species establishment and survival on tailing high in sodium salts. Sixty species, including 30 grasses, 10 forbs and legumes, and 20 shrubs and trees (Table 28.3) were planted on raw tailing. The slopes were fertilized at the rate of 200 lbs./acre, 100 lbs./acre, and 50 lbs./acre of nitrogen, phosphorus, and potassium respectively.

Plant density of seeded grasses was determined by count on a 0.0001 (one ten thousandth) acre plot. Overall shrub and tree success was based upon the values obtained from vegetative density transects of the tailings slope compared with those of the surrounding desert.

Very few plant species successfully germinated and established growth on the first plantings probably due to the extremely alkaline-saline tailing slopes. Much better results were obtained from a second and third planting due primarily to better seedbed conditions, higher amounts of fertility, and lower salt and sodium content. The most successful plants were the saltbush species, desert broom, creosote, acacia, and native cacti. After the second seeding, good success was obtained with all the species evaluated.

TABLE 28.3
Grasses, Forbs, and Legumes Tested on Tailing Slopes

Scientific Name	Common Name	Plant Density (count/0.0001/acre)
Andropogon caucasicus	Blue stem	.88
Arctotis stoechadifolia	African daisy	.77
Atriplex canescens	Fourwing saltbush	12.0
Atriplex lentiformis	Quail bush	6.1
Atriplex semibaccata	Australian saltbush	8.2
Bouteloua curtipendula	Side oats grama	6.21
Chrysanthemum segetium	Chrysanthemum	1.16
Cynodon dactylon	Giant bermuda grass	50.7
Digitaria sanguinalis	Crab grass	4.25
Eragrostis chloromelis	Boer lovegrass	1.14
Eragrostis currula	Weeping lovegrass	9.12
Eragrostis lehmanniana	Lehmans lovegrass	10.19
Eragrostis superba	Welman lovegrass	1.69
Eschscholtzia segetium	California poppy	.97
Hordeum vulgare	Barley	16.22
Lolium perenne	Perennial rye	11.21
Lolium spp.	Annual rye	12.15
Lupinus angustifolius	Blue lupine	2.15
Panicum antidotale	Blue panicgrass	54.23
Pennisetum ciliare	Buffelgrass	100.76
Sorghum sudanense	Sudan grass	14.35
Sporobolus cryptandrus	Sand dropseed	35.45
Voltus corniculatus	Birdsfoot trefoil	.16

TABLE 28.4
Plant Density 15 Weeks After Seeding

Common Name	Seedling/ 0.0001/acre	Common Name	Seedling/ 0.0001/acre
Buffelgrass	87.6	Australian saltbush	8.2
Blue panicgrass	52.1	Quail bush	6.1
Giant bermuda grass	50.7	Blue lupine	0
Sand dropseed	34.2	Yellow sweetclover	0
Fourwing saltbush	12.6	Birdsfoot trefoil	0

The south facing slopes of tailing ponds at Cyprus Pima Mining Company are warmer, and when the tailing are both saline and alkaline, special problems are presented. A study was designed to test a number of drought and saline-alkali tolerant species under these conditions.

A south facing 1.5:1 tailing slope covering approximately eight acres and 35 feet high was hydroseeded with several species on May 5, 1972. The hydroseeding mixture contained seeds, 1,500 lbs./acre of silva-fiber, 200 lbs./acre of nitrogen, 100 lbs./acre of phosphorus, 50 lbs./acre of potassium, and 1,000 gals. of water per acre. The tailing slope was sampled on August 28, 1972.

Buffelgrass had the greatest density, followed by blue panic grass, giant bermudagrass (NK 37), and sand dropseed (Table 28.4). Four-wing saltbush provided good plant cover while quail bush and Australian saltbush were low in densities compared to the other two *Atriplex* species. None of the legumes germinated in this particular study. Australian saltbush, usually a good performer in the arid environment, failed. Since establishment, buffelgrass has maintained exceptional growth and is providing excellent cover on the area.

TRANSPLANTED CACTI

Transplanting cacti is one method of rapidly establishing plant cover. Aesthetics, as well as resistance to rodents, drought, dessication, and sunscald are some advantages of cacti. A large front-end loader, a trunk-mounted crane, and a forklift were used to transplant cacti on the tailing slope area. One purpose of the operation was to salvage native cacti from the tailing ponds that were being enlarged.

Large quantities of the smaller, more easily handled, cacti were transplanted with the front end-loader and forklift. The end-loader was used to scoop up cholla, prickly pear, and barrel cacti. They were transported to the tailing slopes on a flat-bed truck. The forklift was used to lift the strongly rooted species of yucca and ocotillo. A rope was placed around the base of the plants and attached to the forklift for lifting. The truck-mounted crane was used to remove giant saguaro. A webbed belt and thick cloth padding was wrapped around the saguaro at three points to prevent breakage. Once the plant was secure, the roots were dug out and the plant was transported to the tailing slope by crane. Fill material consisting of composted manure, tailing material, and fertilizer was compacted around each

Fig. 28.2. Transplanted desert plants.

plant by driving the wheels of the motorized equipment around the base. All transplanted material survived (Fig. 28.2).

It was important to instruct the equipment operators in proper handling to insure success. Instructions included time of planting, care in excavation, proper plant selection, and necessary initial fertilization and watering practices.

TAILING AND AGRICULTURE

Agricultural soil contains plant nutrients derived from the breakdown of the original "parent" materials. It also contains organic matter, some in the form of humus and some in the form of microorganisms. These three basic factors, essential in forming a soil material, dictated the approach taken by Cyprus Pima Mining Company in reclamation and revegetation of mining wastes.

The establishment of vegetation on tailing slopes can be made possible by an understanding of the physical and chemical problems associated with these areas. An area of tailing, unprotected and exposed to summer and winter heat in the arid southwest, will quickly lose all available moisture in its topmost layer and easily reach temperatures in the hot summer months of 100°F to 110°F. A moderate wind under these conditions can quickly cover freshly emerged plant material with tailing sand.

An important characteristic of tailing material is its homogeneity. This characteristic allows the plant to develop a full root system, a fundamental factor in the growth of healthy plants. Although tailing material presents some unusual problems not normally associated with agricultural soils, they can be defined and corrected to lead to normal agricultural development. It is possible that tailing waste could have more productive potential than many common soils in the desert, but some problems must first be remedied. It is necessary to:

1. Study the chemistry of the tailing material to define the growth deterrents, deficiencies and chemical imbalances.
2. Determine the treatments and amendments necessary to provide an environment suitable for plant growth.
3. Identify plant species best adapted to the adverse growing conditions of tailings slopes.
4. Evaluate laboratory findings under actual field conditions.

The special techniques developed at Cyprus Pima Mining Company to establish vegetation on the tailing slopes includes the following:

1. Heavy or fine grading to smooth up the deep gullies formed on the sides of the tailing slopes.
2. Dragging a spiked chain over the surface of the slope to break up areas where iron sulfides have created a crust, to roughen up the excessively compacted areas and to smooth out irregularities.
3. Grading a two-foot lip on the leading edge of the slope to insure against runoff water eroding away the surface of the slope.
4. Grading cross-dikes adjacent to the slope in order to collect excessive water from rainfall and prolong irrigation.
5. Applying compost manure at five tons per acre to provide for a better soil condition.
6. Compacting barley straw at 2.5 tons per acre with a "sheepsfoot" roller to provide adequate mulching to the tailing slope.
7. Pre-irrigating with applications of fertilizer injected into the irrigation line.
8. Hydroseeding a mixture of desired seeds with various rates of wood fiber to the tailing slope area.
9. Hydraulic spraying of sewage effluent at 1,000 gallons per acre to act as a soil conditioner.
10. Additional irrigating with liquid fertilizer added to insure germination and seedling emergence.

TEMPORARY TAILING SLOPE STABILIZATION

There are instances where a tailings slope is left unstabilized before permanent vegetation can be established. During these periods, temporary stabilization is necessary to reduce wind and water erosion.

Annual grasses function particularly well in temporary stabilization because they quickly germinate and have a high growth rate, and the seed is commercially available and inexpensive. Winter barley *(Hordeum vulgare)* or annual rye *(Lolium spp.)* can be used during the winter growing season, and sudan *(Sorghum sudanese)* or milo-maize *(Sorghum vulgare)*, during the summer growing season. The annual grasses also produce food and cover for birds, game and small animals. At the time of permanent stabilization, the slopes will already be relatively stable and covered with natural organic material.

When seeding a tailing slope, even for temporary stabilization, it is essential to provide an adequate seedbed and fertilizer. Irrigation is also required to obtain adequate plant densities.

In the spring and fall of 1972, a temporary stabilization experiment was initiated to test various annuals on a raw tailing slope. The objectives of the study were to evaluate the effectiveness of annual species, and the amount of surface mulch produced and to determine root production.

Four species of annuals were hydroseeded on a 50 x 50 foot plot during the growing season. Irrigation and fertilization at 200 lbs./acre of nitrogen was applied as needed. Production data were obtained by the harvest method in which all vegetation within a 0.0001 acre quadrat was clipped to ground level. Root production was evaluated by flotation and sieving known volumes of tailing samples (Table 28.5).

TABLE 28.5
Annual Grass Production in One Growing Season

Species	Plant Density (plants/ 0.0001/acre)	Plant Cover (percent)	Above Ground Production (lbs./acre)*	Root Production in Upper 35 cm. (lbs./acre)*
Winter barley	828	89	2088	5722
Annual rye	388	64	394	3015
Milo-maize	628	82	1150	4755
Sudan grass	520	75	689	1215

*Dry weight.

Good germination was obtained on the test area within a few days after hydroseeding. Effective erosion control was obtained within two weeks. Severe windstorms passed through the experimental testing area during March and August without significant erosion damage to the tailing slopes. Interestingly, heavy seeding rates limited production. Barley produced the most plant material for the winter growing season and annual rye produced the least. Milo-maize produced the most plant material for the summer growing season. Sudan grass produced the least vegetative cover.

Results from this study indicate that tailing slopes can be quickly and economically stabilized with annual grasses. Further research is necessary, however, to determine optimum species, seeding rates, fertilizer rates, and seeding dates.

WATER, FERTILIZER AND HYDROSEEDING

Timing is important in the application of seed and fertilizer. The use of a hydroseeder lends itself to successful results at low cost. Hand seeding reduces the chance of success and costs more. However, it is the best alternative for small areas where hydroseeding cannot be done conveniently. Hydroseeding has been used extensively in Arizona along highways, railroads, canal banks, and for tailing slopes, where the fastest possible methods of seeding under adverse conditions are needed. Hydroseeding permits conditioning, seeding, fertilizing, and mulching at high speed, and it is a rather simple operation in which the seed, wood fiber and water are all combined. The materials are slurried, then pumped through a nozzle and sprayed on the side of the slope.

Overburden or chemical stabilizers on the tailing slopes have not been used at Cyprus Pima because of the cost and because of poor results obtained from their use. Mulch materials, such as straw and fibrous materials, appear to give better results. Tests have shown that mulch applied to the tailing slopes conserves moisture, stabilizes surface temperatures, lessens the erosive power of rain, and helps the growth of bacteria and microorganisms. Mulching is also necessary if the seed and fertilizer are to be exposed to adverse weather conditions for a considerable period of time. If operations can be arranged to minimize the adverse conditions, mulching can be eliminated from the planting technique. However, environmental irregularities in the climate, either too high a temperature or freezing snow, require mulch.

Since the beginning of reclamation work on tailing ponds in 1968, tests have shown that where annual precipitation is 11 to 12 inches, it is essential to supply irrigation water during the growing season for the first two years. Once plants have adequately covered the slope area, irrigation can be gradually withdrawn, to allow the plants persisting in the area to rely on natural rainfall for continued growth and reproduction.

Irrigation water can be used to advantage to initiate and maintain growth in low rainfall areas. However, on tailing pond slopes, unless the irrigation equipment is carefully supervised, it should not be used because of the danger of washouts. With irrigation, grasses such as Bermuda grass and buffelgrass can quickly be established to provide an initial cover. They also act as a nurse crop for slower growing, more vulnerable leguminous plants such as alfalfa, lupine and clover. Legumes also thrive in the favorable micro-climate provided by other grasses such as Lehmans lovegrass and weeping lovegrass.

Fertilization is essential to supply sterile tailing with the necessary and proper balance of nutrients during the early growth period. Once second generation plants have been successfully established, it is expected that a fertility cycle will be developed that should require no further addition of commercial fertilizers.

Most of the nitrogen in complete fertilizers is in organic form, either natural or synthetic. Many complete fertilizers contain nitrogen in the ammonia form. A few contain nitrogen in the nitrate form. All nitrogen must come either from the air, from organic matter, or from fertilizers. If organic matter added to the tailing is high in carbon compared to nitrogen, soil organisms working to digest the high carbon material may compete with the plants for nitrogen. Therefore, it is necessary to add additional nitrogen to the organic matter to maintain a proper carbon-nitrogen ratio.

Tests on tailing material have indicated that in nearly all cases major and minor plant nutrients are completely lacking. Unless sufficient fertilizer is supplied to promote good growth, failure is certain. This is particularly true during early establishment. Once the plants have developed an extensive root system, they should survive without further attention drawing the required nutrients from decomposed plant material. Three or more fertilizer applications may be necessary on newly exposed tailing material during the growing season. Large applications are unnecessary if the plant root system is not sufficiently developed to take them up. Fertilizers are also ineffective if there is insufficient soil moisture to bring the nutrients into solution for use by the plant.

Because tailing material is essentially sterile, it can be an excellent base for controlling the plant production by selected fertilizing materials. For example, after the tailing slopes have been seeded, it is desirable to develop a strong root system as quickly as possible. Therefore, the initial fertilizer application should have a low balance of nitrogen but should be high in phosphorus and potash. As growth progresses, the balance should be raised to induce top growth. Fertilizer injected into the irrigation systems has worked well.

Fertilizer rates used on tailing slopes should be those normally applied to poor soils. On the Cyprus Pima tailing, two applications were made at different times for a total of 200 pounds per acre.

CHEMICAL PROPERTIES OF TAILING

Chemical analyses were made of the tailing material at each site to assess the available nutrients and to determine what metals were present in concentrations that could be phytotoxic. However, the tailing materials differed so much from normal soils that few interpretations could be made. For example, because there is no organic matter present in tailing material, the analysis indicated that there is no buffering capacity to a change in pH. Therefore, a relatively small amount of acidic or basic material may affect the pH readings much more than in a normal soil. Organic matter tends to absorb excess anions and cations up to the point of saturation. In this way organic matter acts as a "buffer," which in normal soils eases the shock effect of the addition of chemical materials. The lack of organic material in tailing makes it necessary to evaluate test results on a different basis than for normal soils. Due to the absence of buffering capacity, it is usually necessary to evaluate the salt content by conductivity tests. The pH rating of the tailing will initially dictate what vegetation might be suitable.

In tailing material pH values over 9.5 are common. This suggests a serious salinity problem. Calcium and magnesium are also high, and yet conductivity readings are often low, indicating that the salts of these elements are present only in small quantities. Thus, although the material contains

TABLE 28.6

Analysis of Copper Tailing From Various Tailing Ponds

Tailing Pond	pH*	Organic Matter (per-cent)	Bulk Density (g./cm.³)	Total Soluble Salts (ppm)	(ppm)								
					EC$_e$†	P‡	K	Cu	Fe	Mn	Zn	Mg	NO$_3$—N
3	7.85	0.31	1.98	3428	3.8	2	88	144	869	56	29	729	2
4	7.85	0.28	1.40	3861	3.6	3	97	138	755	44	24	716	3
5	8.24	0.51	1.86	4959	3.1	3	79	155	915	65	32	814	2
6	8.02	0.36	1.82	4812	2.7	1	106	162	929	71	31	821	3
7	8.9	0.20	1.86	4851	3.0	2	135	158	988	69	38	835	3

*Saturation paste.
†Mm/hos/cm, saturation extract.
‡Bicarbonate soluble.

a large potential source of calcium and magnesium, they are not in a form that immediately affects the tailing soil solution (Table 28.6).

Tailing material is saline because of the salty water used for transport to the ponding area. The water is recycled, which further tends to concentrate salts. This salinity could be sufficient to prevent the growth of all but the most salt tolerant plants.

The salinity of freshly deposited tailing material is fairly uniform, but this condition changes rapidly as the material dries out. In a poorly drained tailing area, salt may move up and down in the profile in response to evaporation and precipitation. During the late spring and summer months, the salts accumulate at the surface with greatest concentration on the higher areas of the tailing slope. Salt accumulations are greatest by late fall. If a tailing slope is scarified and leveled and has good drainage, salinity is reduced markedly by the winter and summer rains. Salt can also be reduced in a short time by irrigation.

MOISTURE AND THE SELECTION OF PLANTS

Temperature and precipitation are of great importance to plant establishment. In areas of limited rainfall, the slope and aspect of the tailing berm is important and may be the limiting factor to growth. Some species may grow on eastern and northern slopes, but not on western or southern slopes. Timing is important when planting summer annuals. By taking advantage of moisture which may be available only in fall and spring, it may be possible to develop a ground cover without extensive irrigation by using drought-tolerant plants. The selection of plants which will adapt to the severe conditions with tailing waste material must be based upon:

1. The chemical composition of the tailing material.
2. Local rainfall, wind humidity and temperature.
3. The exposure or aspect of the tailing slope.
4. The availability of the plant material and its ability to persist in the particular tailing material.

Irrigating plants in tailing soil material is critical to revegetation programs. The frequency of irrigation depends upon a number of interrelated factors such as the needs of the particular plant and its age, the season, the weather, the nature of the soil and the method of application. Irrigation simply on a routine basis may subject various plants to drought or drowning.

In irrigating tailing, it is important to wet the entire root zone. When free water reaches the point where it fills the interspaces between soil particles, the supply of oxygen to the roots is reduced. Root elongation stops and nutrient absorption is hindered. If the deficiency continues, oxygen producing toxic organisms may thrive, and beneficial bacteria may die out. Plants vary in their ability to resist these conditions, but many of the choicest reclamation plants absolutely require air in the soil. To maintain a desirable water:air ratio, it was found important not to overirrigate and to maintain the tailings near field capacity.

UTILIZING ORGANIC MATTER

Organic material is broken down by organisms that require nitrogen. If these organisms cannot get sufficient nitrogen from the organic material itself, they will get it from the soil, robbing roots of whatever nitrogen is available. At least 1.5 percent nitrogen is necessary for organic matter decomposition. Leaves, peat moss, plant compost, and manure usually contain sufficient nitrogen (but manure may also contain salts). Sawdust or ground bark, if thoroughly decomposed or if fortified with nitrogen, will not cause nitrogen depletion.

Trace elements, needed only in infinitesimal quantities, are usually present in most tailing, but nitrogen, phosphorus, and potassium must be supplied. In unimproved soils, nitrogen comes from decomposing organic material, which is in very short supply in dry western soils and completely absent in tailing. Nitrogen is used in large quantities by plants and is easily lost by leaching and depleted by soil organisms. As a result, it must be added, usually at least twice a year.

Nitrogen from decaying organic material is not directly available to plants. It must be converted by soil organisms (fungi, molds, bacteria) first into ammonia, then into nitrites, and finally into nitrates which are taken up by plant roots. These micro-organisms, being plants themselves, need a certain amount of warmth, air, water, and nitrogen to do their job. Any soil amendments which improve aeration and water penetration will also improve their efficiency.

The addition of organic matter to tailing increases water holding capacity, decreases the frequency of irrigation, insulates the soil surface, increases germination and seedling emergence, decreases the erosive power of runoff and rainfall, and provides additional nutrients for plant growth. Annual crops, such as cereal grains, sorghum, perennial crops, and native vegetation are an effective way to provide organic matter. They should be incorporated into the soil when they have the greatest biomass. Increasing the water holding capacity by addition of plant material also decreases the danger from marginal toxicity of the large quantities of elements normally found in mine tailing.

PLANT SELECTION

One of the first requirements of the revegetative stabilization or reclamation program is to prevent erosion due to wind or water and to obtain stabilization of the tailing pond slopes. It is necessary to use plants that have the ability to root quickly while protecting others which root more slowly, but which will also withstand adverse conditions and give a ground cover that requires minimum maintenance.

The mixture of plant material should include plants with a diversity of growth habits, sufficiently broad to handle all possible environmental conditions. Seeds of fast germinating, short-lived plant species are used in combination with slower germinating, long-lived drought-tolerant species, and/or winter hardy plants as required by the elevation and exposure of the tailing slope. If moisture is supplied fairly quickly by irrigation or natural rainfall, the plants will develop a significant amount of growth over the entire tailing slope within a few days. On the Cyprus-Pima tailings, the fast germinating plants do not survive beyond the second or third year since they are not truly drought tolerant. Therefore, it was essential to include indigenous plant species in each season's planting program. By doing this, the native species gradually take over while the short-lived species fade out.

It has been suggested that trees and shrubs be used for both stabilizing tailing slopes are improving aesthetics. However, until the tailing slope material has been "knitted" by the grass and legume roots, providing a more amenable condition for tree growth, it is strongly suggested that this type of planting be deferred. Trees planted on extensive barren tailing slopes have done little to prevent erosion, and under ideal circumstances barely succeed in surviving.

Leguminous plants such as alfalfa and clover must be inoculated with suitable bacteria which fix nitrogen from the atmosphere. Once established, they perpetuate microbiological production on the tailing areas.

HEAVY METAL PHYTOTOXICITIES

Heavy metal toxicity is a major factor limiting plant growth on tailing slopes. Copper toxicity has been noted in both legumes and grasses. Many of the legumes evidence a reddish orange, stunted condition shortly after emergence. Within a few days, they lose most of their green color, turn a pale white, and die. Sorghum seeds germinate, but develop a stunted root system. The sorghum plants that did emerge had severe leaf curling and a very dark green color. Dwarfing and root malformation was evident in all the species tested in tailing with high copper concentrations. Laboratory analyses of plant material indicated high levels of copper in both the roots and leaves, with the roots accumulating the most.

There were high concentrations of Mg in the tailing samples tested. Plants tested in the laboratory and field under manganese excess exhibited severe chlorosis. Desert tobacco leaves showed early states of manganese toxicity by the uneven distribution of chlorophyll. High concentrations of iron, zinc, and magnesium were also present in the tailing material. However, toxicity symptoms of zinc, iron and magnesium have not been identified. Because the solubility of these metals is directly associated with pH, lowering the pH by liming could control the solubility of the heavy metals.

CONCLUSIONS

Tailing materials do not behave as normal agricultural soils. Tailing slopes are low in water holding capacity, devoid of plant nutrients, polluted with excessive salts and heavy metals, and subject to wind and water erosion. These adverse characteristics must be altered before tailing can be adequately stabilized with vegetation.

Tailing material is highly variable in physical and chemical characteristics. Differences exist between types of milling operations and among different ages and exposures of tailing. In preparation for treatment, each site must be evaluated to identify the nature and extent of the problems.

Earlier stabilization efforts used agronomic crops. Domestic species were first used and later exotic species were tried. The efforts were successful, but recent observations indicate that native desert species have invaded the planting sites and may eventually develop permanent vegetation. Plants used for vegetative stabilization should be selected from the standpoint of hardy species adapted to the climate of the arid southwest, and the local environmental conditions of the tailing pond areas.

Even though the copper ore tailing material on which this study is being conducted is particularly poor for plant growth, excellent stands of grass species were grown to full maturity in one growing season with proper irrigation and fertilization. This type of stabilization has offered greater protection against wind without physical or chemical treatment.

The primary goal of vegetative stabilization is to quickly establish an economically sound, self-perpetuating plant community that stabilizes the tailing and enhances the aesthetic value of the area. The overall success of vegetative stabilization at Cyprus Pima Mining Company has shown that such a goal is feasible and is being accomplished at a reasonable cost.

PART VI

PLANT SPECIES
FOR DISTURBED LANDS

ALMOST ALL OF THE POTENTIAL SURFACE MINE LANDS in the southwest are in shrub, desert, or grassland ecotypes. Consequently, most of the revegetation research by 1975 had been toward the establishment and maintenance of perennial grass and shrub species. A great number of species have been tested and many more possibilities remain. The great diversity of available plants is encouraging because of the equally great variability of mined materials from site to site in the southwest and the unique problems of each. The selection of the most suitable plant species or combination of species depends upon the particular adverse properties of the spoil material, overburden, or tailing in addition to factors of local environment and economics.

Once a mine site has been successfully revegetated, which requires two or more years in the southwest, grazing may become a principal use of the area. Phil R. Ogden discusses the economics and the requirements necessary for range livestock grazing to be compatible with mining on range lands. Shrub species suitable to rangelands and problem areas in semiarid regions

278   Part VI: Plant Species for Disturbed Lands

are explored by Eamor C. Nord. A. Perry Plummer discusses plant species suitable for stabilizing land disturbed by man, since they are also suitable for stabilizing mine wastes. He provides an excellent shopping list of adaptable species. The success in establishing shrubs and grasses on copper mine wastes in southern Arizona is described by Wendell G. Hassel. The success of operational revegetation programs on coal mine spoils is discussed by Alton F. Grandt. Some of the species discussed are from more humid areas, but their characteristics can be extrapolated to western species in many cases.

29. Range Management and Surface Mining

Phil R. Ogden

EXTENT OF RANGELANDS

Range is a type of land which may be distinguished from urban, forest, or cultivated lands by its vegetation and the uses for which the land is best adapted. Rangeland supports a natural vegetation of grasses, grass-like plants, forbs and/or shrubs; and because of this vegetation, grazing by herbivorous animals has been a major use.

The Forest-Range Task Force (1972) combined plant communities as outlined by Küchler (1964) into ten range ecosystems for the western United States. These broad range ecosystems are: sagebrush, desert shrub, southwestern shrubsteppe, chaparral-mountain shrub, pinyon-juniper, mountain grasslands, mountain meadows, desert grasslands, annual grasslands, and alpine. This list serves to illustrate the great diversity of rangeland.

Approximately 81% of the land in Arizona is rangeland and another 6% supports western forest ecosystems which also are grazed (Forest-Range Task Force 1972). Ninety percent of the land in Nevada supports range ecosystems. The probability is high, therefore, that mining in the southwestern U.S. will be located on rangeland, and that activities during mine development, operation, and reclamation will have to be coordinated with other uses of the rangeland associated with the mined property.

LIVESTOCK GRAZING AS A USE OF RANGELANDS

Livestock grazing has been a use on southwestern U.S. rangelands since 1598 (Humphrey 1958), and livestock grazing in the late 1800s and early 1900s has greatly influenced the present appearance and condition of western ranges. Livestock grazing has been a major use of rangelands for so long that the term range management is often taken to be synonymous with livestock management. Range, however, is the land resource; management of this land is more than livestock management. The need in surface mining and reclamation is to recognize all uses and coordinate these uses as much as possible to obtain the best use of the land resource. Livestock grazing will be emphasized as a use which should be considered in planning.

ECONOMICS OF LIVESTOCK PRODUCTION

Since rangelands are variable in productive capability, it is convenient to discuss economics and production on an animal unit basis, then to relate this production to the land resource as the acres required to support the animal unit within the various range ecosystems. An animal unit (AU) is defined as 1000 pounds of live weight or about equal to a cow and calf (Stoddart and Smith 1955). Replacement heifers average about 0.8 AU, a bull as 1.25 AU, and a horse as 1.25 AU. Thus, a range livestock cow-calf operation of 300 mother cows with 45 replacement heifers, 15 bulls, and 5 saddle horses would be a 360-AU operation. The costs and production are distributed among the 360 total animal units to express on an animal unit basis.

Individual livestock operations vary greatly in costs and returns depending on size, land ownership, management, quality of range forage and livestock, and current prices. One example can help only to provide some perspective to the economics of livestock production. The estimated income and expenses shown in Table 29.1 are for a generalized southwestern U.S. ranch in the early 1970s. The ranch was assumed to be the 360-AU operation described above with 85% of the 300 mother cows producing calves for market. Steer calves averaged 425 pounds at weaning and sold for $40/cwt, and heifers weighed 400 pounds when sold at $35/cwt. Cull cows and bulls averaged 800 pounds and sold for $30/cwt. Capital investment in livestock was $300/AU, and $200/AU was invested in corrals, fences, water developments and other improvements.

The $10 shown in Table 29.1 as return to investment in land and for management was not adequate to pay a reasonable return to management and land at the inflated values of the early 1970s. Land costs were often in excess of $600 in the southwest for enough land to support one animal unit. Value of rangelands was determined in part by uses other than expected returns to grazing. Such uses as to fulfill a desire to own land, investment for urban development, personal outdoor recreation, and mining all contributed to this inflated land value.

A low return to land investment by livestock production is often cited as a reason to eliminate grazing as a use on western rangelands. This, however, is a short-sighted approach and is comparable to not processing secondary minerals found in a mine because they won't pay the full cost of the mining operation. The $125 gross product value per animal unit shown

TABLE 29.1

Estimated Income and Expenses Expressed on an Animal Unit Basis for a Generalized Cattle Ranch in the Southwestern U.S. in the Early 1970s

Source	Amount Per Animal Unit
Gross value of product	$ 125
Annual operation costs*	−60
Depreciation on improvements	−15
Net Ranch Income	$ 50
Return on capital investment in improvements and livestock (8% x $500)	40
Return to management and land investment	$ 10

*Annual operation costs include: supplemental feed, grazing fees, hired labor, taxes, transportation costs, maintenance and repairs, insurance, veterinary services and other annual expenses.

in Table 29.1 has a multiplying factor of about 2.25 to 3.00 times in a local economy (Kearl 1974) and certainly this is a major consideration in many southwestern U.S. communities.

OTHER VALUES OF RANGE LIVESTOCK PRODUCTION

There is another consideration also for livestock grazing on rangelands, and this is in relation to conservation of energy and the world food levels. With the generalized southwestern U.S. ranch of 360 AU discussed previously, each animal unit contributed about 300 pounds of live weight meat to market annually. This contribution of protein from the range is with a relatively low fossil fuel input, the energy converted to meat coming mostly from digestion of cellulose from range forage. This is an energy source that would not otherwise be available for human use.

The development of livestock watering ponds and wells on rangeland may be beneficial for wildlife and recreationists and can be aesthetically pleasing. These water developments are often constructed and maintained from livestock operating costs and do not have to be supported by the public through taxes of income from sources other than the land on which the improvements are put. Livestock operation costs may also include maintenance of roads and trails which make some lands accessible to public use.

GRAZING CAPACITIES WITHIN WESTERN RANGE ECOSYSTEMS

To visualize more readily the effect of a mining operation on range livestock production in various parts of the southwestern U.S., the animal unit data as discussed above may be converted to acres required to support an animal unit in the various western range ecosystems (Table 29.2). An animal unit in 1970 was produced on 100 acres of sagebrush rangeland, 240 acres of pinyon-juniper range, or about 60 acres of desert grassland. With a large number of acres required to support an animal unit, the actual number of livestock displaced by a surface mining operation might be relatively small, but the range area purchased to obtain control of the mine location may cover an extensive area and require planning and good management to maintain or improve this range resource. The range sites and

TABLE 29.2

Grazing Capacity as Average Number of Acres Per Animal Unit Per Year for Western Range Ecosystems, 1970 (Forest-Range Task Force, 1972)

Ecosystem	Acres Per Animal Unit	Ecosystem	Acres Per Animal Unit
Sagebrush	100	Mountain grasslands	40
Desert shrub	400	Mountain meadows	10
Southwestern shrubsteppe	200	Desert grasslands	60
Chaparral-mountain shrub	170	Annual grasslands	10
Pinyon-juniper	240	Alpine	60

condition of the vegetation and soils on these sites determine the opportunities for coordinating various land uses including livestock grazing on rangelands associated with specific mining operations.

RANGE SITES AND RANGE CONDITION

A range site includes units of rangeland of similar climate, soil, and topography with a potential to grow a specific plant community which will respond similarly to management. Within a single precipitation zone, a loamy bottomland site may be expected to produce differently and require different management from a sandy bottomland site or a loamy upland site.

When past mismanagement, adverse weather, fire, or some other factor has caused a deterioration of the potential plant community growing on a specific site, the condition is deteriorated to less than potential for the site. A site may be in poor, fair, good, or excellent condition depending on the current situation compared to the potential for the site.

A loamy upland range site with 16 inches of precipitation may require 40 acres or less per animal unit when in good condition, but when in poor condition this site would require in excess of 100 acres to support an animal unit. Favorable forage species decrease in vigor and abundance and weedy species increase with deterioration of range condition. Also, the habitat value for wildlife for the two conditions would be very different. The poor condition may be favorable habitat for some wildlife species, and the excellent condition would be better habitat for other wildlife species. Watershed and recreational values also are influenced by range condition.

Much of the southwestern U.S. rangeland is in a deteriorated range condition. This has several implications for mining companies. First, mining operations often purchase a ranch to gain control of mining property. As they consider alternatives for management of the property beyond what is actually taken up by the mine and associated operations, they often find that the remaining grazing land is inadequate to maintain an economical size for the grazing unit. One solution is that range improvement by brush clearing, plowing, seeding, burning, good management, and so forth to improve range condition may be utilized to improve the grazing capacity enough that grazing lost by the mine location can be regained and the livestock operation can be operated on the remaining land.

The deteriorated condition of many western rangelands, much of which was initiated in the early 1900s, has also caused the general public to be biased against livestock grazing. Therefore, there often is pressure to eliminate grazing from rangelands in favor of recreation or wildlife without considering all consequences or values for coordinated use. Livestock grazing will not receive public support unless it can be shown that good management will make grazing compatible with other uses or land values.

REQUIREMENTS FOR RANGE LIVESTOCK GRAZING TO BE COMPATIBLE WITH OTHER USES OF RANGELANDS

Only through good management techniques can livestock grazing be compatible with other uses of rangelands, and whether a mining company hires a manager and runs its own livestock operation or the company leases the grazing rights, these conditions for good management should be met.

The livestock operation should be large enough to provide economy of size. This size would be over 300 animal units for most situations. Adequate quality of forage must be provided to livestock all year and especially at breeding time to get a good calf or lamb crop as the salable product. The livestock operation manager must be aware of good animal husbandry techniques.

From the standpoint of the range vegetation, the first requirement for proper range livestock grazing is that there be a proper number of animals grazing an area so that the utilization of forage plants is proper in amount. Distribution of use over space also is important and may require fencing, water development, riding, salting, and trails. Use of desirable plants on a range generally should not exceed about 50% of the current year's growth. In the southwestern U.S. with its variable weather, flexibility in livestock numbers grazing a range is an important attribute of any livestock grazing plan. Persons responsible for lease agreements should be particularly aware of this need. Enough herbage should be left to protect the soil, replenish organic matter to the soil, and enable the plant to carry on its life functions. The life functions of individual range plants which should be provided for to maintain or improve range condition are: 1) seed production, 2) seeding establishment, 3) root growth, 4) shoot growth, and 5) carbohydrate reserve production and storage.

Proper livestock grazing which can be compatible with other rangeland uses does not just happen. It must be planned for and tailored to the specific range sites, to the livestock, and to the requirements of the plants growing on the sites. Properly managed livestock grazing can usually provide income from the land, help finance water and road developments on the land, and continue to provide meat production. Mining companies should consider a properly managed livestock operation as a possible use which can be compatible with other uses on southwestern U.S. rangelands associated with mining operations.

REFERENCES

Forest-Range Task Force. 1972. The nation's range resources — A forest-range environmental study. Forest Resource Report no. 19. Washington, D.C.: USDA Forest Service.

Humphrey, Robert R. 1958. The desert grassland. *The Botanical Review* 24(4): 193–252.

Kearl, W. Gordon. 1974. Returns to rangelands. *Journal of Range Management* 27:413–416.

Küchler, A. W. 1964. *Potential Natural Vegetation of the Conterminous United States.* Special publication no. 36. New York: American Geographical Society.

Stoddart, L. A. and A. D. Smith. 1955. *Range Management.* 2nd ed. New York: McGraw-Hill Book Company, Inc.

30. Shrubs for Revegetation

Eamor C. Nord

To HELP REDUCE FIRE HAZARDS in the chaparral zone of the southwestern United States, plant scientists are searching for vegetation growing under arid conditions that is durable, low growing, and slow burning. They have found more than a score of shrub species that can be established and grown satisfactorily under semiarid-to-arid conditions. Many shrubs tested to date show promise for rehabilitating disturbed areas, such as deteriorated big-game ranges, outdoor recreation sites, mining wastes, and spoil banks.

The search for suitable plant material has taken two broad approaches. One is to seek out domestic as well as foreign sources of such materials (Franclet, *et al.* 1971; Leiser and Nord 1976). Another is to develop methods for propagating plants from seed or vegetative material, followed by testing in greenhouses, nursery, and field planting (Nord and Goodin 1970; Nord, *et al.* 1971a, 1971b; Radtke 1974). Near its Riverside laboratory, the Pacific Southwest Forest and Range Experiment Station is testing selected shrub species in 1- to 2-acre plots of drill seedings and by transplanting.

CRITERIA FOR SELECTING PLANTS

Plants to be used primarily for reduction of fire hazards must meet more stringent requirements than are necessary for most other uses. For this purpose, the plants should fulfill most of these criteria:

1. Low volume — This requirement considers the amount of both woody and herbaceous material produced over a period of years — not just current production.

2. Adaptability — Plants must be adaptable to dry chaparral sites, and preferably to a moderately broad range of elevations, exposures, temperatures, and soils.

3. Growth form — The best are low growing, prostrate shrubs that creep along the ground.

4. Reproduction — Species or varieties which reproduce vegetatively as well as by seed. Preferred plants are those that can be established on

Adapted in part from a forthcoming U.S.D.A. Forest Service Research Paper, by Eamor C. Nord and Lisle R. Green, to be published by the Pacific Southwest Forest and Range Experiment Station, P.O. Box 245, Berkeley, California 94701.

Fig. 30.1. Antelope bitterbrush was established mostly from rodent cached seed on a cut slope in Mono County, California.

wildland sites by direct seeding; other planting methods are much more costly.

5. Root systems — Plants with deep, many-branched and fast-growing root systems are preferred.

6. Relative flammability — Many factors are involved, but plants that contain high moisture content in the foliage, preferably over 75 percent, during the summer and dry season are preferred. This high content is frequently correlated with mineral or ash content of the foliage.

7. Palatability — A desirable characteristic except during the establishment stage when grazing, browsing, or clipping by rabbits may destroy the young plants.

None of the plants tested to date have met all of these criteria. However, creeping or Sonoma sage *(Salvia sonomensis)* fits many requirements for plants intended to reduce fire hazards. This plant provides a dense, low ground cover and has proved that it can reduce the rate of fire spread within annual vegetation (Phillips, *et al.* 1972). Some other shrubs have not been able to reduce fire hazards because of their high fuel volume and upstanding growth habit. As these plants mature, they accumulate dead fuel that is flammable and likely to burn readily during summer and early fall. However, some of these other plants, such as the massive saltbushes *(Atriplex)* and bitterbrush *(Purshia)*, have been tested for use as browse and in outdoor recreation site development (Nord 1965; Magill and Nord 1963). They have proved to be well suited for rehabilitation of deteriorated rangeland and — to some extent — raw cut-fill slopes and spoil banks (Figs. 30.1, 30.2).

Fig. 30.2. Allscale saltbush was established from seed cast on fill slope in Inyo County, California. The saltbush stand developed within 4 years after roadway was completed.

PLANTING ON DISTURBED WILDLAND SITES

In experiments, some 30 shrub species have shown promise for planting on fuelbreak and — to some extent — on other disturbed semiarid-to-arid wildland sites. This group included 17 low growing, 4 semiprostrate or medium height, and 9 upright or tall shrubs. A few plants within each of these groups have demonstrated a rather wide range of adaptability to different soils extending from near the coast to 5,000-ft. elevation (Table 30.1). A few of the shrubs were readily established in the same manner as grasses by direct seedings as well as from transplants (Nord, *et al.* 1971b).

Creeping sage was established by direct seedings, fresh stem cuttings, and rooted transplants. Generally, more than 50 percent of the plantings survived after initial first-year establishment. Plants grew rapidly upon most of the sites where they were planted in southern California. Lavender-cottons *(Santolina)*, rockrose *(Cistus)*, and some saltbushes *(Atriplex)* also adapted well to many different sites and conditions. Certain saltbush, bitterbrush, and creeping sage plants were successfully established from direct seedings upon properly prepared seedbeds and, in some instances, upon highway cut and fill slopes (Figs. 30.3, 30.4). Several other plants tested were suitable for use to reduce fire hazards but were usually restricted to limited sites or other conditions (Table 30.2). In many instances, introduced species such as saltbushes and green galenia *(Galenia pubescens)*

TABLE 30.1

**Shrub and Subshrub Plants With Relatively Wide Adaptability
To Soils and Climatic Conditions**

LOW GROWING — Usually under 1 foot tall:
  Creeping or Sonoma sage *(Salvia sonomensis)*
  Gray santolina or lavendercotton *(Santolina chamaecyparis)**
  Green santolina or lavendercotton *(S. virens)**

SEMIPROSTRATE — Usually under 2 feet tall:
  Descanso rockrose *(Cistus crispus)**

TALLER SHRUBS — Generally over 3 feet tall:
  Saltbushes: Fourwing *(Atriplex canescens)*, allscale or desert *(A. polycarpa)*, quail-
      bush *(A. lentiformis)*, Brewer's *(A. 1. var. Breweri)*
  Bitterbrush: Desert *(Purshia glandulosa)* and antelope *(P. tridentata)*
  Rockroses: Gum *(Cistus ladaniferus)** and purple *(C. villosus)**

*Introduced species.

Fig. 30.3.   Fourwing saltbush on divider strip and adjoining fill slope developed from drill or broadcast seedings made during 1967 to 1968 as Interstate Highway construction between Cabazon and Palm Springs, California, was nearing completion.

Fig. 30.4. Fourwing saltbush seeded by helicopter in early 1970 on fresh terraced cut and fill slopes along Highway 60 east of Riverside, California, produced the best stands among shrub species and grasses.

TABLE 30.2
Shrub and Subshrub Plants With Limited Adaptability

LOW GROWING — Usually under 1 foot tall:

Ione manzanita *(Arctostaphylos myrtifolia)*
Caucasian artemisia *(Artemisia caucasica)**
Sweet sagebrush *(A. ludoviciana* var. *incompta)*
Saltbushes *(Atriplex cuneata)*, Gardner's *(A. gardneri)*, shadscale *(A. confertifolia)*, flat-topped or annual *(A. inflata)*, Muller's *(A. Mulleri)**, Australian *(A. semibaccata)*
Prostrate baccharis *(Baccharis pilularis)*
Ceanothus prostrate or squawcarpet *(Ceanothus prostratus)* and Mason's *(C. masonii)*
Green galenia *(Galenia pubescens)** — probably toxic
Myoporum *(Myoporum parvifolium)**
Newberry's penstemon *(Penstemon Newberryi)*
Rosemary *(Rosmarinus officinalis)** — very flammable

TALLER SHRUBS — Usually over 2 feet tall:

Saltbushes: Mediterranean *(Atriplex halimus)**, silver *(A. rhagodioides)**, Torrey's *(A. Torreyi)*

*Introduced species.

from Australia, the Mediterranean Region, or South Africa did not tolerate cold temperature. Accordingly, they were limited to lower elevations (usually under 2,000 ft.) where they are not likely to be affected by killing frosts.

The *Atriplex* and *Cistus* genera merit special mention as these genera contain many species that are useful for revegetating disturbed sites in semi-arid-to-arid climates of the world. They contain a vast, largely undeveloped storehouse, including more than 100 perennial *Atriplex* and a score or more *Cistus,* with plants ranging from low growing to relatively massive sizes. Many plants in both genera provide excellent ground cover and have proved highly effective for stabilizing raw cut or fill slopes, especially along highways in southern California. Some species reproduce vegetatively from stem layering along prostrate branches, or from root sprouts, as well as from seed (Nord, *et al.* 1969, 1971a, 1971b). Good quality seed of some species in each genus is generally available commercially or by collecting from natural stands or plantings, and at a relatively reasonable price as compared to seed of most other shrub species. For the most part, seed can be readily processed so that it germinates rapidly when growing conditions are satisfactory.

Several *Atriplex* provide ideal browse or habitat for grazing animals and game species, especially upland game birds, and some species can be harvested as fodder that is both wholesome and nutritious for domestic animals (Franclet, *et al.* 1971; Goodin and McKell 1970). Although both *Atriplex* and *Cistus* contain many plants that are drought tolerant, there are several *Atriplex* species that are outstanding in this respect. They can withstand either torrid or frigid temperatures, and soils that range from excessively to poorly drained.

Some *Atriplex* species contain endotrophic mycorrhizae (Alden 1975; Williams, *et al.* 1974), enabling the plants to grow satisfactorily on soils of very low fertility. Practically all the *Atriplex* species have a relatively high mineral content — (up to 25 percent) — in the foliage (Chatterton 1970; Nord and Countryman 1972; Philpot 1970). Consequently, these plants have a high osmotic potential that, together with the unique trichomes or balloon-like vesicles on the leaves (Black 1954; Mozafar and Goodin 1970), enables them to absorb and retain moisture effectively from the ground or from the saturated atmosphere (Wallace and Romney 1972). The foliage maintains relatively high moisture throughout the year, and consequently is less flammable and not as susceptible to burning as most chaparral species during most of the year. The main shortcoming of *Atriplex* is the susceptibility of young plants to damage and destruction by small animals, especially rabbits and squirrels. Although some strains within a species appear to be less vulnerable to such depredation,* no effective and practical means has been developed to overcome this problem in wildland plantings.

---

*Nord, E., and R. Stallings. Unpublished data on file at Pacific Southwest Forest and Range Experiment Station, Riverside, California.

DESCRIPTIONS OF SELECTED SHRUBS

A few shrubs that are useful for planting on disturbed sites and have shown considerable promise for planting on a relatively wide range of conditions between the coast and about 5,000-ft. elevation are described here. Sources for seed for many species will have to be developed or enlarged before extensive seedings can be made. Normally only a few of these plants are available at nurseries; therefore, it is generally advisable to make arrangements at least one year ahead so that seed can be obtained and plants propagated and ready when needed.

Fourwing saltbush (Atriplex canescens). Fourwing saltbush is widely distributed on dry plains and hillsides throughout much of the western United States and Mexico. It is common in the desert and semidesert areas from near the coast and inland to about 7,000-ft. elevations. It grows on a variety of soils where pH ranges from 6.0 to 7.8 or higher. Plants range up to 7 feet tall and 10 feet or wider across the crown, depending on the ecotype and site. Flowering, fruiting, and growth characteristics can vary greatly and a number of climatic or geographic strains have been recognized and described (Hanson 1962; Nord, *et al.* 1971b). Because of its deep and widely spreading root system, drought tolerance, cold hardiness, and ease of establishment, this plant is especially desirable for stabilizing disturbed soils (Plummer, *et al.* 1963). The plants provide excellent browse for livestock and big game and habitat for upland game birds. To maintain a healthy productive stand that does not burn readily, any buildup of excessive foliage or fuel that otherwise is likely to develop as the plants mature should be removed periodically.

Fourwing saltbush can be readily established on disturbed sites by direct seeding or from potted transplants. Seed from a nearby source or at least a similar climate is preferable to seed grown under other conditions. It should be dewinged and sown from ½ to 1 inch deep. The best time for planting in southern California is in late winter or early spring (Nord, *et al.* 1971b). Seeding 2 to 4 pounds dewinged seed per acre is usually sufficient when drilling into the ground. However, up to 8 or 10 pounds per acre have been broadcast on disturbed areas. This seed is generally available commercially.

Newly planted areas should be protected against rabbits throughout at least the first year, but thereafter, the plants become less susceptible to damage by clipping or browsing. A small mesh fence, such as 1-inch poultry wire netting about 30 inches high, is usually adequate to protect plantings from rabbits. After plants are well established and crowns are 2 feet or taller, they resprout vigorously after mowing or burning. Occasional mowing to about a 6-inch stubble height, grazing with livestock, or prescribed burning reduces accumulation of old or dead fuel and stimulates regrowth, much of it laterally, and helps to maintain the plants in thrifty and vigorous condition.

Allscale or desert saltbush (Atriplex polycarpa). This intricately branched shrub has relatively fine foliage up to 5 feet tall and equally as wide across the crown. Found mostly in basins, usually below 5,000-ft. elevations throughout Southwestern United States, it develops best on deep, well-drained soils that are usually basic in reaction with a pH from 7.5 to 8.0 or higher. It is not as cold tolerant as fourwing saltbush but is more drought-hardy, often growing with only 4 to 5 inches total annual precipitation, most of which comes during late fall or winter.

The foliage of allscale is nearly as palatable and nutritious as four-wing saltbush for livestock and big game. And the plants provide excellent cover and nesting for quail. The plants resprout profusely after repeated mowing or clipping. Allscale is similar in many respects to fourwing salt-bush in ease of establishment and seed production, but is more restricted in its adaptability. Seed, which is usually available commercially, does not require scarification or any special processing as it germinates rapidly and usually within a few days in favorable conditions. Possible seed losses by rodent depredation, or whether a repellent may be necessary for plantings, have not been determined. However, young plants are susceptible to damage and destruction, requiring protection from rabbits at least during the first year.

Plants may grow to 1 foot or taller with a root system up to 4½ feet deep and twice as wide during the first year. This shrub has been planted and grown successfully on a variety of soils where pH ranges from 6 to 8 or higher (Nord, *et al.* 1971b). On sites where it is adapted, it may be included with mixtures of other saltbush species. Occasional mowing or grazing by livestock not only maintains plants in better condition, but also helps new seedlings to become established where the ground is disturbed by trampling or by equipment operation.

Rockrose (Cistus *sp.*). Several shrub members of this genus from the Medi-terranean Region were introduced and tested in the United States, starting in the late 1940's. A few species were found to be quite versatile in use — especially for stabilizing roadsides and other disturbed areas and to some extent for wildlife cover and landscaping. Although species vary, they are generally quite drought tolerant and moderately cold hardy. They can grow satisfactorily with as little as 15 inches of annual precipitation and where minimum temperatures do not drop below about 10°F. They are adapted to a wide range of soils, but do best where soil is deep, well drained, mod-erately coarse to medium textured, and have a pH between 5.5 to 7.7. Most species are sensitive to too much water and bad drainage.

Rockrose plantings, practically all from rooted transplants, have made satisfactory growth on a wide range of sites extending from near the coast to 5,000-ft. elevation, on different soils and climatic conditions throughout southern California. Rate of survival for rooted transplants usually exceeds 60 percent. Growth on brush-cleared areas has been satisfactory. Most species tested are low in palatability to livestock and seem relatively immune to rabbits. Therefore, they have an advantage over many other shrub species that are more susceptible to these agents.

Many of the rockrose species have conspicuous, attractive flowers that appear during May and June. Seed of the species grown in southern Cali-fornia ripens during midsummer to early fall, when the capsules or seed heads can be stripped or plucked off the plants and dried, and the seed separated and cleaned. Seeds generally have high viability, but require treatment to speed up germination. An effective and practical means to increase and speed up germination is to soak the seed for 2 minutes in boiling water (Kenneth R. Montgomery, personal communication 1975). Sufficient seed has not been available, and means to improve germination have not been tested sufficiently to recommend direct seedings under field planting conditions. At least some of the species can probably be established by direct seedings if seed is treated to increase and hasten germination, care is exercised in selecting the planting sites, and good agronomic practices are followed in such plantings.

The mineral or ash content ranges from 4.7 to 10.4 percent; the amount varies according to the rockrose species and the time of the year. Moisture content of the foliage usually drops below 70 percent during late summer and fall; consequently, mature stands are susceptible to burning under extreme burning conditions. Except for the low growing Descanso rockrose *(Cistus crispus)* that is generally less than 2 feet tall, most rockroses are not recommended for plantings intended primarily to reduce fire hazards. They can, however, serve effectively for rehabilitating disturbed sites, and for other useful purposes throughout a wide range of sites — especially in the southwest (Juhren and Montgomery 1976).

Descanso rockrose (Cistus crispus). This plant was initially grown at the University of Washington Arboretum (Mulligan 1953). A form of rockrose, morphologically similar to *C. crispus* but of uncertain origin, was set out in the Arboretum field plot at the Descanso Gardens in about 1966. This accounts for the name "Descanso Rockrose." This shrub is attractive, aromatic, semiprostrate, and evergreen, usually 1½ to 2 feet tall and spreads by means of stem layers along the decumbent branches to form a crown several feet wide (Fig. 30.5). The light green, compact foliage provides a dense cover that excludes most herbaceous plants from developing below the crown. During late spring and early summer, plants have numerous attractive, bright rose-lavender flowers. Flowering occurs about 2 to 3 weeks later and persists for a longer period than for most other *Cistus* species.

The plant is relatively drought tolerant and can grow satisfactorily with as little as 15 inches annual precipitation. It is well adapted to a wide range of soils varying from moderately coarse to medium textured between pH 5.5 to 7.7. Survival from potted plants has generally exceeded 70 percent. They often grow from 6 to 12 inches the first year, spreading outward thereafter at a rate of about 1 foot per year. Plantings made in 1974 on fuelbreaks at two southern California sites showed the first year's survival and growth rate as summarized in Table 30.3.

The main difference in size of the plants beginning the second year was due to a winter freeze at the Pisgah Peak site. However, most plants had resprouted by early spring.

Number of seeds per pound averages about 605,000, or 1,330 per gram. Seeds have high viability — ±85 percent — but are dormant. Germination is improved either by a 2-minute soak in boiling water or by drying at 150° or 185°F. for the same period. These treatments during a 75-day

Fig. 30.5. Rockrose *(Cistus)* species can grow rapidly on undeveloped or degraded soils to provide good ground cover. Descanso rockrose *(C. crispus)*, left and center, is a semiprostrate shrub usually less than 2 feet tall that spreads to form a dense crown several feet across. Purple rockrose *(C. villosus)*, right, is an upstanding shrub 4 to 5 feet tall.

test yielded over 85 percent germination as compared to 35 percent or less for untreated seed (Kenneth R. Montgomery, personal communication 1975). Sufficient seed has not been available for field plantings. However, several hundred seedling plants were planted at coastal and inland areas to determine suitability of this plant for extensive use on brush-cleared areas, and to provide a better source of seed in the western United States.

This plant, because of its low growth habit and fuel volume, should burn with less heat output than most chaparral vegetation and therefore could serve to reduce fire hazards. It also shows good promise for rehabilitating disturbed sites, for landscaping, and for other useful purposes throughout a wide range of sites.

TABLE 30.3

Comparison of Two Descanso Rockrose Plantings

Site	Elevation (ft.)	Survival (percent)	Height (in.)
Laguna Beach, Orange County	500	92	15
Pisgah Peak, San Bernardino County	5000	71	4

Fig. 30.6. Gum rockrose *(C. ladaniferus)* is a tall, evergreen shrub with numerous willow-like stems usually 5 to 6 and occasionally up to 10 feet tall, that forms a loose, open crown. Plants bear numerous attractive white, bowl-shaped flowers during late spring and early summer.

Gum rockrose (Cistus ladaniferus). This tall, dark evergreen shrub has numerous willow-like stems, usually five or six, and in some instances grows to 10 feet tall with a loose, open crown (Fig. 30.6). Gum rockrose derives its name from the gum-resin that accumulates, making the leaves, young twigs, and seed capsules very sticky. During late spring and early summer, plants bear numerous, attractive white, bowl-shaped flowers with a basal cluster of rich stamens. Some varieties have prominent crimson spots on the petals.

This plant is hardy and seems well adapted to a wide range of sites from near the coast to about 5,000-ft. elevation. It grows best on deep, well drained soils. It is sensitive to excessive watering and bad soil drainage. Gum rockrose has been planted for stabilizing cut and fill slopes and other disturbed areas and to some extent for landscaping throughout southern California since the early 1940s. Plantings from rooted transplants have generally had high survival — often 80 percent or higher — and on favor-

TABLE 30.4
Survival and Growth of Gum Rockrose (North Mountain Experimental Area, California)

Site	Elevation (ft.)	Annual Precipitation (in.)	Survival (percent)	Growth Height (in.)	Crown Diameter (in.)
Wolfskill plot, NMEA	2600	12	94	20	24
Nixon plot, NMEA	3500	18	98	56	40
Vista Grande, San Bernardino N.F.	5000	23	88	17	37

able site and moisture conditions, have grown up to 1 foot a year during the first few years. Rates of survival and growth of gum rockrose over a 4-year period at three sites on or near the North Mountain Experimental Area (NMEA) in Riverside County, California are given in Table 30.4.

The results and subsequent observations of these and other plantings show that this rockrose requires at least 15 inches annual precipitation to grow satisfactorily. Although the plants withstand somewhat colder temperatures, the stem tops were repeatedly frozen back at the Vista Grande site, where temperatures often drop to about 10°F.

Seed is partially dormant and requires treatment to germinate. Soaking seed for 2 minutes in boiling water is equally as effective for this as with most other *Cistus* seeds. Direct seedings have not been used to any extent with this species. Where large numbers of seedlings are needed, the most satisfactory method is to seed in flats or nursery beds. Cuttings treated with a hormone will root when kept under intermittent mist. Rooted plants, preferably in pots 2 to 3 inches square by 8 to 12 inches deep can be transplanted during late winter or very early spring when the ground is still wet and preferably before new growth commences.

Gum rockrose is seldom grazed or browsed and, therefore, usually does not need special protection from animals. It can be used in conjunction with other shrubs for soil stabilization and for landscaping in brush-cleared areas where it is adaptable. It is not recommended for use to reduce fire hazards because of its upstanding habit. And the large amount of fuel it produces is susceptible to burning when the foliage dries out during late summer and fall.

Purple rockrose (Cistus villosus). This plant is an upstanding, moderately dense evergreen shrub generally 3 to 4 feet tall with rounded crowns 2 to 4 feet in diameter (Fig. 30.5). Foliage is dark green and slightly aromatic, and plants bear showy lavender flowers during late spring and early summer. It is equally as hardy and adapted to most of the same conditions as gum rockrose. It is admirably suited and has been planted extensively for erosion control and to some degree for landscaping. Plantings established 15 to 20 years ago in southern California have persisted throughout the years, but plants stagnate beginning in 8 to 12 years (Kenneth R. Montgomery, personal communication 1975). Thereafter, dead fuel accumulates and as the foliage dries out during late summer and fall, the plants are susceptible to burns.

Under favorable conditions, purple rockrose produces seed within 2 years. Seed of this species is available commercially. There are about 525,000 seeds per pound, or 1,150 per gram. Seed is partially dormant but germination can be increased or hastened by: (1) boiling water soak for 1 minute, (2) dry heat at 180°F. for 2 minutes, and (3) gibberellic acid soak in 200 ppm for 2 hours (Kenneth R. Montgomery, personal communication 1975).

Purple rockrose has naturalized to a limited extent in some areas, but direct seedings on wildland sites have resulted in no more than sparse stands. This lack of more abundant growth may be due to the slow germination and seedling development from untreated seed, or because of insufficient site preparation or adverse soil moisture conditions. Potted transplants such as are carried by some commercial nurseries have shown good survival, and have grown satisfactorily at several places in southern California. Plantings in 2- by 2- by 8-inch bands at three sites on or near the North Mountain Experimental Area in Riverside County, California averaged 70 percent or higher survival, and on two out of three sites were growing satisfactorily 8 years after they were planted. The high mortality that occurred between the second and fourth year at the lower elevation site and dieback of the plant tops at the higher elevation site indicate that this *Cistus* is not well adapted to either drier conditions where annual precipitation is less than about 15 inches or where temperatures may drop below 15°F. Except for moderate clipping of young plants by deer, mostly during the first year, this plant is not greatly affected by grazing or other destructive agents.

The primary use of purple rockrose should be for planting on disturbed sites, such as road cut and fill slopes to protect the soil and to reduce reencroachment of flash fuels, and for landscaping.

Creeping or Sonoma sage (Salvia sonomensis). This sage is an aromatic, semiprostrate, suffrutescent plant found in the chaparral zone along the Sierra Nevada and Coast Range in California at elevations below 6,000 feet. It grows on relatively shallow, moderately acid to neutral soils where annual precipitation averages 15 to 40 inches and temperature ranges from 15° to more than 100°F. It can withstand some frost and drought, but does not tolerate alkaline soils (Pratt, *et al.* 1971).

Creeping sage plants provide excellent protection to soil and have value for fire abatement (Phillips, *et al.* 1972). It forms mats often several feet across and crowds out much of the flashy fuel, primarily annuals, because of its low, dense foliage that is not more than 8 to 10 inches tall (Fig. 30.7). It is not grazed or clipped to any extent by animals and therefore has a distinct advantage over more palatable shrub species. It is adaptable to many site conditions throughout the chaparral zone.

Plants have been established successfully by direct seedlings, fresh stem cuttings, and by transplants. Once established, it spreads up to 1 foot outward per year, and roots profusely along the prostrate branches. Survival of plants grown from seed, stem cuttings, and transplants has been consistently high; usually more than 70 percent of the plants have lived after

Fig. 30.7. Creeping or Sonoma sage is a low growing plant that spreads primarily by layering along the branches to form dense mats several feet across. Plants shown from 8-year-old seeding at North Mountain Experimental Area, Riverside County, California, are over 15 feet across the crown.

the first year's establishment. Plants are easily propagated from stem cuttings taken at almost any season of the year without benefit of hormone treatment (Nord and Goodin 1970), although best rooting in field plantings has developed from "hardened" stem or branch sections taken during the late winter. Cuttings to be planted in the field should be at least 15 inches long and set almost that deep with soil firmly tamped around the stem.

The seed — of which there are about 500,000 per pound — is available only in limited amounts from a few commercial seed dealers. Otherwise, it must be collected from natural stands during mid-May to early June. Seed retains high viability for 5 years or longer when stored in airtight containers at refrigerator temperatures, with or without the recommended gibberellic acid treatment. Satisfactory stands have resulted from planting seed treated by soaking in gibberellic acid at 200 ppm for 1 to 2 hours to break dormancy (Nord, *et al.* 1971a), and planted at ½ to 1 inch depth during the late winter to early spring while there was sufficient moisture present in the soil for plants to become established.

Creeping sage does not resprout from either roots or stems after burns, but new stands may develop from seedlings that appear following fires. It is highly susceptible to weed killers and, thus, would need to be re-established if it were sprayed with herbicides. Such spraying often takes place in fuelbreaks and along roadways in some areas. Moisture content of foliage goes below 60 percent during the late summer and early fall and foliage will then burn if there is much dead litter to carry the fire.

Gray and green lavendercotton or Santolina (Santolina chamaecyparissus L. *and* S. virens). These two plants, native to the Mediterranean region, are both low to semierect, compact, aromatic evergreen subshrubs with numerous small erect to semidecumbent branches and fine, porous foliage (Fig. 30.8). They have a fibrous root system that penetrates quite deeply and widely into the soil. Green lavendercotton plants, up to 30 inches high and 4 to 5 feet wide, are considerably larger than gray lavendercotton, which may grow to be 18 inches high and up to 3 feet across the foliar crown in good growth situations. However, both species are smaller — commonly about 1 foot — upon dry sites.

Both species, but particularly gray lavendercotton, are drought tolerant, hardy and versatile, and can be grown successfully from the low desert (Mathias, *et al.* 1961, 1968) throughout most of the chaparral zone in California. Once established, they persist under relatively harsh conditions. In the field, they spread slowly, usually 1 to 2 feet per year, by layering along the decumbent branches. Initial establishment has been only by

Fig. 30.8.   Green and gray (center) lavendercotton or Santolina are low, compact evergreen subshrubs that tolerate harsh site conditions. Their dense cover and deeply penetrating and wide root system provide good soil protection.

transplants grown from stem cuttings under intermittent mist. I know of no instance where lavendercotton has been established on field sites by direct seedings, nor have any natural seedlings been observed. Munz and Keck (1959) report that green lavendercotton has occasionally escaped and become established in parts of central and southern California. These plants provide good protection to soil as they spread to form dense cover that crowds out most annuals. Plants are not known to be grazed or clipped to any great extent by animals.

Rooted plants grown from stem cuttings have had consistently high survival rates and have made satisfactory growth on a variety of cleared brushland and highway cut-fill and other disturbed areas in the coastal, San Gabriel, and San Jacinto Mountains up to 5,000-ft. elevations. When planted where weedy competition was removed immediately preceding planting, 75 percent or more of potted plants were growing actively 5 years after planting without supplemental irrigation or other maintenance. However, in Los Angeles County some stands have faded out within 5 years (Arthur M. Arndt, personal communication, August 24, 1975).

Flowering heads form beginning the first year. Stem layering that develops during the second year is more frequent on gray than on green lavendercotton plants. The mineral or ash content ranges from 5 to 12 percent, and fuel moisture varies considerably according to the season and between the two species. The foliage dries out sufficiently during late summer and fall so these plants are susceptible to burning. The fine foliage provides a porous fuel that resembles grassland fuels somewhat; thus, wildfires will most likely burn rapidly when fuel moisture is low.

The chief value of these two lavendercottons is their ability to grow in dry situations, and their usually low fuel volume. Green lavendercotton in particular has low fuel volume. Plants of both species are generally available at commercial nurseries.

REFERENCES

Alden, E. F. 1975. Endomycorrhizae enhance survival and growth of fourwing saltbush on coal mine spoils. USDA Forest Serv. Res. Note RM-294. Fort Collins, Colorado: Rocky Mountain Forest and Range Exp. Stn.

Black, R. F. 1954. The leaf anatomy of Australian members of the genus *Atriplex*. 1. *Atriplex visicaria* Howard and *A. nummularia* Lindl. *Australian Journal of Botany* 2:269–286.

Chatterton, N. J. 1970. Physiological ecology of *Atriplex polycarpa:* Growth, salt tolerance, ion accumulation, and soil-plant-water relations. Ph.D. dissertation on file at University of California, Riverside.

Franclet, A., H. N. Le Houérou and others. 1971. [The Atriplex in Tunisia and North Africa.] FO:SF/TUN 11 — Technical Report 7, 271 p. (FAO of UN). [Translation from French.]

Goodin, J. R. and C. M. McKell. 1970. *Atriplex* spp. as a potential forage crop in marginal agricultural areas, pp. 158–161 in *Proc. XI International Grasslands Congress*. Univ. of Queensland Press.

Hanson, C. A. 1962. Perennial *Atriplex* of Utah and the northern deserts. M.S. thesis on file at Brigham Young University, Provo, Utah.

Juhren, M. C. and K. R. Montgomery. Long-term responses of introduced shrubs in southern California. (Manuscript in preparation.)

Leiser, A. T. and E. C. Nord. Caucasian artemisia: A promising ground cover from the USSR. (Manuscript submitted to *Horticultural Science.*)

Magill, A. W. and E. C. Nord. 1963. An evaluation of campground conditions and needs for research. USDA Forest Serv. Res. Note PSW-4. Berkeley, California: Pacific Southwest Forest and Range Exp. Stn.

Mathias, M. E., W. Metcalf, M. H. Kimball and others. 1961. *Ornamentals for Low-elevation Desert Areas of Southern California.* California Agric. Exp. Stn. Ext. Serv. Circ. 496.

Mathias, M. E., W. Metcalf, M. H. Kimball and others. 1968. *Ornamentals for California's Middle Elevation Desert.* California Agric. Exp. Stn. Bull. 839.

Mozafar, A. and J. R. Goodin. 1970. Vesiculated hairs: a mechanism for salt tolerance in *Atriplex halimus* L. *Plant Physiology* 45:62–65.

Mulligan, B. O. 1953. A *Cistus* collection in Seattle. *Journal of the California Horticultural Society* 14(4):131–141.

Munz, Philip A. and David D. Keck. 1959. *A California Flora.* Berkeley: University of California Press.

Nord, E. C. 1965. Autecology of bitterbrush in California. *Ecological Monographs* 35:307–334.

Nord, Eamor C., Donald R. Christensen and A. Perry Plummer. 1969. Atriplex species (or taxa) that spread by root sprouts, stem layers, and by seed. *Ecology* 50(2):324–326.

Nord, E. C. and C. M. Countryman. 1972. Fire relations, pp. 88–97 in *Wildland Shrubs — Their Biology and Utilization.* USDA Forest Serv. Gen. Tech. Rep. INT-1. Ogden, Utah: Intermountain Forest and Range Exp. Stn.

Nord, Eamor C., and J. R. Goodin. 1970. Rooting cuttings of shrub species for planting in California wildlands. USDA Forest Serv. Res. Note PSW-213. Berkeley, California: Pacific Southwest Forest and Range Exp. Stn.

Nord, Eamor C., Louis E. Gunter and Stewart A. Graham, Jr. 1971. Gibberellic acid breaks dormancy and hastens germination of creeping sage. USDA Forest Serv. Res. Note PSW-259. Berkeley, California: Pacific Southwest Forest and Range Exp. Stn.

Nord, Eamor C., Patrick F. Hartless and W. Dennis Nettleton. 1971. Effects of several factors on saltbush establishment in California. *Journal of Range Management* 24(3):216–223.

Phillips, Clinton B., Louis E. Gunter, Grant E. McClellan and Eamor C. Nord. 1972. Creeping sage — a slow burning plant useful for fire hazard reduction. Sacramento: California Div. For. Fire Control Notes 26.

Philpot, C. W. 1970. Influence of mineral content on the pyrolysis of plant or plant materials. *For. Sci.* 16(4):461–471.

Plummer, A. P., S. B. Monsen and D. R. Christensen. 1963. Restoring big-game range in Utah. Utah Division of Fish and Game Publ. 68-3.

Pratt, Parker F., Eamor C. Nord and Francis L. Bair. 1971. Early growth tolerances of grasses, shrubs, and trees to boron in tunnel spoil. USDA Forest Serv. Res. Note PSW-232. Berkeley, California: Pacific Southwest Forest and Range Exp. Stn.

Radtke, K. 1974. Low-fuel plant project report, July 1, 1973–June 30, 1974. U.S. Forest Service and Los Angeles County Fire Department, Coop. Agreement no. 21-253. (Unpublished report on file at Pacific Southwest Forest and Range Experiment Station, Riverside, Calif.)

Wallace, A. and E. M. Romney. 1972. Radioecology and ecophysiology of desert plants at the Nevada Test Site. TID-25954. Springfield, Virginia: National Technical Information Service, U.S. Dept. of Commerce.

Williams, S. E., A. G. Wollum and E. F. Aldon. 1974. Growth of *Atriplex canescens* Pursh., Nutt. improved by formation of vesicular-arbuscular mycorrhizae. Las Cruces, New Mexico: J. Series 487, New Mexico Agric. Exp. Stn.

31. Revegetation of Disturbed Intermountain Area Sites

A. Perry Plummer

THE ABILITY OF PLANT SPECIES TO ADAPT to a site is an all-important requirement for successful revegetation, whether the goal is stabilizing disturbed areas or producing improved cover and greater forage for game or livestock. Good establishment qualities are essential to adaptation. Other attributes to be taken into account are soil building qualities, cover, growth rates, availability of seed and planting stock, and tolerance to grazing. Precipitation or available moisture, aspect, slope, and the nature of the soil substratum are important factors of the habitat to be reckoned with in making decisions on what to plant. Persons making the selection not only must have a good understanding of the site and climate, but also of the capabilities and attributes of plant species.

As part of a cooperative research project with the Utah Division of Wildlife Resources on restoration of depleted ranges, a large number of species have been tested on disturbed areas such as roadcuts and fills, mine spoils, gullies, and seriously eroded hillsides. Several scientists are engaged in these programs, and their experiences are drawn upon in this paper.

A number of shrubs, grasses, trees, and forbs are useful for revegetating bare or eroded areas by direct seeding or transplanting. While there are new species and variants to be discovered or developed for particular sites, a sufficient number of adapted ones are known for fairly adequate revegetation of most sites, although seeds and transplanting stock may not be available for some species.

Because of the demand for information regarding shrubs, most emphasis is placed on this class of plants. Also, the need for forbs to revegetate disturbed areas has prompted more research than in the past. While some trials are continuing with grasses, this work is limited because good background knowledge has been developed for many of them over the past 50 years (Hafenrichter, *et al.* 1949, 1968; Anderson, *et al.* 1953; Hanson 1965; Plummer, *et al.* 1955, 1968; Farmer, *et al.* 1974).

Development of some of the important information in this chapter was made possible by funds provided by the Federal Program for Wildlife Restoration Project W-82-R.

Much remains to be learned in making refinements to provide improved plants as well as to assure their success through improved techniques. No doubt, there will be improvement in the future by breeding and selection to make plants more productive and efficient. However, the problems are urgent, and it is not possible to wait for the consummation of improvement programs.

Vegetation Establishment

Seven major purposes for establishing vegetation on disturbed areas are to:

1. Stabilize the site against water and wind erosion to reduce hazards of silt pollution and dust.
2. Prevent undesirable plants from gaining a competitive foothold.
3. Furnish cover and food for wildlife and livestock.
4. Beautify the area with appropriate cover plants.
5. Establish vegetation to screen unsightly backgrounds.
6. Provide shade and a pleasant environment for people.
7. Provide additional area for crops where climate and soil are satisfactory.

Planting Techniques

In restoration work on disturbed areas, direct seeding, nursery stock, wildings from naturally occurring stands, container stock grown in greenhouses and cold frames, and cuttings have been tried. All have a place, which varies with environment. Techniques are mentioned briefly because of their importance to adaptation. Transplanting techniques may vary somewhat by species because certain plants may be better established through one approach than another. For instance, oldman wormwood, chee reedgrass, and shadscale saltbush have not been successfully established from seed, but establish well from vegetative plantings. Wormwood is planted directly by stem cuttings or by rooted cuttings, chee reedgrass by rhizomes, and shadscale by wildings.

Because of the difficulty of operating large power machinery on many disturbed areas, pulaskis, shovels, planting bars, hand augers, and similar tools have been used in transplanting. On steep slopes, hand techniques proved best, particularly on sites of less than one acre. Plants grown in containers in greenhouses or cold frames for 16 to 30 weeks have been readily established by removing them from the carton and planting them individually as bare-root stock in a prepared hole. These seedlings often survive better than 1- and 2-year-old nursery stock. They seem to be fresher, and new root growth starts more quickly.

Of course, planting the entire contents of a carton helps to insure establishment on harsh sites and is a widely recommended procedure (Ferguson and Monsen 1974). When using this plan, only one or at most a few seedlings should be grown in a container. Plants and soil are generally removed from the container at planting time. Cardboard milk cartons are suitable containers. Other, more sophisticated types of containers have been used to advantage.

Direct seeding has been employed most often for planting herbaceous species. However, on severe sites, vegetative plantings may be warranted (Monsen, in press). Seeds of rapid growing native shrubs have often been included in mixtures with grasses and forbs in direct seeding.

Because of the problem involved in using drills on irregular disturbed areas, broadcasting is the usual seeding technique. Broadcasting proved more acceptable than drilling, because the resulting stands have a more natural appearance. Also, the planting depth is better suited to the variety of different-sized seeds found in mixtures of species. Aerial seeding has been found to be superior to hand broadcasting, particularly on large and irregular areas, because seeds can be more uniformly dispersed (Plummer *et al.* 1968). A helicopter provides an efficient means of broadcasting seed on small areas. Fixed-wing planes, appropriately equipped, work well on large areas.

Pipe harrowing or chaining provides a suitable way to cover the seed and provide a seedbed where this is needed (Plummer *et al.* 1968). Ample application of seeds is not waste, but helps to insure a stand. If savings are to be made, they should not be at the expense of seed unless seed is scarce and difficult to obtain. Where annual precipitation is in excess of 12 inches and soils are sterile, application of 40 to 60 pounds of available nitrogen and 100 pounds of available phosphorus is usually helpful. Need for other nutrient minerals can be determined from soil analysis.

Mulch is recommended. The types that do not blow or wash away are the most effective. A tar-straw mulch is effective for large disturbed areas, but should be thin enough to avoid smothering emerging seedlings. Mulch also aids in holding moisture in the soil, particularly on steep slopes. Pits, basins, or furrows also help to reduce runoff and store a greater amount of moisture.

Transplanting is usually done in early spring while the ground is still wet from accumulated winter moisture. In contrast, direct seeding is best done in the late fall or winter to take advantage of accumulated moisture. There are exceptions. In the southwest where summer and fall rains are dependable, warm season species are best planted just prior to the summer rains (Anderson *et al.* 1953). In high mountain areas, seeding is best done in early summer (Plummer and Fenley 1950). Transplanting in high mountains is best done in late fall, just before winter snows.

Selecting Species

Adapted species are listed in Table 31.1, at the end of this chapter. They are rated from 1 to 5 (very poor to very good) for eight characteristics that make them useful on disturbed sites. Their adaptation to acid and alkaline soil also is indicated. Some species have a much greater range of adaptation than others. For each species, vegetal zones are indicated in approximately the order of how well suited the species is to each. Admittedly, judgments made for some are much better than for others. The amount of study and background of experience has varied among species.

Any firsthand knowledge gained about plants in a specific vegetal, soil, or climatic area is helpful in making a good choice of species for revegetation purposes. A knowledge of the attributes of plants is helpful to select the most suitable. This usually requires observation by trial planting under the type of environment to be encountered in the field. Often this is more of an art than a science. A "green thumb" is helpful.

Trials take time. Consequently, the experience of others furnishes a good beginning. As a result of developing knowledge, the best plant today may not be the best tomorrow, but one must proceed on the basis of available information. Checking species against the ratings of attributes in Table 31.1 should be of assistance in selecting those for particular problem areas.

Some Attributes to Consider

Differences in the mature stature of plants are important. These differences may be traceable to the environment, particularly site quality. This is more evident in woody than herbaceous plants because of their inherently larger size. Herbaceous plants are similarly affected by environment, but differences are not so striking.

Where there is abundant moisture and deep fertile soil, Siberian elm, Russian olive, boxelder, bigtooth maple, silver buffaloberry, honeylocust, or Rocky Mountain juniper will be trees (Fig. 31.1). On a severe site, they may grow and persist as low- or medium-height shrubs. Although water requirements are higher, the same is true of streamside trees such as thin-

Fig. 31.1.   A 12-year-old Russian olive transplant maintaining itself well as a shrub along with grasses and forbs on steep south facing roadcut.

leaf alder, water birch, Scouler willow, and narrowleaf cottonwood. They become shrubby when growing some distance from the stream.

In addition to environmental aspects, genetically controlled factors are also important. Expressed genetic differences in populations of a species between geographical and altitudinal areas sort out into ecotypes or biotypes (Clausen, et al. 1940, 1941).

Usually ecotypes have evolved over long periods of time as a result of natural selection. Consequently, they are generally well adapted to the sites on which they grow, and usually to similar kinds of areas. Hence, it can usually be assumed that ecotypes, native in a particular habitat and climate, would also adapt to a similar climate or vegetal type.

When plants are established in environments quite different from their natural sites, there is a question of how long they may survive. For instance, a planting of fourwing saltbush from a southern desert shrub source died out after 4 years when established in a colder northern desert type 2,000 feet higher in elevation. However, the northern shrub ecotype was successful in the southern desert shrub type, and has persisted as well as local ecotypes. It is generally true that cool climate ecotypes of woody plants will adapt to warmer climates better than warm climate ecotypes will adapt to colder areas. Because aboveground parts of herbaceous species die down to the ground and are not markedly affected by cold temperatures, there is somewhat greater latitude in planting them outside of their natural environment.

Over broad areas, genetically small, medium, and large ecotypes of a species are generally found on harsh, moderate, and favorable sites respectively. Similar genetic size differences may also relate to the length of growing seasons, such as those resulting from differences in elevation. The smaller ecotypes are found where the growing season is shortest, and the larger ones where it is longest (Clausen, et al. 1940). The genetic differences in size are readily seen when ecotypes of a species are compared (Fig. 31.2). However, luxuriance of growth may be dramatically modified by rainfall during the growing season.

Generally, ecotypes and biotypes of a species are best adapted and can be depended on to reproduce on sites similar to those on which they developed. A rule of thumb to follow is to plant species, subspecies, or ecotypes in an environment similar to those on which they occur natively.

Other Considerations

Aside from the protective effect of vegetation on road banks, there is special need for vegetation low enough so that game or livestock cannot jump out of concealment into the path of motor vehicles. Along some roads, it may be desirable to establish plants that are not sought by animals that might otherwise distract motorists.

There are many plants of low stature, but not many fulfill the requirement of low selectivity to animals. Species of snowberry and creeping barberry achieve this fairly well on foothill and mountain ranges. American or wild licorice is a sod-forming herbaceous legume of low palatability that shows promise for this purpose through the sagebrush and mountain brush types. These plants establish well from seeds or transplanting materials.

Fig. 31.2. Two even-aged bitterbrush plants of different ecotypes. Plant on the right is an aggressive crown spreader more suitable for stabilizing loose soils.

On drier sites, such as where shadscale saltbush occurs, low rabbitbrush, matchbrush, and Louisiana sagebrush provide good cover and fit these dual objectives fairly well. This phase of revegetation of disturbed areas deserves more attention. Since it is native, why not plant shadscale? For reasons unknown, we have not been successful in seeding this saltbush. However, we have successfully transplanted young wildings from one place to another.

Other species in addition to shadscale have excellent attributes for stabilizing disturbed areas, but artificial establishment methods have not been found. An important species in this category is galleta grass. This sod-forming grass grows through a fairly wide range of climate and soils in both the northern and southern desert shrub types (West, et al. 1972). It is frequently listed as a species suitable for stabilizing disturbed sites in dry desert areas where it occurs naturally. Consequently, bid invitations are often circulated for its seed. Since it is a poor seed producer, bids are rarely made. At best, it is a difficult grass to establish by seeding or sodding. Roundleaf buffaloberry and inland saltgrass are similarly difficult to establish from seedings.

An Assortment of Plants Is Best

Planting trees or shrubs with forbs and grasses is desirable where they are adapted naturally to provide full and adequate cover under varying microenvironments, as well as to control soil slippage. The deep roots of shrubs and trees are more effective than forbs and grasses in preventing mass slippage and erosion. In addition, an assortment of shrubs, trees, forbs, and grasses provides a better blending of foliage with adjacent landscapes. Forbs and grasses serve as a protective understory cover between trees and shrubs.

Grasses and forbs that spread vegetatively, such as by rhizomes, help immensely in stabilizing ground surface. Plants of this type provide smaller openings than bunch-type vegetation. The smaller the bare openings, the better the infiltration and control of runoff. If they are equally adaptable, intermediate wheatgrass, bluestem wheatgrass, Louisiana sagebrush, and Pacific aster, all sod formers, are superior to crested wheatgrass, Indian ricegrass, arrowleaf balsamroot, and sticky geranium. However, if the site has considerable variability with a tendency to dry and slick spots, it would probably be wise to plant all of them.

As a minimum, it is generally desirable to have woody and herbaceous cover in combinations, unless special conditions are to be met, such as the need for uniform height growth which only one or two herbaceous species can furnish. It is not suggested that attempts be made to establish trees where they do not occur or will not grow; but generally there are shrubs, forbs, and grasses that will grow together on even the driest sites. The appropriate selection of these is of prime importance.

OBTAINING PLANTING STOCK

A critical need exists for areas from which desirable and adapted plants can be readily obtained for transplanting (Plummer 1970). Typical nurseries that demand intensive care are not always the answer, but they can be of help. Some acreage where grasses, forbs, and some shrubs are maintained to provide pieces of sod and portions of crowns for transplanting would serve a useful purpose. In some instances, seed may be obtained from the same plantings.

Some commercial nurseries in the west sell at least some stock of native and exotic trees, shrubs, and herbs suited to certain types of disturbed wildland areas. A few specialize in natives. There are also individuals who specialize in collecting seeds and wildings of native plants. Generally, they do this on a contract basis. Even though this is a growing enterprise, availability of adapted species, particularly of natives, continues to be much too limited. Forest Service nurseries and Soil Conservation Service plant material centers are developing materials for expanded use, but their service to private enterprise has been limited. The recently published *Seeds of Woody Plants in the United States* provides pertinent information relative to seeds of a great many of the species referred to in this paper (U.S. Dept. of Agriculture 1974).

To get the needed plant materials assembled requires planning 2 to 5 years ahead. Unfortunately, this is generally not done.

SOME IMPORTANT SHRUBS FOR USE ON DISTURBED AREAS

Sagebrush and rabbitbrush (Sunflower family). Many shrubby sagebrush species, subspecies, and ecotypes cover a vast array of sites on western wildlands. With a few exceptions, most of them are native to the West, and range from high to low in palatability. These shrubs are capable of providing planting stock and seed for revegetating disturbed areas which may range from wet to dry, sandy to heavy clay, highly alkaline to acid, and from sea level to high mountains.

Most noteworthy are the members of the big sagebrush complex which comprises the shrubs that characterize major western ranges. Four subspecies of big sagebrush are recognized. They are: (1) basin big sagebrush found mainly on lowland plains and valleys on basic soils; (2) Wyoming big sagebrush occurring over dry sites in valleys and foothills on a wide range of soils varying from acid to basic; (3) mountain (or Vasey) big sagebrush usually occupying somewhat higher elevations, but with some ecotypes occurring on lowland ranges where the soils are moderately basic to fairly acid; and (4) subalpine (or Rothrock) sagebrush occurring only in the higher mountains on slightly basic to fairly acid soil. There are variations within all subspecies (Hanks, *et al.* 1973; and McArthur, *et al.* 1974).

Other important closely related species that are lower-growing are black sagebrush, alkali sagebrush, silver sagebrush, Bigelow sagebrush, scabland (or stiff) sagebrush, and pigmy sagebrush. Some distantly related species are spiny sagebrush, fringed sagebrush, and sand sagebrush.

All of these native species establish well as transplanted nursery and container stock, and as wildings. While seeds are small, seedlings establish quickly from broadcasting. Seeds need to be cleaned only to a 10 percent purity. Usually hammermilling the collected seed heads is sufficient. Young plants grow rapidly. They reproduce in 2 to 5 years depending on how favorable the environment may be.

On the basis of experience, ecotypes within subspecies of big sagebrush and black sagebrush (Fig. 31.3) show the greatest promise for stabilizing

Fig. 31.3. Four-year-old transplants of black sagebrush, left; and basin big sagebrush, right. Both are excellent for stabilization.

the widest range of sites. However, sand and Bigelow sagebrush appear to be useful on sandy soils, at least within areas of their natural occurrence. Ecotypes of silver sagebrush are useful for stabilizing sites of above-average moisture, including fairly wet sites. The most erratic species in establishment are spiny and pigmy sagebrush. These smaller-statured species grow on severely depleted areas on desert ranges and juniper-pinyon sites. (See also Table 31.1.)

Oldman wormwood (Fig. 31.4) is an introduced sagebrush from Europe that shows important ability for stabilizing subsoil sites. Planters merely stick 12- to 20-inch cuttings into the ground in the early springtime while the soil is still wet from the winter snowmelt (Plummer 1974). Stems from the past year's growth are usually best for making cuttings, but those from older stems root well if the soil remains moist. In some instances on severe sites, rooted cuttings have been more successful than freshly made cuttings. This sagebrush is valuable because of the microenvironment it provides for the establishment of other vegetation that later replaces it. In the intermountain west, Oldman wormwood has only sparingly produced good seed. Seedlings have occasionally been found.

Rabbitbrush species, like those in the sagebrush complex, show excellent promise for stabilizing disturbed sites. Seed-producing plants develop rapidly from seedlings. The plumed or feathered seeds are carried by the wind. Where they land on disturbed places or where competition is lack-

Fig. 31.4. Oldman wormwood plants, established 4 years previously from cuttings, stabilize a roadcut in juniper-pinyon type.

Fig. 31.5. Mountain rubber rabbitbrush stabilizing disturbed opening in aspen after broadcast planting.

ing, they have a propensity for good establishment. The rabbitbrush species is commonly separated into rubber rabbitbrush and low rabbitbrush. Rubber rabbitbrush is tall-growing and composed of many subspecies and ecotypes. These may vary in height from 2½ to 7 feet. Low rabbitbrush ordinarily varies in height from 1 to 2 feet. Variation of ecotypes within subspecies is great.

Attention has been given to four subspecies of rubber rabbitbrush because of their wide occurrence in the west through a broad range of vegetal and climatic zones from low valley bottoms to higher mountains. (Hanks, *et al.* 1975.)

Mountain rubber rabbitbrush with grey-green foliage (Fig. 31.5) grows at the higher elevations and may form part of the lower subalpine vegetation. It extends well down onto foothill ranges to mix with white rubber rabbitbrush and green rubber rabbitbrush. The latter two, however, are most common on well-drained foothills and also grow out into valley and plain expanses where they mix to a considerable degree with alkali (or threadleaf) rubber rabbitbrush. Alkali rubber rabbitbrush is green, but can be easily distinguished from green rubber rabbitbrush by its threadlike leaves; green rubber rabbitbrush leaves are usually ¹⁄₁₆- to ⅛-inch wide.

Most rabbitbrush forms tend to grow on basic soil, but mountain rubber rabbitbrush may occur on acid soil as do some forms of white rubber rabbitbrush. Mountain rubber rabbitbrush is the most palatable of the subspecies followed in order by white, green, and alkali rubber rabbitbrush. However, there is great variation in this quality. All are aggressive spreaders within their zones of occurrence. Plants can be easily established by a light broadcast seeding. Collected seed needs only hammermilling before broadcasting. A purity of 8 to 10 percent is sufficient. In a seed mixture of grasses,

Fig. 31.6.   Wideleaf rabbitbrush is excellent for revegetating any disturbed area on mountain ranges.

forbs, and other shrubs, rabbitbrush should comprise no more than 20 percent of the mixture. Mixtures can be sown by aircraft.

On salt-bearing areas, alkali rabbitbrush is preferred, but on acid soil, white or mountain rabbitbrush is best. Spreading rabbitbrush, which primarily occurs in the upper Colorado River drainage, increases aggressively by underground root stocks. It has high tolerance to alkalinity, attains a height of 2 to 6 feet, and frequently occurs in borrow pits. Consequently, it is highly promising as a stabilizer for alkaline disturbances.

Several low rabbitbrush species show particular usefulness on drier desert sites. However, Douglas rabbitbrush and wideleaf rabbitbrush (Fig. 31.6) show good adaptation to higher elevations. In contrast, desert (small) rabbitbrush does best on desert sites. However, any low rabbitbrush growing in the vicinity or on a similar vegetal type can be readily established on disturbed areas by broadcasting. All forms grow well from transplanted nursery stock or wildings. Rapid growth of all rabbitbrush species makes them good complements to perennial grasses and forbs on disturbed areas which require rapid revegetation.

Saltbushes (Goosefoot family). Woody saltbushes (or chenopods) are important plants in alkaline soils over the world. While limited in number of genera and species, they cover a greater expanse of arid ranges than any other group of plants. Natural hybridization between species occurs widely (Drobnick and Plummer 1966). Fourwing saltbush is the principal species in this group (see Table 31.1).

Other saltbushes will play an important part in the future in revegetating salt-bearing lands as more is learned. Important ones may be cuneate

Fig. 31.7. Prostrate kochia showing rapidly developing ground cover by natural seeding on disturbed area.

saltbush (or Castle Valley clover), mat saltbush, Gardner saltbush, shadscale saltbush, spineless hopsage, spiny hopsage, prostrate kochia, winterfat, black greasewood, inkbush, and seepweed. The latter is a suffrutescent saltbush form that along with inkbush shows unusual ability to grow in highly alkaline areas. All saltbushes grow fairly readily from transplants, and most establish well from field plantings. Because of germination problems, a few are difficult to establish by direct seeding. Unfortunately, shadscale and mat saltbush, two of the most salt-tolerant types, do not establish readily from direct seeding. Prostrate kochia (Fig. 31.7), a species introduced from Russia, is proving an excellent low shrub for stabilization of disturbed sites on basic soils. The species is ½-foot to 2½-feet high; under favorable conditions it reproduces in 1 year.

Bitterbrush, cliffrose and other shrubs (Rose family). A fairly large number of woody species in the rose family show special promise on disturbed areas. Some are closely related and hybridization among these is relatively common. Most important among them are antelope bitterbrush, desert bitterbrush, cliffrose, and Apache plume. They all show good promise where annual precipitation is 11 inches or more. Apache plume (Fig. 31.8), cliffrose, and desert bitterbrush grow well in the upper southern desert shrub type, and warmer parts of the northern desert shrub type. Antelope bitterbrush grows best in the northern shrub type and also grows well through the juniper-pinyon and mountain brush zones.

Desert bitterbrush, Apache plume, and some populations of antelope bitterbrush are fire tolerant, but cliffrose is not. The fire-tolerant forms of antelope bitterbrush are usually the creepers.

Fig. 31.8.   Apache plume spreads aggressively
by root sprouting and seeding on sterile soils.

A symbiotic nodule-forming organism on roots of many antelope bitter-brush populations appears to fix atmospheric nitrogen (Krebill and Muir 1974). Similar organisms have been found on cliffrose and presumably occur on other shrubs in this group of plants. This symbiotic relationship may explain why shrubs can develop good growth on sterile sites. (See also chapter 30.)

True mountain-mahogany (Fig. 31.9) and curlleaf usually grow on rocky soils and open hillsides in the juniper-pinyon and mountain brush types. Both grow on sterile soils. Once established, they are highly persistent where they are adapted. Curlleaf holds its leaves through the winter and may attain a height of 15 to 20 feet; true mountain-mahogany is deciduous and usually attains a height about one-half that of curlleaf. True mahogany is highly fire tolerant; curlleaf is not. Littleleaf mahogany, a dwarf form of curlleaf, has much narrower leaves. Numerous intermediate forms are found, and all hybridize with true mountain-mahogany.

Wildrose (or woods rose) is found in nearly all life zones from the juniper-pinyon to the high subalpine. The species exhibits large variations in height growth, underground spread, and spininess. While absence of thorns makes this rose more acceptable for planting, presence of thorns is probably desirable on disturbed areas to reduce unwanted grazing. This spreader is readily established from seed as well as by transplanting. All forms of wildrose are fire tolerant and should find an important place in the revegetation of disturbed areas.

Saskatoon serviceberry and Utah serviceberry also have a fairly wide range of adaptation in juniper-pinyon and mountain brush types and are

Fig. 31.9.   A mature true mountain-mahogany. The
species establishes well on disturbed areas.

fire tolerant. Both can be readily established by direct seeding and transplanting. They vary in height from 2 to 15 feet. Berries on Utah serviceberry persist; those on Saskatoon serviceberry do not. Ecotypic variation
from area to area is usually encountered (Blauer, *et al.* 1975).

Anderson peachbrush and desert peachbrush show remarkable adaptation through a large range of sites from the southern and northern desert
shrub types well into mountain brush and ponderosa pine types. Both have
established better from direct seeding than from transplanting. Anderson
peachbrush spreads aggressively underground. Both are rapidly developing
shrubs and show promise for disturbed areas.

Bessey (or sand) cherry, western common chokecherry, bitter cherry,
and squawapple are useful for stabilization purposes through the juniper-
pinyon, mountain brush, and ponderosa pine types. Nanking (or bush)
cherry, blackthorn (or sloe), and Peking cotoneaster (introduced species)
appear to have similar adaptation. Once established, all are highly persistent and furnish good cover.

Snowberry and related species (Honeysuckle family). Snowberry species are
among the important native shrubs with special value for stabilization purposes. Mountain snowberry and desert (or longflower) snowberry comple-

Fig. 31.10.   Mountain snowberry growing on a raw slope.

ment each other very well through a wide range of vegetal and precipitation zones and soils (Plummer 1970). Because of their above- and below-ground spreading characteristics, they are outstanding for disturbed areas. Mountain snowberry (Fig. 31.10) is best planted on sites in and above the mountain brush zone where there are 11 or more inches of precipitation; desert snowberry can be used to advantage where there may be as little as 9 inches of precipitation in some sagebrush and juniper-pinyon types. These species establish readily from transplants and seeds. Wildings as well as bare root and container stock also establish well. Common snowberry is an aggressive species in timber types as is western snowberry in the northern plains. In initial trials, both species have shown good promise and should have an especially important place in northerly areas.

Blueberry and redberry elder show special usefulness on disturbed mountain areas. Best success with these has been with nursery stock. Tatarian honeysuckle and common lilac, two exotics, show similar promise. All are fire tolerant.

Wild lilac (or ceanothus, Buckthorn family). Shrubs of the Buckthorn family grow through a wide range of coarse igneous soils. In adaptation, these shrubs are in marked contrast to the saltbush shrubs, all of which require alkaline soils. Martin and desert ceanothus (Fig. 31.11) show ability to

Fig. 31.11.   Martin ceanothus covering a steep exposed roadcut.

grow in both basic and acid soils. Except for Martin ceanothus, most of the ceanothus species hold their leaves fairly well through the winter. In a few instances, there may be a limited association of desert and Martin ceanothus with fourwing saltbush on limestone soils.

Shrubs in the buckthorn family can perform an important service in the stabilization of neutral and acid soils. Except for desert ceanothus, they are best suited to the mountain brush and conifer timber types where average annual precipitation exceeds 14 inches. Because of the difficulty and expense in harvesting seeds of shrubs in this family, transplanting of nursery and container stock is generally resorted to. Seeds of some species may stay in the ground for many years, then after chaining or burning heavy overstory they germinate and establish new plants. In some juniper-pinyon areas in eastern Nevada, Martin ceanothus and desert ceanothus established stands where no one could recall having seen them before. The additional moisture and sunlight made available by eliminating the tree competition stimulated germination of seed from shrubs that had long since vanished.

Broadleaf snowbush ceanothus is probably the most widely used shrub in this family. However, Martin ceanothus has a wider range of adaptation. It grows best on disturbed areas in the mountain brush, aspen, and high mountain conifer types. Several other species are promising. Squaw carpet, a low spreading form, shows special usefulness on disturbed areas. Some species bear root nodules. Consequently, they fix nitrogen in the soil to bene-fit their own growth as well as that of nearby vegetation (Hellmers and Kebleher 1959).

Legumes (Legume family). There is a wide assortment of shrubby legumes, native and exotic, to choose from. Because of associated symbiotic rhizobium soil bacteria, most legumes manufacture their own nitrogen from the air. Many grow in the southwest and are largely comprised of mesquite and catclaw species. Some have important merit for stabilizing disturbed sites in the dry, semitropical areas. New Mexico locust, by virtue of its large number of ecotypes, shows good promise for a wide range of sites from the warm lowland to cool mountains. False indigo is a promising shrub in the juniper-pinyon and mountain brush types. Black locust has also demon-strated good adaptation over temperate climate sites where annual pre-cipitation is 12 inches or more. This species may be a small tree where there are more than 15 inches of moisture. Honeylocust shows similar adap-tation and may be a shrub or tree depending upon precipitation. Siberian peashrub and bladdersenna, introduced legumes, show good adaptation over a wide range of sites. Bladdersenna appears to be particularly well suited to the sterile soils derived from shales. Most shrubby legumes can be trans-planted or seeded, but establishment by transplanting nursery stock is more certain.

Other shrubs. A large number of native shrubs representing several other families show high promise for stabilizing disturbed areas. Important native shrubs include Gambel oak, mountain maple, bigtooth maple, boxelder, western virginsbower, golden currant, redozier dogwood, common juniper, Nevada ephedra, green ephedra, New Mexican forestiera, skunkbush sumac, Rocky Mountain sumac, Scouler willow, purple sage, silver buffaloberry, and mountain ash. Additional introduced shrubs include matrimony vine, bittersweet, and five-stamen tamarisk. These are considered in Table 31.1 along with native shrubs.

TREES

Some trees are useful in revegetation work, particularly on roadcuts where deep-rooted vegetation is required to hold slopes from mass slippage. Where disturbed sites are within timber types, the nearby native species are recommended. Like shrubs, trees are immensely variable with respect to eco-types and biotypes. Planting nursery or container stock has been superior to direct seeding.

Conifers appropriate for disturbed sites include subalpine fir, white fir, Douglas-fir, Colorado blue spruce, Engelmann spruce, lodgepole pine, limber pine, ponderosa pine, two-leaf pinyon pine, singleleaf pinyon, Ari-zona cypress, Rocky Mountain juniper, and Utah juniper. The introduced Scotch pine is a conifer that should find good adaptation on disturbed sites.

Useful broadleaf deciduous trees are thinleaf alder, waterbirch, quaking aspen, narrowleaf cottonwood, Fremont poplar, black poplar, netleaf hack-berry, and eastern hackberry. If removed from ground water, most are shrubby, but nevertheless useful, on disturbed areas having good moisture in the spring. Netleaf hackberry is an exception and is highly useful in stabilizing disturbed areas in the drier juniper-pinyon and mountain brush types where it tends to be a brushy shrub. It establishes readily from direct seeding or transplanting nursery or container stock.

FORBS

An increasing number of forbs are showing good adaptation for stabilization over a wide range of sites. Important among these are alfalfa, chickpea milkvetch, sicklepod milkvetch, Pacific aster, sweetclover, bramblevetch, Utah sweetvetch, German (or common) iris, bouncing bet, small burnet, gooseberry globemallow, western yarrow, blueflax, Palmer penstemon, showy goldeneye, oneflower helianthella, Louisiana sagebrush, tarragon sage, and absinth sage (or common wormwood). With the exception of common iris, these can be seeded directly. However, on harsh sites it sometimes pays to resort to transplanting as well as direct seeding. This is especially helpful in establishing rhizomatous species such as Louisiana sagebrush, Pacific aster, tarragon sage, western yarrow, and chickpea milkvetch. Once established, these species spread both vegetatively and by seed. Showy goldeneye, blueflax, and yellow sweetclover establish well from seed, but are relatively short-lived. Yellow sweetclover is a biennial; showy goldeneye and blueflax persist from 3 to 5 years but they reseed and are good natural increasers on disturbed areas.

German iris (Fig. 31.12) can be planted by merely dropping rhizomes in 3- to 4-inch pits made with a hoe or shovel and replacing the soil in the pits. It grows well on disturbed areas from northern desert shrub types into high mountain areas and from alkaline to acid sites. If this species fails to establish and grow, the site can be regarded as severe.

Fig. 31.12.   German iris in
a wildland setting.

Range types of alfalfa grow well from the lowland big sagebrush types to openings in timber and aspen. The seeds of these and other legumes should generally be inoculated unless it is known that rhizobium bacteria are in the soil. Alfalfa and sweetclover are useful and in some instances may be overused because seeds are readily available from commercial sources.

GRASSES

Grasses are highly useful in conjunction with shrubs and forbs in stabilizing disturbed areas. Because of demonstrated adaptation and availability of seed, introduced grasses have a paramount place in most revegetation programs. They are aggressive in establishment and most are good soil stabilizers. Livestock and big game show a fairly high preference for them.

Outstanding introduced bunchgrasses are Fairway crested wheatgrass, Standard crested wheatgrass, hard fescue, tall wheatgrass, Russian wildrye, meadow foxtail, mountain rye, tall oatgrass, and orchardgrass. The latter two have wide use as understory species in shade such as under aspen and many shrubby plants such as Gambel oak (Plummer, *et al.* 1955). Noteworthy is an orchardgrass (P.I. 109072) from Russia that shows marked ability to grow on disturbed sites in foothills as well as on higher mountain sites.

Tall wheatgrass and Russian wildrye adapt well on alkaline areas. Russian wildrye shows outstanding qualities of establishment and persistence on desert sites. This grass has persisted on some salt desert shrublands for more than 30 years where average annual precipitation has been about 6 inches. Although tall wheatgrass shows good qualities of adaptation to alkaline areas, it requires about double the minimum precipitation. Both grow well in association with shrubby chenopods such as black greasewood and shadscale.

Fairway crested wheatgrass and standard crested wheatgrass are highly useful grasses on the foothill and valley ranges in the big sagebrush and juniper-pinyon types. Fairway has extended adaptation on disturbed places in the mountain brush and is fairly shade tolerant. Hard fescue shows everwidening adaptation to disturbed places on foothill lands in the juniper-pinyon, sagebrush, and higher mountain areas on both acid and basic soils. Meadow foxtail is unsurpassed in adaptation on disturbances on high alpine and subalpine sites in mountainous areas. The species is weakly rhizomatous but is primarily a bunchgrass. It is an exceptional natural increaser when once established.

Mountain rye, a short-lived but rapidly developing perennial, is outstanding for stabilizing disturbed sites in the mountain brush and ponderosa types. It is not highly competitive so slower developing perennials in seed mixtures including it establish and eventually control the site. The annual winter rye can be used in a similar way in the same types but it may be too suppressive to include perennials. Low amounts of seed (10 to 15 pounds per acre) are suggested in mixtures with perennials.

Intermediate wheatgrass, pubescent wheatgrass, smooth brome, quack-grass, Kentucky bluegrass, and Canada bluegrass are highly useful introduced sod formers. The latter two have particular value in shady places; the wheatgrasses are best suited to sunny sites. There are several named selections of them.

Smooth brome forms are usually of either southern or northern climatic origin in Europe and Asia. Southern types are more aggressive sod formers than northern ones in the low and warmer climates on sites in foothills and valleys of the intermountain area (Plummer *et al.* 1955). Northern selections, however, are best suited to the cooler temperatures of the higher mountains in alpine, aspen, and subalpine ranges. Because of the more aggressive sod-forming attributes, southern races are better for stabilization of disturbed areas than northern ones at all elevations. The Lincoln selection is a generally good one for this purpose. Smooth brome is excellent for growing in association with meadow foxtail and mountain lupine in high mountains. Intermediate and pubescent wheatgrass are essentially variants of the same species and have best adaptation in the juniper-pinyon and mountain brush types. Luna pubescent wheatgrass appears to be one of the best strains of several for disturbed areas. A rhizomatous crested wheatgrass (P.I. 109012) is also showing good cover characteristics on disturbed sites in foothill areas.

Quackgrass is an excellent sod former for disturbed areas over a wide range of sites from alkaline to fairly acid. Because it is a pernicious plant on croplands, care must be taken that it is not planted near crops.

Kentucky bluegrass and Canada bluegrass are also useful in high mountain ranges, particularly where low-growing sod formers are desired. They are suited to mountain ranges from mountain brush to subalpine and alpine. Canada bluegrass appears to do better on acid or neutral soil. Kentucky bluegrass, although more versatile than Canada bluegrass, grows better on limy soils. Once established, both are good natural increasers.

While there are a number of native grasses which can perform well on disturbed areas, lack of seed for most precludes extensive direct seeding. The most useful native bunchgrasses for disturbed areas are Indian ricegrass, sand dropseed, mountain brome, slender wheatgrass, bluebunch wheatgrass, basin wildrye, and bearded wildrye. Native sod-forming grasses finding increased use on disturbed sites are alkali sacaton, creeping wildrye, bluestem wheatgrass, streambank wheatgrass, and Salina wildrye.

Indian ricegrass adapts to a wide assortment of sites in northern and southern deserts as well as in the big sagebrush, juniper-pinyon, and southerly exposures of the mountain brush types. It is very important that adapted ecotypes be used. Often the seeds have hard-seedcoat dormancy. This can be overcome to a considerable extent by treating them with sulfuric acid to a moderate intensity (Plummer and Frischknecht 1952). Sand dropseed and bottlebrush squirreltail are useful on many similar disturbed sites where Indian ricegrass is adapted, but squirreltail is also adapted to disturbed sites in high mountain areas. For good growth, sand dropseed requires some summer moisture.

Bluebunch wheatgrass prefers well-drained soils in the big sagebrush and pinyon types. Bearded and beardless forms are the same species (Daubenmire 1939). Whitmar, a beardless strain selected and developed by the Soil Conservation Service, is widely adaptable and is in commercial production. This grass can be used to advantage with crested wheatgrass. Moun-

Fig. 31.13. Basin wildrye has a wide range of adaptation.

tain brome and slender wheatgrass, two rapidly developing bunchgrasses, quickly establish cover from direct seeding on mountain areas. They are good in mixtures with shrubs, forbs, and slower developing grasses in disturbed areas.

Basin wildrye (Fig. 31.13), a large and robust grass, has adaptation from valley areas well up into openings in aspen. It is especially desirable for establishing cover on disturbed sites. One ecotype (P5797), discovered by the Soil Conservation Service, shows unusual ability to establish well and adapt to a wide range of sites from saline valley bottoms to the higher mountain slopes. This ecotype is being propagated for seed and should appear on the market in at least limited amounts.

Bluestem (or western) wheatgrass and alkali sacaton, sod-forming natives, are well suited for stabilizing disturbances in moderately alkaline areas. While they may grow together in the northern desert shrub type, alkali sacaton is superior in drier climates of the southern and northern desert shrub types. Both usually grow well where big sagebrush prevails and the soil is somewhat alkaline. Alkali sacaton does better in climates with some summer precipitation. Both are more tardy in establishment by direct seeding than many other species, but once established, they are highly persistent. Streambank wheatgrass resembles bluestem in sod-forming characteristics but has finer leaves. Stands are usually more readily established from direct seeding. Seeds of the Topar selection developed by the Soil Conservation Service are available. Two other sod formers, creeping and Salina wildrye, show considerable promise on alkaline areas and should find important places in the future. Salina wildrye grows well in rocky well-drained soil, whereas creeping wildrye does better in the heavier, less well-drained sites.

TABLE 31.1
Species Adaptation Attributes and Adaptation to Disturbance Areas

Common and scientific name	Established by seed	Established by transplants	Seed production and handling	Natural spread (seed)	Natural spread (vegetative)	Growth rate	Soil stability	Adaptation to disturbance	Soil preference Acid (AC) Alkaline (AL)	Type or community to which species is adapted*
SHRUBS AND TREES										
Acacia, catclaw *Acacia greggii*	2†	4	3	4	3	3	3	3	AL/AC	SD
Alder, thinleaf *Alnus tenuifolia*	3	4	3	4	4	4	4	2	AC/AL	BS, JP, MB, A
Apache plume *Fallugia paradoxa*	3	3	4	4	4	5	5	3	AL/AC	JP, SD, ND, MB
Apricot, Siberian *Prunus armeniaca*	2	4	3	0	1	4	3	3	AL/AC	MB, JP
Aspen, quaking *Populus tremuloides*	0	4	2	0	4	4	5	4	AL/AC	A
Barberry, creeping *Berberis repens*	3	3	3	4	4	5	5	4	AL/AC	MB, JP, A, SA, ND
Birch, water *Betula occidentalis*	2	3	4	3	2	3	4	3	AC/AL	MB, JP, A, BS
Bitterbrush, antelope *Purshia tridentata*	4	4	4	2	3	4	4	4	AC/AL	JP, MB, ND, SD
Bitterbrush, desert *P. glandulosa*	4	4	4	3	3	3	4	3	AL/AC	JP, SD, ND, MB
Blackthorn (or sloe) *Prunus spinosa*	2	5	4	1	4	4	4	4	AL/AC	MB
Bladdersenna *Colutea arborescens*	4	5	5	4	2	5	4	3	AL/AC	MB, A, JP

*JP — Juniper-pinyon   ND — Northern desert shrubland   IS — Inland saltgrass   AL — Alpine
MB — Mountain brush   SD — Southern desert shrubland   A — Aspen   WM — Wet meadows
BS — Big sagebrush   SS — Salt shrubland   SA — Subalpine   C — Chaparral

†1 Very poor; 2 Poor; 3 Medium; 4 Good; 5 Very good.

TABLE 31.1 — cont.

Species Adaptation Attributes and Adaptation to Disturbance Areas

Common and scientific name	Established by seed	Established by transplants	Seed production and handling	Natural spread (seed)	Natural spread (vegetative)	Growth rate	Soil stability	Adaptation to disturbance	Soil preference Acid (AC) Alkaline (AL)	Type or community to which species is adapted*
SHRUBS AND TREES — cont.										
Boxelder *Acer negundo negundo*	4	5	5	4	2	4	4	3	AL/AC	MB, A, JP
Buffaloberry, roundleaf *Shepherdia rotundifolia*	0	1	2	1	2	2	4	3	AL	JP
Buffaloberry, silver *S. argentea*	2	3	3	3	3	3	4	3	AL	JP, ND, IS, WM
Ceanothus, cuneate *Ceanothus cuneata*	4	4	3	3	2	3	4	4	AC	MB, JP
Ceanothus, Mohave desert *C. greggii*	2	4	2	3	3	3	4	5	AC/AL	ND, SD, C
Ceanothus, Martin *C. martini*	4	4	2	3	3	3	4	3	AL/AC	MB, JP, A
Ceanothus, snowbrush *C. velutinus*	2	4	2	3	4	4	5	4	AC/AL	MB, A
Ceanothus, squaw carpet *C. prostratus*	2	3	1	3	3	3	4	4	AC	MB
Cherry, Bessey *Prunus besseyi*	3	5	4	3	4	4	3	3	AC/AL	MB, JP, A
Cherry, bitter *P. emarginata*	2	4	3	3	5	4	5	4	AC/AL	MB, JP, A
Cherry, ground *P. fruticosa*	3	5	3	3	3	5	5	4	AL/AC	JP, MB
Cherry, Nanking *P. tomentosa*	4	5	4	3	3	4	4	4	AL/AC	JP, MB
Chokecherry, black *P. virginiana melanocarpa*	3	4	4	3	5	4	5	5	AC/AL	MB, A, JP

Cinquefoil, bush *Potentilla fruticosa*	2	5	3	3	3	3	3	3	AC/AL	WM, A, SA, MB
Cliffrose, Stansbury *Cowania mexicana stansburiana*	3	4	4	4	3	4	3	3	AL/AC	JP, MB, ND
Cotoneaster, Peking *Cotoneaster acutifolia*	2	4	3	2	2	3	3	3	AL/AC	MB, JP
Cottonwood, black *Populus trichocarpa*	2	4	2	3	3	4	4	4	AL/AC	MB, A, BS
Cottonwood, Fremont *P. fremontii*	2†	4	2	3	4	3	4	4	AL/AC	BS, JP, ND
Cottonwood, narrowleaf *P. angustifolia*	1	4	2	3	4	4	4	2	AL/AC	MB, A, BS
Currant, golden *Ribes aureum*	4	5	4	5	5	4	4	4	AL/AC	MB, JP, WM
Currant, squaw *R. cereum inebrians*	2	3	2	2	3	2	3	3	AC/AL	MB, SA, A
Currant, sticky *R. viscosissimum*	2	3	2	3	3	2	4	3	AC/AL	SA, A, MB
Cypress, Arizona *Cupressus arizonica*	4	3	4	3	3	4	3	3	AL/AC	MB, JP
Dogwood, redozier *Cornus stolonifera*	3	4	5	3	4	4	4	4	AL/AC	BB, A, BS
Douglas-fir *Pseudotsuga menziesii menziesii*	1	3	4	1	0	3	3	3	AC/AL	SA, MB
Elder, blueberry *Sambucus cerulea*	3	5	5	3	5	5	4	4	AC/AL	MB, ND, A
Elder, redberry *S. racemosa pubens microbotrys*	3	3	5	2	5	5	3	5	AC/AL	SA, A, MB
Elm, Siberian *Ulmus pumila*	4	5	5	5	2	5	4	4	AL/AC	JP, MB, BS
Ephedra, green *Ephedra viridis*	4	3	2	2	3	2	3	3	AL/AC	ND, JP, MB, SD
Ephedra, Nevada *E. nevadensis*	4	3	2	2	3	2	3	3	AL	ND, JP, SS

*JP — Juniper-pinyon
MB — Mountain brush
BS — Big sagebrush

ND — Northern desert shrubland
SD — Southern desert shrubland
SS — Salt shrubland

IS — Inland saltgrass
A — Aspen
SA — Subalpine

AL — Alpine
WM — Wet meadows
C — Chaparral

†1 Very poor; 2 Poor; 3 Medium; 4 Good; 5 Very good.

TABLE 31.1 — cont.

Species Adaptation Attributes and Adaptation to Disturbance Areas

Common and scientific name	Established by seed	Established by transplants	Seed production and handling	Natural spread (seed)	Natural spread (vegetative)	Growth rate	Soil stability	Adaptation to disturbance	Soil preference Acid (AC) Alkaline (AL)	Type or community to which species is adapted*
SHRUBS AND TREES — cont.										
Fir, subalpine *Abies lasiocarpa*	1	3	4	3	0	2	4	3	AC/AL	SA, A
Fir, white *A. concolor*	3	4	5	4	0	4	4	4	AL/AC	SA, MB
Forestiera, New Mexican *Forestiera neomexicana*	4	5	4	3	1	4	4	3	AL/AC	MB, JP, ND
Greasewood, black *Sarcobatus vermiculatus*	3	3	3	3	2	4	3	3	AL	SS, ND, SD, IS
Hackberry, eastern *Celtis occidentalis*	3	4	4	2	0	3	4	4	AL/AC	MB, JP
Hackberry, netleaf *C. reticulata*	3	4	3	3	2	4	4	4	AL/AC	MB, JP
Honeylocust *Gleditsia triacanthos*	2	4	5	3	1	3	2	3	AC/AL	JP, MB
Honeysuckle, bearberry *Lonicera involucrata*	1	4	2	2	3	3	3	2	AC/AL	A, SA, MB
Honeysuckle, tatarian *L. tatarica*	3	5	4	3	3	4	3	3	AL/AC	MB, JP, ND
Hopsage, spineless *Grayia brandegei*	3	3	4	3	1	4	4	2	AL	SS, ND, JP
Hopsage, spiney *G. brandegei*	3	3	3	3	1	4	3	3	AL	ND, JP, MB
Indigo, false *Amorpha fruticosa*	3	4	4	3	3	4	4	3	AC/AL	JP, MB, ND, BS
Juniper, common *Juniperus communis*	2	3	2	2	3	2	4	4	AC/AL	MB, A, SA

Juniper, Rocky Mt. *J. scopulorum*	2	3	4	3	3	3	4	AC/AL	MB, A, JP, ND
Juniper, Utah *J. osteosperma*	1	2	3	3	3	3	3	AC/AL	JP, MB, ND
Kochia, prostrate *Kochia prostrata*	5	5	5	5	2	5	4	AL	SS, BS, ND, SD, IS
Lilac, common *Syringa vulgaris*	2†	4	3	2	4	3	3	AL/AC	MB, JP, ND
Locust, black *Robinia pseudoacacia*	3	4	4	4	4	3	4	AC/AL	MB, JP, ND
Locust, New Mexico *R. neomexicana*	3	3	3	4	4	4	2	AL/AC	MB, JP, A
Maple, Amur *Acer ginnala*	4	5	4	3	3	3	3	AC/AL	MB, JP, A
Maple, bigtooth *A. grandidentatum*	4	4	3	3	3	2	5	3	AL/AC	MB, A, JP
Maple, mountain *Acer glabrum*	4	4	3	4	4	3	3	4	AC/AL	MB, A, JP
Matchbrush (or broom-snakeweed) *Gutierrezia sarothrae*	3	5	3	5	5	3	5	AL	JP, BS, ND, SD
Matrimonyvine *Lycium halmifolium*	1	5	1	0	5	5	4	AL/AC	JP, MB, BS
Mesquite *Prosopis juliflora*	4	3	4	5	5	5	2	AL	SD, C
Mt.-Mahogany, curlleaf *Cercocarpus ledifolius*	3	2	3	4	2	3	3	AL	MB, JP
Mt.-Mahogany, littleleaf *C. intricatus*	2	2	3	3	1	2	2	AL	JP, MB, ND, SD
Mt.-Mahogany, true *C. montanus*	2	3	3	3	3	3	3	AL	MB, JP, ND, A
Nightshade, bittersweet *Solanum dulcamara*	3	4	3	4	4	3	4	AL/AC	MB
Oak, Gambel *Quercus gambelii*	3	2	3	4	4	5	2	AL/AC	MB, JP, A

*JP — Juniper-pinyon
MB — Mountain brush
BS — Big sagebrush

ND — Northern desert shrubland
SD — Southern desert shrubland
SS — Salt shrubland

IS — Inland saltgrass
A — Aspen
SA — Subalpine

AL — Alpine
WM — Wet meadows
C — Chaparral

†1 Very poor; 2 Poor; 3 Medium; 4 Good; 5 Very good.

TABLE 31.1 — cont.
Species Adaptation Attributes and Adaptation to Disturbance Areas

Common and scientific name	Established by seed	Established by transplants	Seed production and handling	Natural spread (seed)	Natural spread (vegetative)	Growth rate	Soil stability	Adaptation to disturbance	Soil preference Acid (AC) Alkaline (AL)	Type or community to which species is adapted*
SHRUBS AND TREES — cont.										
Peachbrush, Anderson *Prunus andersonii*	3	5	4	3	3	4	5	5	AL/AC	ND, JP, MB, SD, C
Peachbrush, desert *P. fasciculata*	5	3	4	3	4	5	4	4	AL	ND, JP, MB, C
Peashrub, Siberian *Caragana arborescens*	3	5	4	3	3	4	3	4	AL/AC	MB, JP, ND
Penstemon, bush *Penstemon fruticosus*	3	4	4	4	4	4	4	5	AC	MB, A
Pine, Jeffrey *Pinus jeffreyi*	2	3	4	2	0	3	4	3	AC	MB
Pine, limber *P. flexilis*	1	3	4	3	0	3	4	4	AL/AC	MB, A, SB
Pine, lodgepole *P. contorta*	1	3	4	4	0	4	5	5	AC/AL	A, SA, MB
Pine, ponderosa *P. ponderosa*	3	4	4	4	0	4	5	4	AC/AL	MB, A
Pine, Scotch *P. sylvestris*	3	4	4	4	0	4	5	4	AL/AC	MB, A
Pinyon, two-leaf *P. edulis*	3	3	4	5	0	3	3	4	AL/AC	JP, MB, BS
Pinyon, one-leaf *P. monophylla*	3	3	3	5	0	3	3	4	AL/AC	JP, BS
Plum, American *Prunus americana*	2	4	4	3	4	3	4	4	AL/AC	MB, JP, WM
Rabbitbrush, alkali rubber *Chrysothamnus nauseosus consimilis*	4	4	3	4	5	5	5	5	AL	IS, ND, SS, BS

Species	1	2	3	4	5	6	7	8		
Rabbitbrush, desert *C. stenophyllus*	5	4	3	5	2	3	3	3	AL	ND, JP, SD
Rabbitbrush, low Douglas *C. viscidiflorus viscidiflorus*	4†	4	5	5	3	5	5	5	AL/AC	MB, JP, ND
Rabbitbrush, low wideleaf *C. viscidiflorus lanceolatus*	3	4	5	4	4	3	3	3	AL/AC	MB, A, SA, BS
Rabbitbrush, mountain rubber *C. nauseosus salicifolius*	4	4	4	4	3	5	5	5	AL/AC	MB, JP, A, SA, BS
Rabbitbrush, green rubber *C. nauseosus graveolens*	4	4	5	4	3	5	5	4	AL	BS, JP, ND
Rabbitbrush, white rubber *C. nauseosus albicaulis*	5	4	4	4	3	5	5	4	AL/AC	BS, JP, MB, ND
Rabbitbrush, spreading *C. linifolius*	3	3	4	3	5	4	5	5	AL	ND, BS, JP, MB
Rose, woods *Rosa woodsii ultramontana*	4	4	4	3	5	3	4	5	AL/AC	MB, JP, A, SA
Russian-olive *Elaeagnus angustifolia*	3	4	5	4	3	4	4	4	AL/AC	MB, JP, IS, SD
Sage, purple *Salvia dorrii carnosa*	3	4	4	3	0	4	3	4	AL/AC	JP, ND, SD, MB
Sagebrush, alkali *A. longiloba*	3	5	3	4	3	4	5	4	AL	ND, BS
Sagebrush, big *Artemisia tridentata tridentata*	4	5	4	5	3	5	4	5	AL	ND, JP, MB, SB
Sagebrush, Bigelow *A. bigelovii*	3	4	3	4	4	4	3	3	AL	ND, JP, SD
Sagebrush, black *A. nova*	4	5	3	5	3	4	3	4	AL	ND, JP, MB
Sagebrush, bud *A. spinescens*	1⁻	3	2	3	2	5	4	3	AL	ND, SS, JP
Sagebrush, fringe *A. frigida*	3	5	3	4	3	5	4	3	AL/AC	ND, SD, JP, SA
Sagebrush, low *A. arbuscula*	3	5	3	4	3	5	4	4	AC/AL	MB, A, JP

*JP — Juniper-pinyon   ND — Northern desert shrubland   IS — Inland saltgrass   AL — Alpine
MB — Mountain brush   SD — Southern desert shrubland   A — Aspen   WM — Wet meadows
BS — Big sagebrush   SS — Salt shrubland   SA — Subalpine   C — Chaparral

†1 Very poor; 2 Poor; 3 Medium; 4 Good; 5 Very good.

TABLE 31.1 — cont.

Species Adaptation Attributes and Adaptation to Disturbance Areas

Common and scientific name	Established by seed	Established by transplants	Seed production and handling	Natural spread (seed)	Natural spread (vegetative)	Growth rate	Soil stability	Adaptation to disturbance	Soil preference Acid (AC) Alkaline (AL)	Type or community to which species is adapted*
SHRUBS AND TREES — cont.										
Sagebrush, mountain A. tridentata vaseyana	4	5	3	5	1	5	5	5	AC/AL	A, MB, BS, SA
Sagebrush, pygmy Artemisia pygmaea	2	4	2	3	2	4	4	3	AL	ND, JP
Sagebrush, sand A. filifolia	2	4	4	4	3	4	4	3	AL	BS, JP, ND, SD
Sagebrush, scabland A. rigida	3	4	3	3	0	4	4	4	AC/AL	MB, JP
Sagebrush, silver A. cana cana	3	5	4	4	4	5	5	5	AC/AL	MB, A, SA, JP
Sagebrush, spiny A. spinescens	2	3	2	4	0	4	3	4	AL	ND, SS, BS
Sagebrush, subalpine A. rothrocki	4	5	3	4	2	5	5	5	AC/AL	SA, A
Sagebrush, Wyoming A. wyomingensis	4	5	4	5	0	5	5	5	AL/AC	BS, ND, MB
Saltbush, cuneate Atriplex cuneata	3	4	3	4	4	4	4	3	AL	SS, ND
Saltbush, fourwing A. canescens	3	4	5	3	3	4	4	3	AL	SD, ND, JP, MB
Saltbush, gardner A. gardneri	3	4	3	3	4	4	4	3	AL	SS, ND, JP
Saltbush, mat A. corrugata	2	4	3	4	2	4	4	5	AL	SS
Saltbush, shadscale A. confertifolia	1†	2	3	4	2	4	3	3	AL	SS, ND, SD, JP

Salt-tree, Siberian *Halimodendron halodendron*	3	4	4	3	5	4	4	4	AL	SS, ND, JP, MB
Serviceberry, Saskatoon *Amelanchier alnifolia*	3	3	3	3	3	3	4	4	AL/AC	MB, JP, A
Serviceberry, Utah *A. utahensis*	3	3	3	3	3	3	4	3	AL/AC	MB, JP
Snowberry, common *Symphoricarpus albus*	3	5	3	5	5	5	5	5	AC/AL	MB, A
Snowberry, desert *S. longiflorus*	3	5	2	3	5	3	4	4	AL/AC	JP, ND, MB
Snowberry, mountain *S. oreophillus*	3	5	3	4	5	3	5	5	AC/AL	MB, A, SA, JP
Snowberry, western *S. occidentalis*	3	5	3	4	5	5	5	4	AC/AL	MB, A
Spruce, blue *Picea pungens*	2	4	5	2	0	4	4	3	AL/AC	A, MB
Spruce, Engelmann *P. engelmannii*	2	3	4	2	0	3	4	3	AC/AL	SA, A
Squaw-apple *Peraphyllum ramosissimum*	3	4	3	3	3	3	4	3	AL/AC	MB, JP, ND
Sumac, Rocky Mt. smooth *Rhus glabra cismontana*	2	4	4	2	4	4	4	3	AC/AL	MB, JP, ND
Sumac, skunk bush *R. trilobata*	3	5	4	3	3	3	4	4	AL/AC	MB, JP, ND, SD, C
Virginsbower, western *Clematis ligusticifolia*	2	4	4	3	4	4	4	4	AL/AC	MB, JP, BS
Willow, purpleosier *Salix purpurea*	—	4	—	—	3	4	4	3	AC/AL	MB, A, BS
Willow, Scouler *S. scouleriana*	—	4	—	—	3	3	4	4	AC/AL	MB, A, SA
Winterfat, common *Ceratoides lanata*	3	4	3	4	3	4	4	4	AL	ND, JP, SS, MB
Wormwood, Oldman *Artemisia abrotanum*	1	5	—	1	4	5	4	4	AC/AL	MB, A, SA

*JP — Juniper-pinyon
MB — Mountain brush
BS — Big sagebrush

ND — Northern desert shrubland
SD — Southern desert shrubland
SS — Salt shrubland

IS — Inland saltgrass
A — Aspen
SA — Subalpine

AL — Alpine
WM — Wet meadows
C — Chaparral

†1 Very poor; 2 Poor; 3 Medium; 4 Good; 5 Very good.

TABLE 31.1 — cont.

Species Adaptation Attributes and Adaptation to Disturbance Areas

Common and scientific name	Established by seed	Established by transplants	Seed production and handling	Natural spread (seed)	Natural spread (vegetative)	Growth rate	Soil stability	Adaptation to disturbance	Soil preference Acid (AC) Alkaline (AL)	Type or community to which species is adapted*
FORBS										
Alfalfa (Ladak, Nomad, Rambler) *Medicago sativa*	5	4	5	2	3	4	3	4	AL/AC	MB, JP, BS
Alfalfa, sickle *M. falcatus*	4	4	3	3	4	3	4	4	AL/AC	MB, JP, BS, A, SA
Alfileria *Erodium cicutarium*	4	—	1	4	0	4	3	4	AL/AC	BS, MB, SD, JP
Aster, pacific *Aster chilensis adscendens*	4	5	4	4	5	4	5	5	AL/AC	MB, A, SA, BS, IS, WM
Aster, smooth *A. glaucodes*	4	5	4	4	5	4	4	5	AL	MB, A, SA, BS, JP, IS
Balsamroot, arrowleaf *Balsamorhiza sagittata*	4	3	4	4	0	2	3	4	AC/AL	BS, MB, JP
Bouncing-bet *Saponaria officinalis*	5	5	4	3	4	4	3	4	AL/AC	MB, A, BS
Burnet, small *Sanguisorba minor*	5	5	5	2	0	4	2	3	AL/AC	JP, MB, BS
Crownvetch, coronilla *Coronilla varia*	2†	5	3	2	5	3	5	4	AL/AC	MB, JP, A, BS
Eriogonum, cushion *Eriogonum ovalifolium*	2	—	3	3	0	4	3	4	AL	JP, MB, BS
Flax, blue (or Lewis) *Linum lewisii*	5	4	4	5	0	5	3	5	AC/AL	MB, JP, BS, A, SA, WM
Geranium, sticky *Geranium viscosissimum*	4	3	2	4	0	4	3	5	AC/AL	A, SA, MB
Globemallow, gooseberry *Sphaeralcea grossulariaefolia*	3	4	4	4	0	4	3	5	AL	ND, BS, JP, SD

Species									Site	Habitats
Goldeneye, showy *Viguiera multiflora*	5	4	5	0	5	4	5		AL/AC	A, SA, MB, JP
Goldeneye, Nevada showy *V. multiflora nevadensis*	5	4	5	4	4	3	5		AL/AC	JP, BS, ND, SD
Goldenrod, Parry *Solidago parryi*	3	5	3	4	5	5	5		AL/AC	A, SA, MB, JP
Helianthella, oneflower *Helianthella uniflora*	5	5	4	0	4	3	4		AL/AC	MB, A, SA, JP, BS
Iris, German (or common) *Iris germanica*	1	5	—	1	4	5	5		AL/AC	MB, BS, ND, A, SA, SD
Kochia, annual *Kochia scoparia*	5	5	5	0	5	4	5		AL	BS, ND, SD, SS, MB
Licorice, American *Glycyrrhiza lepidota*	4	5	3	5	3	4	5		AL	MB, JP, BS, ND, SS
Lomatium, Nuttall *Lomatium nuttalli*	3	2	4	0	4	4	4		AL/AC	MB, A, SA, BS
Lupine, mountain *Lupinus alpestris*	5	2	5	0	4	3	5		AC/AL	SA, A, MB
Lupine, Nevada *L. nevadensis*	5	2	4	0	4	3	5		AL	JP, BS, MB, ND
Lupine, silky *L. sericeus*	5	2	3	0	4	3	5		AC/AL	JP, BS, MB
Milkvetch, chickpea *Astragalus cicer*	4	5	4	4	4	4	5		AL/AC	MB, A, JP, BS
Milkvetch, Snakeriver plains *A. filipes*	3	4	4	0	4	4	4		AC/AL	MB, JP, BS
Penstemon, Eaton *Penstemon eatonii*	5	5	4	0	4	3	5		AL/AC	JP, MB, A
Penstemon, low *P. humilis*	4	5	4	4	4	4	5		AL/AC	MB, JP, BS, A
Penstemon, palmer *P. palmeri*	5	4	5	0	5	4	5		AL/AC	JP, MB, BS
Penstemon, sidehill *P. platyphyllus*	3	4	5	4	4	4	4		AC/AL	MB, JP, A

*JP — Juniper-pinyon
MB — Mountain brush
BS — Big sagebrush
ND — Northern desert shrubland
SD — Southern desert shrubland
SS — Salt shrubland
IS — Inland saltgrass
A — Aspen
SA — Subalpine
AL — Alpine
WM — Wet meadows
C — Chaparral

†1 Very poor; 2 Poor; 3 Medium; 4 Good; 5 Very good.

[333]

TABLE 31.1 — cont.

Species Adaptation Attributes and Adaptation to Disturbance Areas

Common and scientific name	Established by seed	Established by transplants	Seed production and handling	Natural spread (seed)	Natural spread (vegetative)	Growth rate	Soil stability	Adaptation to disturbance	Soil preference Alkaline (AL) Acid (AC)	Type or community to which species is adapted*
FORBS — cont.										
Penstemon, thickleaf *Penstemon pachyphyllus*	5	5	5	4	0	5	3	4	AL/AC	JP, BS, MB
Penstemon, Wasatch *P. cyananthus*	5	5	5	4	0	5	3	4	AC/AL	MB, A, JP
Russianthistle *Salsola kali tenuifolia*	5	—	4	5	0	5	5	5	AL	ND, SD, BS, MB, JP
Sagebrush, Louisiana *Artemisia ludoviciana*	3	5	3	4	5	4	5	5	AL/AC	MB, JP, A, SA, BS, ND
Sagebrush, tarragon *A. dracunculus*	3	5	3	4	5	4	5	5	AL/AC	MB, A, JP, BS, ND
Seepweed (pickleweed) *Suaeda* spp.	3	4	4	4	0	5	4	5	AL	IS, ND, SD
Sweetvetch, Utah *Hedysarum boreale utahensis*	3†	3	3	4	4	2	3	4	AL/AC	MB, JP, BS, ND
Wyethia, mulesears *Wyethia amplexicaulis*	3	4	4	4	3	3	4	4	AL/AC	A, MB, SA
Yarrow, western *Achillea millefolium lanulosa*	5	5	4	5	5	5	4	5	AC/AL	SA, MB, WM, BS
Yellow, sweetclover *Melilotus officinalis*	5	5	5	5	0	5	3	5	AL/AC	BS, ND, MB, A
GRASSES										
Alkaligrass *Puccinella airoides*	3	4	4	4	0	5	3	4	AL	IS, SS, ND, SD
Bluegrass, big *Poa ampla*	4	5	5	4	0	3	4	4	AC/AL	MB, A, JP, BS
Bluegrass, bulbous *P. bulbosa*	5	5	5	5	4	5	2	4	AL/AC	BS, JP, MB

Bluegrass, Canada *P. compressa*	4	5	5	4	5	4	4	4	AC/AL	MB, A
Bluegrass, Kentucky *P. pratensis*	5	5	5	5	5	4	4	4	AL/AC	MB, A, SA, JP
Brome, cheatgrass *Bromus tectorum tectorum*	5	—	5	5	0	5	3	5	AL/AC	JP, BS, MB, ND
Brome, meadow *B. erectus*	5	5	5	3	0	5	3	5	AC/AL	MB, JP, MB, A, SA
Brome, mountain *B. carinatus*	5	5	5	5	0	5	3	5	AC/AL	A, SA, MB
Brome, smooth (northern) *B. inermis*	4	5	5	3	4	4	5	4	AL/AC	A, SA, MB, AL
Brome, smooth (southern) *B. inermis*	5	5	5	4	5	5	5	5	AC/AL	A, MB, JP, SA, BS
Brome, subalpine *B. tomentellus*	5	3	5	4	0	5	4	3	AL/AC	SA, A, MB
Canarygrass, reed *Phalaris arundinacea*	3	5	4	5	5	4	5	2	AL/AC	WM, SA, A
Dropseed, sand *Sporobolus cryptandrus*	4	4	4	5	0	4	4	5	AL/AC	JP, BS, ND, SD, MB
Fescue, hardsheep *Festuca ovina duriscula*	5	4	4	5	0	4	5	5	AC/AL	MB, A, JP, SA
Foxtail, meadow *Alopecurus pratensis*	5	5	3	5	4	3	5	5	AC/AL	SA, AL, WM
Foxtail, reed *A. arundinaceus*	4	5	3	5	4	3	5	2	AL/AC	SA, AL, WM, A
Galleta, grass *Hilaria jamesii*	Has not been successfully planted.									ND, SD, BS, SD
Muhly, mat *Muhlenbergia richardsonis*	2	4	3	4	4	3	4	3	AL/AC	IS, BS, ND, MB
Needlegrass, green *Stipa viridula*	4	4	4	3	0	4	4	3	AC/AL	MB, A
Needlegrass, letterman *S. lettermani*	4	4	2	5	0	4	4	5	AC/AL	MB, A, SA, JP

*JP — Juniper-pinyon
MB — Mountain brush
BS — Big sagebrush

ND — Northern desert shrubland
SD — Southern desert shrubland
SS — Salt shrubland

IS — Island saltgrass
A — Aspen
SA — Subalpine

AL — Alpine
WM — Wet meadows
C — Chaparral

†1 Very poor; 2 Poor; 3 Medium; 4 Good; 5 Very good.

TABLE 31.1 — cont.
Species Adaptation Attributes and Adaptation to Disturbance Areas

Common and scientific name	Established by seed	Established by transplants	Seed production and handling	Natural spread (seed)	Natural spread (vegetative)	Growth rate	Soil stability	Adaptation to disturbance	Soil preference Acid (AC) Alkaline (AL)	Type or community to which species is adapted*
GRASSES — cont.										
Oatgrass, tall *Arrhenatherum elatius*	5	4	4	4	0	4	4	4	AC/AL	A, MB, SA
Orchardgrass *Dactylis glomerata*	5	5	5	5	0	4	4	5	AC/AL	A, MB, SA, BS, JP
Quackgrass *Agropyron repens*	4	5	3	3	5	4	5	5	AL/AC	MB, JP, A, SA
Reedgrass, chee *Calamagrostis epigeios*	0†	5	—	0	5	5	5	5	AL/AC	MB, A, SA, JP
Ricegrass, Indian *Oryzopsis hymenoides*	3	4	3	4	0	4	4	5	AL/AC	JP, ND, SD, BS
Rye, mountain *Secale montanum*	5	5	4	5	0	5	4	5	AC/AL	MB, JP, A
Rye, winter *S. cereale*	5	5	4	5	0	5	4	5	AL/AC	BS, MB, A
Sacaton, alkali *Sporobolus airoides airoides*	3	5	4	4	4	4	5	4	AL	IS, ND, SD, SS, JP
Saltgrass, inland *Distichlis spicata stricta*	0	3	4	0	4	3	5	4	AL	SD, ND, IS, BS
Squirreltail, bottlebrush *Sitanion hystrix*	4	5	3	4	0	4	3	5	AL/AC	JP, ND, SS, SD, MB, A, SA
Timothy *Phleum pratense*	5	4	5	3	0	4	3	4	AC/AL	A, SA
Wheatgrass, bearded *Agropyron subsecundum*	5	4	4	5	0	5	4	4	AL/AC	A, MB, SA
Wheatgrass, bearded bluebunch *A. spicatum*	4	4	4	3	2	3	5	3	AL/AC	BS, JP, MB
Wheatgrass, beardless bluebunch *A. spicatum inerme*	4	4	5	3	2	3	5	3	AL/AC	BS, JP, MB

Wheatgrass, bluestem (or western) *A. smithii*	3	4	4	2	5	2	5	4	AL/AC	BS, MB, JP, ND
Wheatgrass, crested (Fairway) *A. cristatum*	5	5	5	5	2	4	5	4	AL/AC	BS, JP, MB, ND, A
Wheatgrass, crested (standard) *A. desertorum*	5	5	5	0	4	4	4	3	AL/AC	BS, JP, MB, ND
Wheatgrass, intermediate *A. intermedium*	5	5	4	4	3	5	5	4	AL/AC	JP, MB, A, ND, SD
Wheatgrass, pubescent (or stiffhair) *A. trichophorum*	5	5	3	5	3	5	5	4	AL/AC	JP, BS, MB, ND, SD
Wheatgrass, Scribner *A. scribneri*	2	3	2	0	3	3	2	4	AL/AC	AL, SA, A
Wheatgrass, Siberian *A. sibiricum*	5	5	4	0	4	4	3	3	AL/AC	JP, BS, MB
Wheatgrass, slender *A. trachycaulum*	5	5	5	0	5	5	3	5	AL/AC	A, SA, MB
Wheatgrass, streambank *A. riparium*	4	5	3	5	2	4	5	4	AL/AC	JP, MB, BS
Wheatgrass, tall *A. elongatum*	5	5	5	0	5	5	4	4	AL	IS, ND, JP, MB, BS
Wildrye, blue *Elymus glaucus*	5	5	4	0	5	5	3	5	AL/AC	MB, A
Wildrye, creeping *E. triticoides*	5	4	3	4	2	3	5	4	AL	IS, JP, BS, MB
Wildrye, Basin *E. cinereus*	4	4	4	2	5	4	4	4	AL/AC	JP, MB, A
Wildrye, mammoth *E. giganteus*	3	4	3	4	1	4	4	3	AL/AC	JP, MB, BS
Wildrye, Russian *E. junceus*	4	4	5	0	4	3	4	4	AL	ND, BS, MB
Wildrye, Salina *E. salina*	2	3	4	4	3	2	5	3	AL	JP, ND, BS, MB
Wildrye, sabulosa *E. sabulosus*	3	4	4	4	2	4	4	3	AL/AC	JP, MB

*JP — Juniper-pinyon
MB — Mountain brush
BS — Big sagebrush

ND — Northern desert shrubland
SD — Southern desert shrubland
SS — Salt shrubland

IS — Inland saltgrass
A — Aspen
SA — Subalpine

AL — Alpine
WM — Wet meadows
C — Chaparral

†1 Very poor; 2 Poor; 3 Medium; 4 Good; 5 Very good.

REFERENCES

Anderson, Darwin, Louis P. Hamilton, Hudson G. Reynolds and Robert R. Humphrey. 1953. (Rev. 1957.) *Reseeding Desert Grassland Ranges in Southern Arizona.* Ariz. Agric. Exp. Stn. Bull. 249.

Blauer, A. Clyde, A. Perry Plummer, E. Durant McArthur, Richard Stevens and Bruce C. Giunta. 1975. *Characteristics and Hybridization of Important Intermountain Shrubs. I. Rose Family.* USDA For. Serv. Res. Pap. INT-169.

Clausen, Jens, David D. Keck and William M. Hiesey. 1940. *Experimental Studies on the Nature of Species.* Carnegie Inst. Wash. Publ. 520. Washington, D.C.

Clausen, Jens, David D. Keck and William M. Hiesey. 1941. Regional differentiation in plant species. *Am. Nat.* 75:231–250.

Daubenmire, R. F. 1939. The taxonomy and ecology of *Agropyron spicatum* and *A. inerme. Torrey Bot. Club* 66:327–329.

Drobnick, Rudy and A. Perry Plummer. 1966. Progress in browse hybridization in Utah. Proc. Conf. West. State Game and Fish Comm. 46:203–211.

Farmer, Eugene E., Ray W. Brown, Bland Z. Richardson and Paul E. Packer. 1974. *Revegetation Research on the Decker Coal Mine in Southeastern Montana.* USDA For. Serv. Res. Pap. INT-162.

Ferguson, Robert B. and Stephen B. Monsen. 1974. Research with containerized shrubs and forbs in southern Idaho. *Great Plains Agric. Publ.* 68:349–358.

Hafenrichter, A. L., Lowell A. Mullen and Robert L. Brown. 1949. *Grasses and Legumes for Soil Conservation in the Pacific Northwest.* U.S. Dep. Agric. Misc. Publ. 678.

Hafenrichter, A. L., John L. Schwendiman, Harold L. Harris, Robert S. Mac-Lauchlan and Harold W. Miller. 1968. *Grasses and Legumes for Soil Conservation in the Pacific Northwest and Great Basin States.* U.S. Dep. Agric. Handb. 339.

Hanks, D. L., E. D. McArthur, A. P. Plummer, B. C. Giunta and A. C. Blauer. 1975. Chromatographic recognition of some palatable and unpalatable subspecies of rubber rabbitbrush in and around Utah. *Journal of Range Management* 28:144–148.

Hanks, D. L., E. D. McArthur, Richard Stevens and A. Perry Plummer. 1973. *Chromatographic Characteristics and Phylogenetic Relationships of* Artemisia, *Section Tridentatae.* USDA For. Serv. Res. Pap. INT-141.

Hanson, A. A. 1965. *Grass Varieties in the United States.* U.S. Dep. Agric. Handb. 170.

Hellmers, H. and M. M. Kebleher. 1959. *Ceanothus leucodermis* and soil nitrogen in southern California mountains. *For. Sci.* 5:275–278.

Krebill, R. G. and Joyce M. Muir. 1974. Morphological characterization of *Frankia purshiae* the endophyte in root nodules of bitterbrush. *Northwest Sci.* 48:266–268.

McArthur, E. Durant, Bruce C. Giunta and A. Perry Plummer. 1974. Shrubs for restoration of depleted ranges and disturbed areas. *Utah Sci.* 35:28–33.

Monsen, Stephen B. 1975. Selecting plants to rehabilitate disturbed areas. P. 76–90, in: Campbell, R. S., and C. H. Herbel. Improved range plants. Soc. Range Manage., Denver, Colo., 90 pp.

Plummer, A. Perry. 1970. Plants for revegetation of roadcuts and other disturbed and eroded areas. USDA For. Serv., Intermt. Reg. Range Improv. Notes 15(1):1–8.

Plummer, A. Perry. 1974. Oldman wormwood to stabilize disturbed areas. *Utah Sci.* 35:26–27.

Plummer, A. Perry, Donald R. Christensen and Stephen B. Monsen. 1968. *Restoring Big Game Ranges in Utah.* Utah Div. Fish and Game Publ. 68-3.

Plummer, A. P. and John M. Fenley. 1950. *Seasonal Periods for Planting Grasses in the Subalpine Zone of Central Utah.* USDA For. Serv., Intermt. For. and Range Exp. Stn., Res. Pap. 18.

Plummer, A. P. and Neil C. Frischknecht. 1952. Increasing field stands of Indian ricegrass. *Agron. J.* 44:285–289.

Plummer, A. Perry, A. C. Hull, Jr., George Stewart and Joseph H. Robertson. 1955. *Seeding Rangelands in Utah, Nevada, Southern Idaho, and Western Wyoming.* U.S. Dep. Agric. Handb. 71.

Plummer, A. P., Stephen B. Monsen and Donald R. Christensen. 1966. *Four-wing saltbush — A Shrub for Future Game Ranges.* Utah State Dep. Fish and Game Publ. 66-4.

U.S. Department of Agriculture, Forest Service. 1974. *Seeds of Woody Plants in the United States.* C. S. Schopmeyer (Tech. Coord.) Agric. Handb. 450.

West, Neil E., Russell T. Moore, K. A. Valentine, Lamont W. Law, Phil R. Ogden, Fred C. Pinkney, Paul T. Tueller, Joseph H. Robertson and Allan A. Beetle. 1972. *Taxonomy, Ecology, and Management of* Hilaria jamesii *on Western Rangelands.* Utah Agric. Exp. Stn. Bull. 487.

32. Plant Species for Critical Areas

Wendell G. Hassell

A MAJOR OBJECTIVE OF THE SOIL CONSERVATION SERVICE is to reduce soil erosion wherever it may occur. Soil and water conservation planning and land treatment begin with plants, for plants are indispensable to the life of the land. Plant materials centers (PMC) are established to provide improved plants and cultural methods for conservation uses including economic, aesthetic, wildlife, and other values.

Two broad categories of critical areas in the southwest desert are: (1) mine wastes and (2) disturbed soils such as construction sites.

Critical areas can be defined as "potential sediment-producing, highly erodible or severely eroded areas such as: dams, dikes, levees, cuts and fills along highways and in urban areas, mine wastes, surface-mine areas, and denuded or gullied areas where vegetation is difficult to establish with usual seeding or planting methods" (USDA Soil Conser. Ser. 1974).

Some basic principles of soil stability apply to all problem areas. This is illustrated by referring to the Universal Soil-Loss Equation $A = RKLSCP$, where A is the computed soil loss (sheet and rill erosion) in tons per acre per year; R, the rainfall factor; K, the soil-erodibility factor; L, the slope-length factor; S, the slope-gradient factor; C, the cropping management factor; and P, the erosion-control practice factor (Agricultural Research Service 1965).

The Universal Soil-Loss Equation was developed for use on cropland. Erosion problems can be more serious on construction sites because of the net effect of increasing the degree and length of slope, the erosive surface materials, and the lack of vegetative cover. Cover is a factor that can be manipulated by man. It is thus one of the key factors in restoring and stabilizing disturbed sites.

Table 32.1 shows a comparison of soil loss in relation to slope gradient and cover (mulch). The C factors were determined by methods outlined in SCS Technical Release No. 51, "Procedures for Computing Sheet and Rill Erosion on Project Areas."

[340]

TABLE 32.1
Soil Loss From Sheet and Rill Erosion (Brooks 1974)
Tons/acre/year

Length of Slope (ft.)	Percent Slope								
	20*	20†	20‡	30*	30†	30‡	40*	40†	40‡
20	.1	3.1	14	.3	8.5	38	.7	17	76
40	.2	4.4	20	.4	10.0	46	1.0	24	107
80	.2	5.9	27	.7	17.0	76	1.3	32	145

*95% mulch cover in contact with the surface. C factor = .004.
†40% mulch cover in contact with the surface. C factor = .10.
‡No mulch. C factor = .45.
This table shows a relative comparison of soil loss in relation to slope gradient and cover (mulch). C factor reference is (USDA Soil Conservation Service, Eng. Division)

Mulch or ground cover is an important factor to reduce the detachment by raindrop impact and physical movement of soil particles by flowing water. Also, it affects soil surface microclimate and nutrients for germinating seedlings.

Plants establish best when provided with the essential elements for plant growth in optimum quantities.

Subsurface soils and inert soil materials need to be supplied with fertilizer and micro-organisms to support good plant growth. Sewage sludge, cow manure, and organic mulch material can improve growth conditions.

Climatic conditions affect plant establishment, particularly in the desert southwest. Generally, the rainfall pattern is erratic. Conditions favorable for establishing desert vegetation may come only every four to six years. Therefore, without artificial irrigation, plant establishment is risky even with adapted species on good soils.

On disturbed areas, plants may invade naturally over many years if the soils are stable and do not contain materials toxic to plant growth. First, the weedy annuals come, followed by a succession of plants until eventually the soils are stable and support adapted perennial vegetation. Reseeding operations with adapted plants and improved methods can often shortcut this long revegetation process. Irrigation can be used to accelerate the process of establishing permanent cover.

Plants protect the soil surface and reduce the velocity of water onto and over the soil surface. Plants have other values including economic uses, visual appeal, and wildlife food and habitat.

Perennial vegetation on mine wastes helps reduce the cost of controlling undesirable annual weeds, which rob soil moisture and plant nutrients and then blow off the slope when mature.

A cooperative effort was initiated in 1970 between Cyprus Pima Mining Company and the Soil Conservation Service to evaluate plants adapted for mine tailing revegetative work.

Test trials were on the following soil materials:

1. Overburden — all submarginal grade ore covering copper deposit.
2. Tailing — processed waste material after ore is removed during milling operation.

3. Mixed Tailing — tailing capped with a 6- to 8-inch layer of overburden.

Mulch treatments added to the mixed tailing were:

a. Conweb fiber — 2,000 pounds per acre conweb fiber hydromulched
after seeding.
b. Stubble — from barley grown on tailing slopes approximately 5
inches high.
c. No stubble — no barley grown, bare soil surface.

Slopes were hand raked to break the crust and loosen the surface. Seeds
were broadcast and raked in before sprinkler irrigating. Shrubs were seeded
in late winter, and warm-season grass plots were seeded in early June.

Fertilizer was applied with sprinkler irrigation. The overburden plots
and tailing plots received no water.

ADAPTED PLANTS

Plants suitable for mine wastes generally have to be drought hardy
and adapted to a wide range of soils. Land changes generally increase climatic extremes in relation to plant growth. The following are descriptions
of the outstanding species evaluated on Cyprus Pima Mining Company
copper tailing (Table 32.2):

(1) Four-wing saltbush, *Atriplex canescens*. This plant has established
well on mine tailing and has been very competitive in eliminating annual
weeds such as Russian thistle. Rodents have encroached on revegetated areas
but have not been a problem in maintaining stands of four-wing saltbush
on tailings. Once established, it should require minimum care since it is a
long-lived perennial. (See Tables 30.1 and 31.1.)

(2) Australian saltbush, *Atriplex semibaccata,* is adapted on areas
which receive a minimum annual precipitation of 12 inches. In lower rainfall areas supplemental irrigation or planting in areas which receive runoff
is required. (See Table 30.1.)

Australian saltbush is a short-lived perennial (3 to 5 years), but reseeds
readily. It provides a good low cover on tailing slopes. It spreads about 1.5
feet in one year and responds very well to fertilizer. Australian saltbush has
been established on soils with pH 8.2. Seeding rate should be 8 pounds per
acre pure live seed with some adjustments for seeding method and seedbed
condition. This shrub is moderately competitive with invading grasses and
weeds. An established stand provides good cover.

(3) Ruby sheepbush, *Enchylaena tomentosa,* is a dark-green shrub,
2 feet high, with bright red berries. It is rather slow establishing, but once
established, appears to withstand drought and saline soils. There has been
a noticeable reduction of Russian thistle, *Salsola kali,* on ruby sheepbush
plots when infested with weeds.

Ruby sheepbush is adapted to elevations below 3,500 feet. It provides
attractive cover. Best success for rapid establishment has been from transplanting small plants. However, successful stands have been established from

TABLE 32.2
Condition of Plant Cover on Mine Tailing (third year after establishment)

Common Name	Scientific Name	SCS Accession Number	Irrigated Mixed Tailing			Non-irrigated	
			Stubble	Fiber Mulch	No Mulch	Overburden No Mulch	Tailing No Mulch
WINTER-SEEDED SHRUBS							
Australian saltbush	*Atriplex semibaccata*	P-15653	3*	3	3	—	4
Australian saltbush	*Atriplex semibaccata*	P-15654	2	3	3	4	4
Four-wing saltbush	*Atriplex canescens*	P-15644	2	—	4	3	4
Four-wing saltbush	*Atriplex canescens*	A-16652	2	2	3	2	4
Quailbush	*Atriplex lentiformis*	A-17156	2	—	4	4	5
Ruby sheepbush	*Enchylaena tomentosa*	P-15560	4	2	4	3	5
SPRING-SEEDED GRASSES							
Buffelgrass	*Cenchrus ciliaris*	P 15625	—	—	3	5	—
Wilman lovegrass	*Eragrostis superba*	Palar	—	4	5	3	—
Creeping dropseed	*Sporobolus usitatus*	P-15641	—	3	3	5	—
Nodding panic	*Panicum stapfianum*	A-14156	—	—	5	5	—
Yellow bluestem	*Bothriochloa ischaenum*	P-15626	—	—	5	3	—
Lehmann lovegrass	*Eragrostis lehmanniana*	A-68	—	3	5	4	—
Blue panic	*Panicum antidotale*	P-15630	—	1	3	5	—
Sand dropseed	*Sporobolus cryptandrus*	A-16352	—	—	5	5	—
Mixture†			—	—	5	—	—

*1 excellent; 2 good; 3 fair; 4 poor; 5 failure.

†5 lbs. each of Wilman lovegrass and Creeping dropseed; ½ lb. of Yellow bluestem; 1 lb. each of Lehmann lovegrass and Four-wing saltbush; and 2 lbs. of Blue panic grass.

direct broadcast seedings. Seeding rates of about 20 pounds pure live seed per acre are recommended for broadcast seedings.

(4) Lehmann lovegrass, *Eragrostis lehmanniana,* is an old standard range grass used throughout southeastern Arizona and New Mexico where annual precipitation is 10 to 12 inches or more. Lehmann is a warm-season, short-lived perennial bunchgrass that reseeds readily. It is drought tolerant but not winter hardy. It is generally recommended for elevations below 4,500 feet.

Lehmann lovegrass seed is very small. One pound of seed contains approximately 6.5 million seeds, which means 2 pounds per acre of good quality seed should generally provide adequate stands on a good seedbed. Seeds can be planted in midspring when supplemental irrigation is applied or it can be seeded prior to summer precipitation.

(5) Blue panic, *Panicum antidotale,* is an extremely vigorous sod-forming grass with an extensive root system. It reaches heights of 5 to 7 feet. P-15630 was developed at the Tucson PMC for its apparent cold tolerance compared with other varieties of blue panicgrass. This grass is generally used in basin areas on rangeland.

Although blue panic is adapted to floodplains, it has also shown drought tolerance on slopes. This grass may require additional water on tailing slopes in the lower rainfall areas. Blue panicgrass will grow on rather saline soils and has been used to add organic matter and litter to tailing surfaces. Seeding rates of blue panic on critical slopes should be 4 to 6 pounds pure live seed per acre, depending on seedbed and seeding method.

(6) Buffelgrass, *Cenchrus ciliaris,* is a grass that appears to have good potential on mine tailing at lower elevations. It has shown great potential on south-facing slopes with adequate irrigation and fertilizer.

Test work with other desert shrubs has been done west of Phoenix and east of Yuma for plants adapted to structures and highway slopes. These sites are located in an area with an annual rainfall of 4 to 8 inches. Chances for establishment without supplemental water are very poor at these locations. Rock or gravel mulch is beneficial in conserving moisture on these dry sites.

Based on evaluation plantings conducted by plant materials centers in Arizona, the following species appear to have good potential for use on dry slopes in the southwest:

(1) Globe mallow, *Sphaeralcea spp.,* is a herbaceous perennial, dark green with pubescent stems and leaves. It has an orange flower. It does not establish well from seeds but has persisted on very dry sites once established from transplants. (See Table 31.1.)

(2) White bursage, *Franseria dumosa* (Ambrosia), is a low-desert shrub which forms pure stands over large areas in the Mohave Desert from California to Utah and Arizona and south into Mexico. It has a grayish-green leaf which is lobed or finely divided. This plant is extremely drought tolerant but does not grow where temperature drops below 25 degrees. It is recommended for use at elevations up to 3,000 feet. It appears to withstand southern exposures quite well. Seeds are difficult to collect and germination is relatively low due to the thick seed coat and the multi-seed nature of white bursage. However, transplanted plants have performed well.

(3) Brittlebush, *Encelia farinosa,* is a native shrub common in hot, dry regions of southern Arizona. It has grayish leaves and bright yellow flowers. It can be found in dense stands covering dry, rocky slopes. Plants establish during the winter months when moisture conditions are above normal.

Seed dormancy appears to be a problem with direct seeding. Seed treated with Clorox* (soaked 1 hour) showed improved germination in limited tests. Brittlebush grows at elevations up to 3,000 feet.

(4) Desert saltbush, *Atriplex polycarpa* (Chapter 30 and Table 30.1).

(5) Creosote bush, *Larrea tridentata,* is an extremely drought-tolerant shrub found at elevations up to 5,000 feet. This plant occupies thousands of acres in southern and western Arizona. Seed set is usually poor. Cracking the seed hull with a hammer mill lets water penetrate and improves seed germination.

These plants grow with good vigor once established but getting large numbers of plants for transplanting has been difficult. There is a wide range in growth forms in creosote bush. Some appear better adapted for erosion control than others.

(6) Quailbush, *Atriplex lentiformis* (see Table 30.1). These plants perform best where they receive run-in moisture or where the roots can reach a shallow-water table. Stands are usually open, but individual plants have made good growth on tailings.

Two additional plants for conservation and beautification uses on tailing and disturbed sites in southern Arizona are:

(1) Desert broom, *Baccharis sarothroides,* a bright-green evergreen shrub that has good potential for beautification on mine tailing and construction sites. It is moderately salt tolerant and grows at elevations between 1,000 and 5,000 feet. The seed is hard to collect and clean but it germinates readily. (See also *Baccheris* pilularis, Table 30.1.)

(2) Desert marigold, *Baileya multiradiata,* is an attractive low-growing herbaceous plant. It is adapted at elevations up to 5,000 feet. It has large, showy yellow flowers. Desert marigold establishes easily from seed. This plant appears to tolerate saline soil conditions.

Desert plants have many mechanisms that trigger or hold back seed germination, making direct seeding sometimes difficult with respect to getting a high percentage of germination at one time. Some have hard, waxy seed coats, others have germination inhibitors in the seed coat. These substances must be washed away before germination occurs.

A method of "soil plating" (Soil Conservation Service 1973) with topsoil takes advantage of the accumulation of native adapted seed in the soil. Stockpiling topsoil takes advantage of nature's ready-prepared and aged seed and has been successfully used in establishing desert species. Sprinkler irrigation can be used to start seedling germination and establish-

---

*Trade names are used solely to provide specific information. Mention of a trade name does not constitute a guarantee of the product by the U.S. Department of Agriculture nor does it imply an endorsement by the Department over comparable products that are not named.

ment after a 2- to 4-inch layer of topsoil has been spread evenly over the site to be revegetated.

Information about soils and germination dates for different species should be considered for best success.

Cover, vegetative or otherwise, is needed to reduce erosion. The amount of input to rehabilitation determines the extent of vegetative cover that is maintained on a particular area. Rehabilitation of critical areas should be based on good technical information and knowledge. Cost of continued maintenance should be considered as part of long-range plans. Adequate cover and soil stability are prerequisites for maintaining any land use plan.

REFERENCES

Agricultural Research Service. 1965. Agricultural Handbook 282.

Brooks, Frank. 1974. *Universal Soil Loss Equation — Computer Solutions.* USDA Soil Conservation Service, Technical Service Center. Advisory Agron. PO 9, October.

Soil Conservation Service. 1975. Procedure for Computing Sheet and Rill Erosion on Project Areas. Technical Release no. 51. Engineering Division.

Soil Conservation Service. 1974. Critical Area Plantings. Arizona Technical Guide. Section IV. June.

Soil Conservation Service. 1973. Soil Plating Method of Critical Area Planting. Agronomy Note no. 51.

Williams, D. A. 1966. Plant Materials in Conservation, Search. Pub. of the American Seed Research Foundation. Spring.

33. Species Trials on Strip Mine Areas

Alten F. Grandt

SPECIES ADAPTATION STUDIES on coal surface mined lands have been carried on in the eleven states where Peabody Coal Company operates. These trials are established at a new mine as soon as possible to evaluate the performance of the species on the particular disturbed lands. The eleven states — Ohio, Indiana, Kentucky, Alabama, Illinois, Missouri, Arkansas, Oklahoma, Montana, Colorado and Arizona — represent a wide range in climatic conditions, such as precipitation, temperatures, frost-free period and prevailing wind velocity.

Next to climate, soils are the most important factor influencing adaptation of species. The soil materials of mined lands available for plant growth are changed by mining. This material is a heterogeneous mass whose physical and chemical properties are dominated by the character of the geologic strata overlying the coal.

In many areas, especially on Black Mesa in Arizona, the original overburden contains only a thin layer of soil that is permeable. This soil rests on impervious material that does not permit the limited rainfall to infiltrate. During severe thunderstorms this results in flash floods familiar to such areas. However, when the overburden is broken up during the mining process, this cast overburden or spoil contains voids capable of storing large quantities of water.

Peabody's reclamation grading program at Black Mesa is designed to trap all precipitation possible, which will result in the mined lands acting as a "sponge" to hold runoff water and to release this water to growing plants. The disturbed material averages over fifty feet at this mine. Deep roots of plants such as alfalfa have been traced to over twenty feet.

A system of classification of mined lands is essential in selecting species that are adapted. Such a system will be a guide to determine land use and species adaptation and productivity. However, experience has taught that "the book" on species adaptation on coal surface mined lands has not been written completely. The science of mined lands revegetation is a relatively new field.

The major soil properties that affect species adaptation on mined lands are pH, texture, and slope. Every kind of plant has a suitable soil reaction or pH range for its best growth. Some plants grow best in a rather narrow pH range, while other species tolerate a wide pH range. For example, autumn olive *(Elaeagnus umbellata)* is known to grow in pH ranges from 4 to 8.

Criteria used in selecting species are those whose demands upon the environment appear to meet best the site conditions and include tolerance of pH, lack of tilth, drought resistance, varying nutrient levels, indifference to extreme climatic influences, the ability to develop ground cover and stabilize the soil, and productivity potential. Such plants are classed as legumes, grasses, shrubs and trees.

LEGUMES

Legumes have the greatest potential for providing quick, low maintenance ground cover and productivity for most mined land soils. Where soil tests show a pH between 5 and 8, and plant nutrients such as calcium, phosphorus, and potassium are high, legumes are generally well adapted. All legumes should be inoculated with the proper nitrogen-fixing bacteria. This is important because most mined land soils are low in nitrogen and organic matter and thus require introduction of nitrogen-fixing bacteria specific to the particular legume plant.

1. *Alfalfa* (Medicago sativa) "Queen of the Legumes" is one of the best adapted forage species. It is potentially a long-lived, deep-rooted perennial legume. It is valued as feed for livestock and wildlife, for hay and erosion control. The deep-rooting habit allows it to seek out the moisture available and produce high yielding nutritious forage. In Illinois, original seedings of wilt resistant winter hardy varieties have lasted for 15 years without reseeding. It has done very well in Arizona, Colorado, and Montana in average precipitation of 14 inches without irrigation.

    It is palatable to livestock. This may be a disadvantage when grown for grazing since it will cause bloat in ruminants. Bloating can be controlled by grazing only mature plants or when over 50% of the pasture mixture is grass or nonbloating legumes.

2. *Sweet Clover* (Melilotus sp.) is a widely adapted biennial legume with a high calcium requirement and relatively low available (by soil tests) phosphorus and potassium content. As a biennial, it will reseed itself. It is an excellent pioneer or soil building plant, and is used by livestock and wildlife. Sweet clover is not very palatable to livestock. Its best use on mined lands is as a soil builder, adding nitrogen for grasses.

    Sweet clover weevil has seriously damaged both new seedings and second-year growth. Root rot *(Phytophtora cactoriam)* has killed second-year growth before seed production is established.

3. *Birdsfoot Trefoil* (Lotus sp.) is a long lived perennial legume that grows from 12 to 30 inches high and branches profusely. Its root system is more fibrous than alfalfa. It is a cool season crop, but about 10 days later than alfalfa. It is slow to establish and requires a special bacteria inoculant. The seed is quite small, containing 1,000,000 seeds/pound. It is valued as forage for livestock and wildlife. It does not create bloat problems and will persist for many years even when closely grazed. In

more humid regions, it is seeded alone since the sod forming grass species tend to crowd out the plant.

4. *Lespedeza* (Lespedeza sp.) All species are warm season, drought resistant plants, making very slow early growth. Annual lespedezas, Korean and Kobe, provide late summer and fall pasture. Serecia lespedeza is a perennial, grows taller than the annual species, has larger stems and is more woody. It does well on soils with low plant nutrients. Lespedeza bicolor is a shrub, excellent for wildlife and is adaptable to arid spoils. It is best established by plants, but can be propagated by seeding. Bicolor is a good quail food. Many other species and varieties have been seeded with limited results.

5. *Cicer Milkvetch* (Astragalus cicer) is a moderately long-lived, nonbloat legume. It is used for livestock, wildlife, erosion control, and aesthetics. It is slow to become established on mined lands when compared to alfalfa. Seed must be properly scarified before planting to establish suitable stands. Seed supply has been scarce. It is a relatively new species being tried on mined land. Thus, a limited knowledge of its adaptation is known and proven.

6. *Sanfoin* (Onobrychis viciaefolia) is a pink flowered, short-lived, nonbloat legume. Its major value is as forage for livestock and wildlife. Because of its short life, it is not well suited to erosion control. In trials conducted in Illinois, Colorado, and Montana poor stands were obtained. Seed supply has been limited.

7. *Clovers* (Trifolium sp.) include red clover, alsike, common white clover, crimson, Ladino and other species. Most of the clovers are short-lived perennials, shallow to medium rooted species. They are well adapted to the mined lands in humid to semi-humid regions. They are used by livestock and wildlife and for quick-cover erosion control. Ladino forage is high in protein and minerals. It is very palatable to all classes of livestock, but it is also a bloat forming legume.

8. *Crown Vetch* (Coronilla varia) is a long-lived, winter-hardy perennial legume. It forms a dense cover and spreads by underground root stalks and by seed. It is very slow in establishing itself, probably due to the many hard seeds. Its major use is for aesthetics and erosion control. It is not very palatable to livestock as either hay or forage and is relatively low in production of feed.

GRASSES

Without specific soil amendments grasses become established on mined land soils more slowly than legumes. Undoubtedly the main reason is the lack of nitrogen in the raw soil material. As the legumes grow and add nitrogen to the soil, the grasses become more competitive.

Grasses are important plants in the pasture mixture used for forage and also as plants to be used for erosion control. To establish a good grass cover on mined lands will take a minimum of two years and often longer. Orchard grass and fescue have produced good stands after two full growing seasons.

1. *Bromegrass* (Bromos inermis) (see Table 31.1).
2. *Fescues* (Festuca sp.) (see Table 31.1).
3. *Wheatgrasses* (Agropyron sp.) (see Table 31.1).
4. *Indian ricegrass* (Oryzopsis hymenoides) is a densely tufted perennial bunchgrass. It is a highly palatable cool season grass, adapted to arid and semiarid regions. Its best use probably is for winter grazing.
5. *Bermuda grass* (Cynodon dactylon) is a leafy perennial sod forming warm season grass. It will not tolerate cold temperatures, which limits its northern latitude use. It is propagated by either seed or vegetative sprigs. Because of its vigorous rooting system, it is condemned in some agricultural areas producing row crops such as corn and cotton.

    Bermuda grass requires high nitrogen for superior yields. Most any legumes will make an excellent growth in association with Bermuda, thus decreasing the dependence of nitrogen fertilizer for maintenance. In Peabody's reclamation program, its primary use is in Oklahoma and just recently in western Kentucky.
6. *Big bluestem* (Andropogon furcatus) is a long-lived, warm season, sod-forming grass. It is very slow to become established. Since it is also very palatable, careful management of grazing is required until it is firmly established.
7. *Grama grasses* (Bouteloua sp.) Blue grama is a long-lived, warm season, short grass. It is one of the most common grasses found prior to mining on Black Mesa in Arizona and at the Big Sky Mine in Montana. The seed is very fluffy and difficult to seed with most seeding equipment. The Ezee Flow type broadcast seeder has done a good job of distributing the seed. It is a palatable grass, retaining its feeding value into the winter months. It is drought resistant and tolerant to alkaline conditions.

    Side-oats grama is a long-lived, warm season, bunch grass. It is palatable to all classes of livestock. Seed matures in late summer and can be combined

    Many other grasses have been tested with varying success. Weeping lovegrass is used in Kentucky on low pH (3.8–4.0) soil types. Green needle grass is used in Montana. Switch grass does well in Oklahoma and Arkansas. Ryegrass, wild-rye grasses, Orchardgrass, redtop, timothy, and many others are used under specific conditions.

SHRUBS

Shrubs are low, usually several stemmed, woody plants. They are frequently used for range in arid regions, wildlife food and protection, erosion control, and some for their aesthetic value.

1. *Fourwing saltbush* (Atriplex canescens) (see Table 30.1).
2. *Rubber rabbitbrush* (Chrysothamnus nauseosus) (see Table 31.1).
3. *Autumn-olive* (Elaeagnus umbellata) is a nitrogen fixing, non-leguminous shrub. It is a multiple stemmed, densely branched shrub reaching a height of 15 to 20 feet. Few species adapt to so many sites and climatic conditions, and grow as vigorously as autumn-olive. It is an excellent wildlife shrub.

4. *Bristly locust* (Robinia fertilis) is one of the best plants for erosion control. The pink flowers enhance its value for aesthetic uses. It is a rapid growing legume shrub, spreads by root suckers, provides a dense cover, and lays down a heavy leaf litter. It is adapted to low fertility, acid sites.

Other shrubs used for various and specific purposes include honeysuckle, willow, coral berry, sand cherry and Russian olive. Chinese chestnut, making up not more than 5% of the mixture, has been successful in Indiana and Kentucky.

TREES

Some of the earliest attempts at revegetating surface mined lands were with the planting of trees. Early conservation-minded operators observed trees growing naturally on the disturbed areas and began planting trees. The comment was that "if nothing else will grow, trees will."

As early as 1918, peach and pear trees were planted on rough, ungraded mined lands in Indiana. The first record of forest tree planting on mined lands was in 1926 when 1,400 cottonwoods were planted in Indiana. Research studies of the adaptation of tree species were established by the United States Forest Service in 1946. From these and similar studies, adapted species can be recommended for various soil conditions.

Hardwood Trees

1. *European black alder* (Alnus glutinosa) was first grown on mined lands in the United States by Ohio Reclamation Association on Ohio lands. It is a fast growing, nitrogen fixing tree and is a primary tree species used to establish forest plantings. Some trees have straight stems while other plants branch profusely. It is adapted to all classes of spoils. Survival has been observed at pH as low as 3.5. Thin leaf alder performs well on some western sites (see Table 31.1).
2. *Black Locust* (Robinia pseudo-acacia) is widely adapted to all spoil classes, fixes nitrogen, grows rapidly and gives quick cover. This species is valuable as a nurse crop for forest plantings. By adding nitrogen and organic matter, it improves the soils. This species is affected by locust borer, which results in multiple stem shoots sprouting after the main stem deteriorates (see Honey locust, Table 31.1).
3. *Sycamore* (Platanus occidentalis) is a vigorous, rapid growing species. It is often a volunteer tree on newly mined lands. Seeds are wind borne.
4. *Northern red oak* (Quercus borealis varmaxima) is a slow growing species during the first few years of establishment, but has a good survival record. It is adapted to acid sites. Older plantings pick up in growth rate making it a good commercial forest planting (see gambel oak, Table 31.1).
5. *River birch* (Betula nigra) is adapted to a wide range in soil classes. It is a rapid growing tree species with a good survival record (see Table 31.1).

Other hardwood tree species used under varying use conditions include walnut, yellow poplar, green ash, and silver maple. An acceptable acid site hardwood mixture is the combination of northern red oak, sweetgum, syca-more, riverbirch and European black alder, each comprising 20% of the mixture.

Conifers

1. *Red Cedar* (Juniperus virginiana) is especially adapted to high clay mined land and is a rather slow growing tree species (see western juni-pers, Table 31.1).
2. *White Pine* (Pinus strobus) has acid tolerance, showing good survival and growth on pH of 4.0 and above. In hardwood plantations that are hand planted, the last row is a white pine row, serving as a marker row during planting. It contributes aesthetically to the mixed hardwood forest (see western pines, Table 31.1).

Other conifers planted include pitch pine, Jack pine, Virginia pine, and cypress. Virginia pine is used on extremely acid sites and is used for cover. It is used by wildlife.

INDEX

GENERAL

PLANT SPECIES FOR REVEGETATING DISTURBED LANDS

Forbs